AF736416

Power Systems

Electrical power has been the technological foundation of industrial societies for many years. Although the systems designed to provide and apply electrical energy have reached a high degree of maturity, unforeseen problems are constantly encountered, necessitating the design of more efficient and reliable systems based on novel technologies. The book series Power Systems is aimed at providing detailed, accurate and sound technical information about these new developments in electrical power engineering. It includes topics on power generation, storage and transmission as well as electrical machines. The monographs and advanced textbooks in this series address researchers, lecturers, industrial engineers and senior students in electrical engineering.

Power Systems is indexed in Scopus

More information about this series at http://www.springer.com/series/4622

Mohammadreza Daneshvar • Somayeh Asadi
Behnam Mohammadi-Ivatloo

Grid Modernization – Future Energy Network Infrastructure

Overview, Uncertainties, Modelling, Optimization, and Analysis

Springer

Mohammadreza Daneshvar
Faculty of Electrical & Computer Engineering
University of Tabriz
Tabriz, Iran

Somayeh Asadi
Department of Architectural Engineering
Pennsylvania State University
University Park, PA, USA

Behnam Mohammadi-Ivatloo
Faculty of Electrical & Computer Engineering
University of Tabriz
Tabriz, Iran

Department of Energy Technology
Aalborg University
Aalborg, Denmark

ISSN 1612-1287 ISSN 1860-4676 (electronic)
Power Systems
ISBN 978-3-030-64098-9 ISBN 978-3-030-64099-6 (eBook)
https://doi.org/10.1007/978-3-030-64099-6

© Springer Nature Switzerland AG 2021
This work is subject to copyright. All rights are reserved by the Publisher, whether the whole or part of the material is concerned, specifically the rights of translation, reprinting, reuse of illustrations, recitation, broadcasting, reproduction on microfilms or in any other physical way, and transmission or information storage and retrieval, electronic adaptation, computer software, or by similar or dissimilar methodology now known or hereafter developed.
The use of general descriptive names, registered names, trademarks, service marks, etc. in this publication does not imply, even in the absence of a specific statement, that such names are exempt from the relevant protective laws and regulations and therefore free for general use.
The publisher, the authors, and the editors are safe to assume that the advice and information in this book are believed to be true and accurate at the date of publication. Neither the publisher nor the authors or the editors give a warranty, expressed or implied, with respect to the material contained herein or for any errors or omissions that may have been made. The publisher remains neutral with regard to jurisdictional claims in published maps and institutional affiliations.

This Springer imprint is published by the registered company Springer Nature Switzerland AG
The registered company address is: Gewerbestrasse 11, 6330 Cham, Switzerland

Keywords: Grid modernization, Future multi-carrier energy networks, Renewable energy resources, Uncertainty modelling, Multi-carrier energy systems, Energy management and control.

To:
My mother, Fatemeh, and my father, Ali, for their valuable and unconditional love and support.
—Mohammadreza

Preface

In recent years, reliable and extensive energy networks have fueled the nation's growth from various aspects. However, the current structure of the energy grids does not have the necessary characteristics to meet today's energy demands and beyond. Energy networks such as electricity, gas, heat, and water with multi-carrier energy systems need to be operated harmoniously due to the emergence of new hybrid devices and the development of energy hubs. Given the economic and environmental justifications, the future energy network infrastructure is also expected to develop into a fully equipped one with renewable energy resources (RERs). This means that future energy networks will face a large number of uncertainties on the supply side. To overcome these challenges, future energy network modernization is proposed as a promising solution for securely and reliably delivering greater quantities of zero to low-carbon energy, including handling different types of RERs such as solar and wind power.

On the other hand, modernization of the energy customer's side will not only allow various consumers to effectively and easily participate in the different energy markets for maximizing their economic benefits but also applying the advanced communication protocols and control systems will provide additional choices for the consumers and will improve energy efficiency through intelligent demand-side energy management schemes. This is why modern energy grids will need to be incorporated with the capable technologies that provide satisfactory solutions for implementing various types of RERs for clean energy production and economic benefits by reducing the need for unsustainable and expensive conventional power plants. Feeling this great need for modernizing our future multi-carrier energy networks (MCENs), this book is dedicated to cover the basic requirements in concepts, modeling, optimization, and analysis of future energy grids by developing six chapters.

Chapter 1 provides an overview of the electric power system, the natural gas grid, the district heating network, and the water distribution system. Moreover, this chapter investigates the MCENs, discusses the key issues and their various structures, and scrutinizes the different energy markets.

Chapter 2 deeply investigates the modernization of energy systems from the customer's side. This chapter explores the concepts and systems for customer-side modernization in future modern MCENs. Also, modernization of smart homes, commercial and industrial sectors, and buildings are evaluated in this chapter.

Chapter 3 covers the technical and theoretical analysis of future energy network modernization from various aspects. In this chapter, the energy generation sector, energy storage systems, integration of the decentralized energy production systems in the MCENs, demand-side energy management schemes, and food-energy-water nexus technologies are evaluated from the grid modernization perspective.

Chapter 4 deals with the advanced data management systems in the modernization of future MCENs. In this chapter, big data analytics, smart meters, advanced metering infrastructure, Internet of things, energy cyber-physical systems, energy Internet, machine learning, and deep learning techniques are assessed for possible presence in future modern MCENs.

Chapter 5 studies the energy-trading possibilities in modern MCENs. The possibility of the presence of various energy-trading techniques in the future modern MCENs infrastructure is fully analyzed to provide a good overview of the energy-trading issue in modern grids.

Chapter 6 covers the optimization models and uncertainty management of modern MCENs. In this chapter, mathematical models are presented for different systems of the electric power system, the natural gas grid, the district heating network, and the water distribution system to make their implementation possible. Moreover, different common uncertainty modeling techniques are investigated, along with providing the mathematical models for them. In the end, a case study is considered to examine the effectiveness of the proposed model for future modern MCENs.

This book covers the required theoretical, technical, and practical information regarding the modernization of future energy networks. Almost all the required concepts for switching from the current energy system to future modern MCENs are explained in detail. This book evaluates various aspects of the overall energy networks such as reliability, resiliency, stability, sustainability, and security in the presence of coupled electricity, gas, heat, and water networks. Because future energy network modernization is critical for solving the energy crisis, so ongoing activities considering their impacts are rapidly growing in the energy grids. Therefore, this book can be helpful for researchers and postgraduate students who work in various fields of energy systems, such as designing, planning, scheduling, and operation.

Tabriz, Iran — Mohammadreza Daneshvar
University Park, PA, USA — Somayeh Asadi
Aalborg, Denmark — Behnam Mohammadi-Ivatloo
September 2020

Contents

Chapter 1
Overview of the Grid Modernization and Smart Grids

Abstract Nowadays, the forms of energy generation, transportation, utilization, and application are changing all over the world. Given the significant penetration of intelligent systems in human society as well as modern energy distribution systems globally, smart grids are effectively developed to integrate various innovative technologies aiming to improve energy supply process. The array of these technologies has formed widespread efforts to ensure the satisfying of modern living standards that have led to the grid modernization process. In this regard, modernizing the future multi-carrier energy networks is essential for effectively handling the widespread presence of the numerous advanced technologies and intelligent systems in the network place. This chapter is targeted to analyze the role of electricity, natural gas, heat, and water networks in the modernization of the multi-carrier energy networks. Also, this chapter responds to this key question that why we need grid modernization by evaluating different aspects of the future modern grids. Finally, this chapter focuses on various energy markets and structures to give a superlative viewpoint of the future energy markets with the aim of laying the groundwork for energy market development in the modern power system landscape.

Keywords Grid modernization · Smart grids · Hybrid energy networks · Energy markets · Microgrids · Energy hubs · Prosumers

1.1 Introduction

Nowadays, energy has become a fundamental tool, which plays a critical role in people's lives. The popularity of energy among the people has led to rapid developments in energy systems and has intensified the trend of increasing energy consumption worldwide [1]. In recent decades, this issue has become a serious concern and a basic challenge for the existing power systems [2]. Even though the availability of energy in various types has brought appealing advantages, increasing the dependency of the human activities on it at a high level has posed considerable risks

© Springer Nature Switzerland AG 2021

M. Daneshvar et al., *Grid Modernization – Future Energy Network Infrastructure*, Power Systems, https://doi.org/10.1007/978-3-030-64099-6_1

related to its shut down. Hence, in order to avoid discomforts arising from different failures in the energy supply system, the reliability of energy supply is becoming increasingly important and should be seriously considered in the energy network development schemes. Due to this, smart grids are proposed as the new structure of the power system. A smart grid is an intelligent electric power grid that integrates smart energy measures, energy-efficient resources, renewable energy resources (RERs), and network-based technologies to properly respond to the challenges ahead of the system [3]. Various definitions are provided for smart grids, and almost all of them have a common denominator in expressing their undeniable advantages. Smart grids were firstly defined by the Energy Independence and Security Act (EISA), which explicitly described the features of the smart grids [4, 5]. According to the definition provided by the EISA, smart grids should have specific features, which are listed as follows [5].

- A set of advanced control and digital information technologies for improving the electric power system efficiency, security, and reliability.
- Provision of a full cyber-security system through dynamic optimization of the energy networks operations.
- Integration and deployment of the wide range of distributed energy resources (DERs), especially the clean energy resources.
- Development and engagement of the different demand-side energy management programs such as demand response and energy-efficiency resources.
- Deployment of intelligent technologies such as interactive, automated, real-time (RT) technologies for optimizing the physical operation of devices as well as the customer services for establishing the distribution automation.
- Optimal integration and deployment of the advanced hybrid energy storage systems and peak shaving techniques such as thermal storage air conditioning, hybrid electric vehicles, and plug-in electric systems.
- Incorporation of intelligent appliances and smart customer services.
- Identification of existing barriers for avoiding the adoption of unnecessary and unreasonable technologies and devices throughout the smart grid.
- Provision of additional customer choices to the consumers along with the control options and useful information.
- Development of communication standards as well as the interoperability of the equipment connected to the energy network and other appliances.

Given the official definitions of the smart grids, their features can be summarized as follows: the smart grid is a new intelligent structure of the electric power system that enables various elements for coordinately working for the special goals, provides an additional choice for energy consumers with the aim of increasing their convenience, and cost-efficiently and intelligently integrates all sectors in the energy network including energy producers, transmitters, distributors, and consumers to ensure the reliability, sustainability, and resiliency of the power system with minimum power losses and high levels of power quality, safety, and stability by welcoming the cost-effective and environmentally friendly technologies. Indeed, the smart grid employs capable technologies for monitoring, control, communication, and management of the systems to achieve its goals, which are shown in Fig. 1.1.

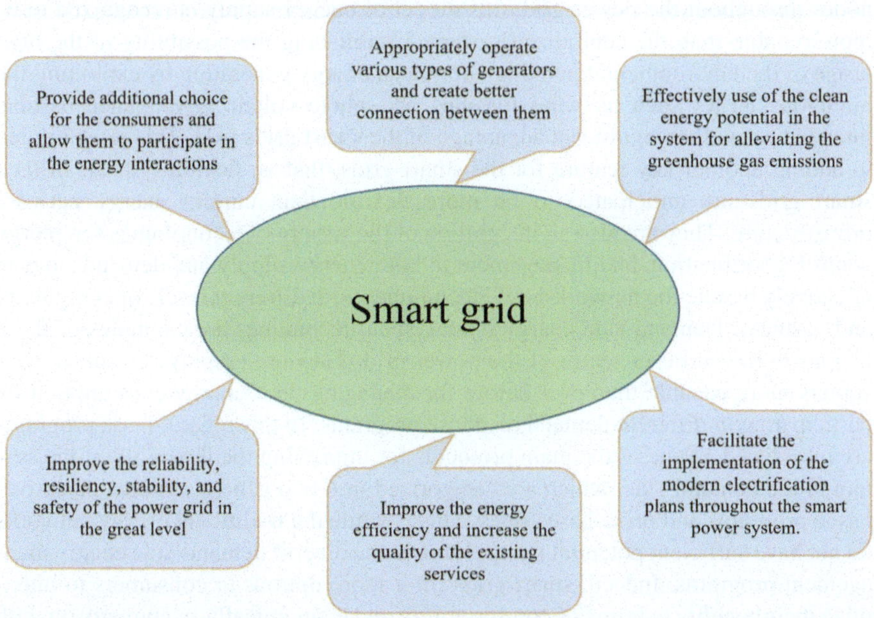

Fig. 1.1 Main goals of smartization of the energy networks

Smart grids typically offer different benefits by applying the automation systems and digital processing to the power grids [6]. Indeed, considering the deep integration of digital technologies with advanced communication protocols has armed smart grids with different capabilities enabling them to bring significant benefits for all parts of the system. Under these conditions, smart grids are taken into account as the full suite for reliably responding to the current challenges and particular concerns ahead of the future energy networks. One of the important focuses of smart grid development plans is reliability. The smart grids use capable technologies for improving the fault detection process with the aim of creating a self-healing system and eliminating the human-based control process by implementing automation systems. Indeed, omitting human interventions makes the system more reliable in managing and controlling energy generation, transmission, and distribution by reducing the vulnerability to attacks, human errors, and natural disasters. Traditional power grids are structured based on centralized power production, which are encircled by the radial-based electricity line models [7]. This is a worthy important issue because most of the current energy networks are engaged with the radial distribution systems, which are more exposed to the power outages and an adequate degree of reliability cannot be guaranteed based on these structures.

Smart grids are developed based on decentralized energy generation mechanisms [8]. Their structures have been designed with the aim of adopting various types of distributed energy resources (DERs), especially clean energy production [9]. The decentralized energy generation paradigm allows economical electrification to the remote area without needing to build the new transmission lines and other related equipment [10]. Moreover, due to providing a large number of energy generation nodes throughout the power grid, this model of energy supply is recognized to be more reliable than the centralized system. Establishing the possibility of the high usage of the environmental potentials for clean energy generation by exploiting the numerous RERs such as wind turbines and photo-voltaic (PV) panels is also intended as another significant advantage of the smart grids [11]. This issue has led to adding another key feature for the smart grids, that is, flexibility [12]. In fact, smart grids are anticipated to be more flexible than current energy network infrastructure. They are also an integration of the variety of technologies for energy control, coordination, and management in both energy supply and demand sides to effectively handle the network-based techniques with different levels of complexity and vastness. Demand-side energy management techniques are examples of these actions in the end-user sector of the power grid. They are employed to make consumers more capable than ever before for managing their energy consumption by participating in different demand response programs. In this regard, these programs are introduced as one of the main protocols for improving the flexibility of the system in the demand-side, which are categorized into two general groups: incentive-based programs and price-based ones. This is while the traditional energy networks do not have sufficient potential for applying to the diverse demand-side energy management programs. Indeed, smart grids offer more options to consumers to maximize their benefits in terms of comfortability and economically in comparison with current networks. The superiority of smart grids over current networks is not only summarized in the reliability and flexibility issues, but also more other important factors exist that need to be considered. In order to clearly identify the other factors, detailed comparison is carried out between smart grids and traditional networks, which is illustrated in Fig. 1.2.

Though smart grids are touted as the advanced infrastructure of the system with numerous advantages, behind these benefits lies the growing challenges. Satisfying most of the benefits indicated in Fig. 1.2 is becoming increasingly more challenging due to different reasons. Important examples of these factors that affect the reliability and insurability of the smart grids are:

- Ever-increasing grid congestion.
- Existence of the numerous devices throughout the system, which increases the complexity of the system and makes the optimization process even more difficult.
- The operation of the grid at its "edge" in more locations has a negative impact on network reliability.
- Larger "carbon footprints" are created by a consolidation of operating various systems.
- Massive utilization of RERs accentuates the volatility and complexity of the smart grids.
- Automation of processes increases system vulnerability to cyberattacks.
- Reliable energy supply is a challenge in the presence of the high share of RERs and the system will need other capable technologies for providing continuous energy supply in the deregulated environment.

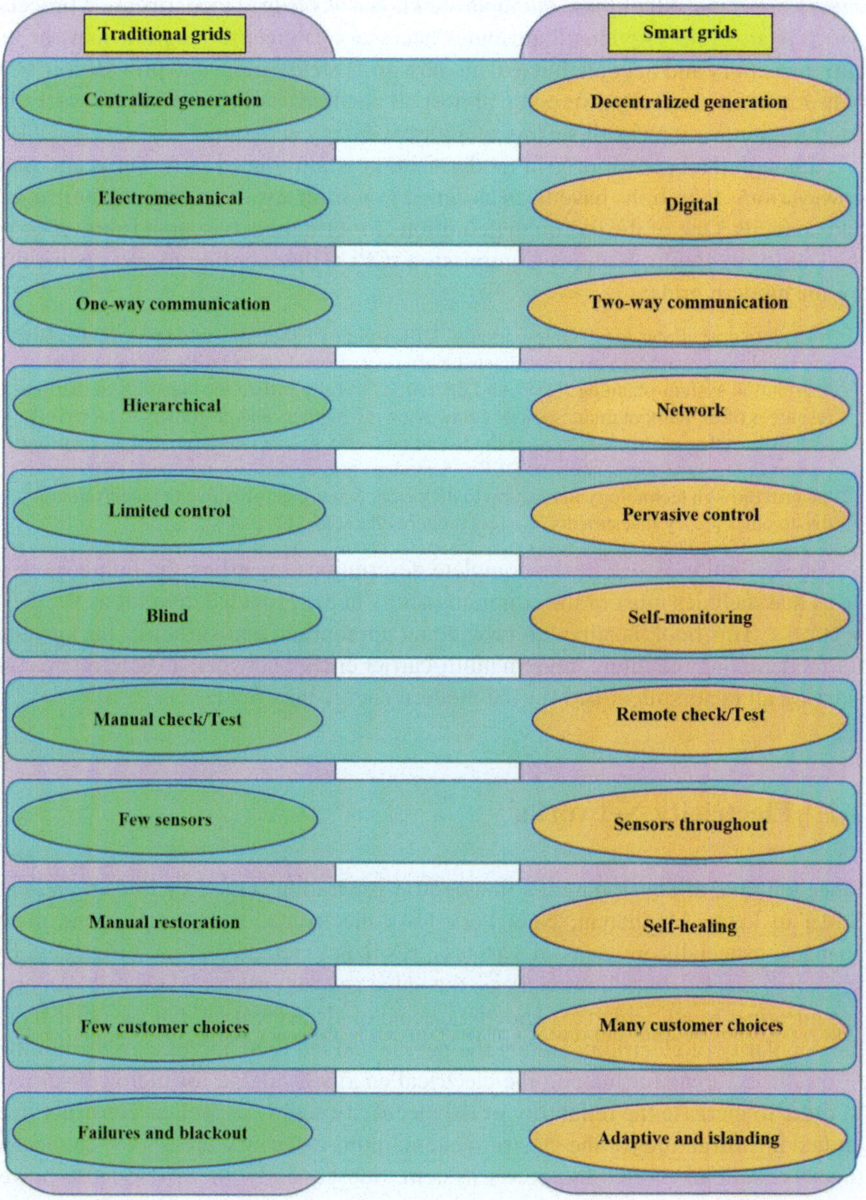

Fig. 1.2 Comparison between features of the traditional energy networks and smart grids

All of the aforementioned factors are only a small important part of the challenges that smart grids face for full implementation. For effectively considering the challenges ahead of the smart grids, a great need is felt for developing technologies, tools, and concepts required to control, protect, analyze, predict, and manage the energy networks. All of these attempts are known as the grid modernization process, which is aimed to provide all attributes necessary to meet the requirements of the current century and beyond. Indeed, modern grids are developed to provide a critical platform to successfully overcome almost all challenges of the energy networks by adopting innovations in all sectors of a global energy economy.

Although the concept of grid modernization is introduced in recent years, until now, various definitions have been assigned to it from various energy organizations and councils. One of the thorough definitions for grid modernization is provided by the California Public Utilities Commission (CPUC). According to this description for the modern grids:

> A modern grid allows for the integration of DERs while maintaining and improving safety and reliability. A modern grid facilitates the efficient integration of DERs into all stages of distribution system planning (DSP) and operations to fully utilize the capabilities that the resources offer, without undue cost or delay, allowing markets and customers to more fully realize the value of the resources, to the extent cost-effective to ratepayers, while ensuring equitable access to the benefits of DERs. A modern grid achieves safety and reliability of the grid through technology innovation to the extent that is cost-effective to ratepayers relative to other legacy investments of a less modern character [13, p. 4.1].

This definition is one of the complete descriptions regarding the future modern grids that includes most of the common factors in the provided definitions for these networks. This book is aimed to provide an appropriate and sufficient background for modernizing the future modern multi-carrier energy networks (MCENs) by considering all factors identified for the modern energy networks.

1.2 Electricity Networks

Since the time of developing the first central generation station by Edison on Pearl Street in lower Manhattan, New York [14], electrification has undergone many changes, especially in terms of vastness, resiliency, reliability, and efficiency. By supplying electricity to limited areas, the value and importance of electrical energy became clear to all, which expanded the scale of electrification. On the other hand, the popularity of electricity among the people had led to an increase in the number of electrified areas for making the electrical energy available for more consumers. In order to increase the reliability of the electricity supply as well as constructing a unified infrastructure for the electric system, multi-electrified areas were connected together and had created the power system. Subsequently, the first electric power system was developed in the alternating current mode in 1886 in Great Barrington, Massachusetts [15]. The electric power system generally consisted of three main parts, that is, energy generation, transmission, and distribution.

The electricity generation part is responsible for energy production by using different types of energy sources and is usually located near the energy centers for easy access to the primary fuel with minimum fuel transmission costs. The earlier power production systems were developed based on the centralized generation system, which relied on central conventional fossil fuel-based energy production units. However, due to the disadvantages of the centralized systems, which created significant challenges and concerns for the power system, the idea of the decentralized power grids received considerable attention more than ever. Some of the important characteristics, applications, advantages, and disadvantages of the centralized and decentralized systems are depicted in Fig. 1.3 [16].

In the electric power system, transmission systems are used for bulky movement of the electricity from the power generation stations to the electric substa-

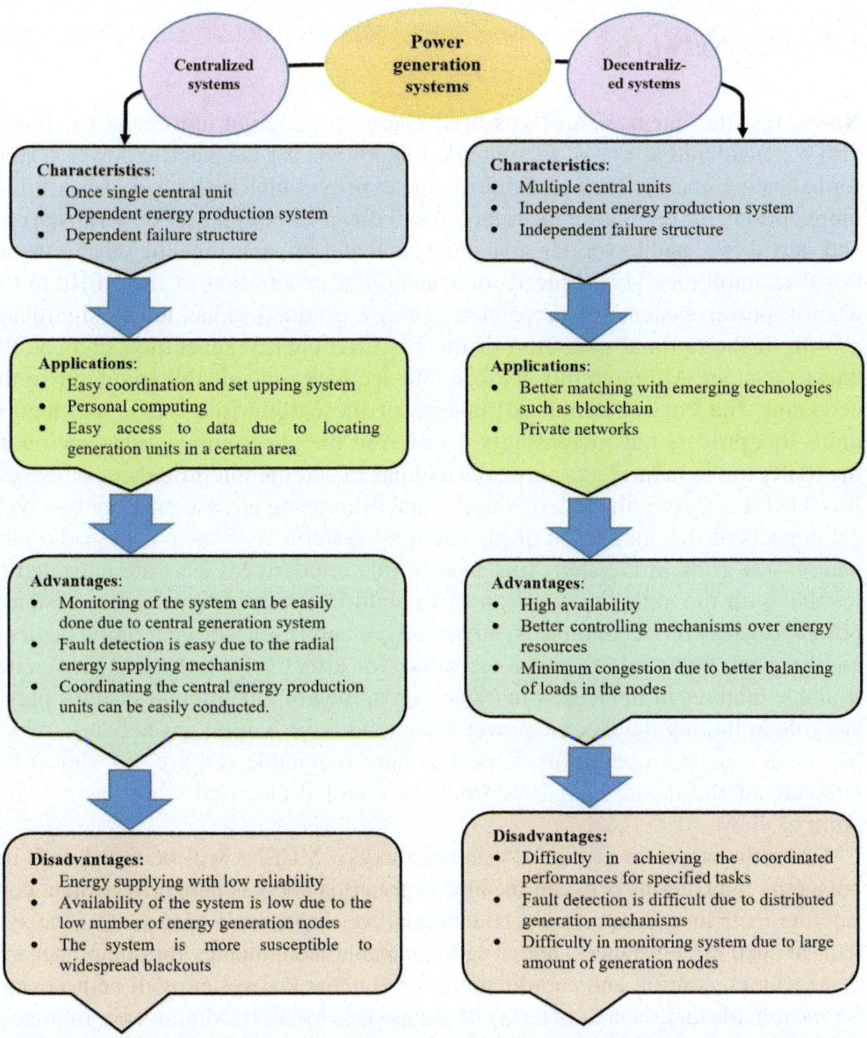

Fig. 1.3 Characteristics, applications, advantages, and disadvantages of centralized and decentralized systems [16]

tions in the distribution part. In this respect, transmission lines play the role of vital arteries for the transmission systems that have been traced across the power system. Finally, distribution systems are taken into account as the final step of electrification that delivers electrical energy from the distribution system to the different consumers.

Given the advantages of the decentralized energy production paradigm, future modern MCENs are preferred to be developed based on the distributed energy generation systems, especially clean energy production resources.

1.3 Gas Networks

Nowadays, the integration of gas-fired energy production units with the power grid is considered as one of the remarkable options for the electric power system for balancing energy. This is primarily due to providing a high level of flexibility, more options for regulating the ramp up and down levels, shorter time for start-up and start-down, and lower negative environmental impacts than the other conventional technologies [17]. Indeed, increasing the penetration of the RERs in the electric power system for more clean energy production has led to significant growth in the optimal operation of the gas-fired energy generation units as the backup system. Although this is taken into account as a valuable trend due to the economic and environmental advantages of the carbon-free energy production units, this process has increasingly transferred the uncertainties associated with the RERs to the natural gas networks and has raised the intermittences of the gas flow [18]. Increasing the risk of venting and decreasing gas pressure are two typical impacts of the integration of the gas-fired systems with the power grid on the natural gas grids in the short run. Due to this, modern MCENs are expected to involve with the automatic version of the multi-carrier energy storage systems, power-to-gas (PtG) technologies, advanced co- and tri-energy generation devices, and linepack potential of the gas networks for effectively mitigating the uncontrollable features of the RERs. In other words, the aforementioned systems play a key role in linking the electric power system with the natural gas network, which has made the interoperability between them inevitable. Figure 1.4 shows the structure of the modern MCENs from the coupled electricity and gas network point of view.

Given the structure in Fig. 1.4, future modern MCENs will be faced with the numerous interactions between the electricity grids and gas networks, which, considering their limitations, would be essential for the optimal operation of the system. In such circumstances, adopting the capable technologies for comprehensive management, control, and coordination of the integrated system will be necessary for the reliable and secure operation of the modern MCENs. Moreover, a great need is felt for the holistic models that can significantly facilitate the entrance of the new multi-carrier systems to the modern MCEN infrastructure with overall consideration of the technical, financial, and environmental aspects.

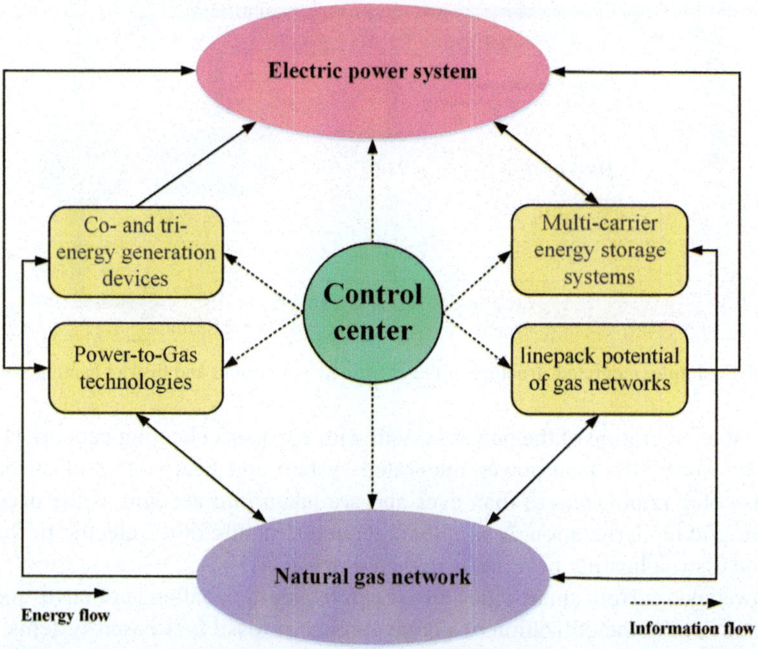

Fig. 1.4 The integrated structure of the electricity and natural gas networks in the future modern MCENs

1.4 Heat Networks

Nowadays, clean energy production continues to flourish and is forming a more cost-effective and sustainable electricity industry [19]. This is while future modern grids are targeted to be developed as sustainable infrastructures with a high or full share of the clean energy production units. However, new strategies, devices, and technologies are required to solve the unpredictable challenges of the RERs for maintaining a dynamic power balance in the system. The development of new high-performance heat pumps and electric boilers has created increasingly competitive conditions for electrical heating devices in terms of reducing environmental emissions as well as operating costs. The minimization of the operation costs and greenhouse gas emissions are usually taken into account as the main factors in optimizing the electric power system. Moreover, these factors are expected to play a key role in developing the modern energy grids in the future. In this respect, given that the heat storage is scalable and cheap, heating systems will play an active role as one of the cost-effective devices in the modern future MCENs with a high share of the RERs aiming to balance energy in real time [20]. On the other hand, the electric power system is affected by utilization of a large fraction of domestic electrical energy for heating energy production in vast areas, especially in the winters. These facts not only indicate the synergistic effects between electrical and heating energy but also

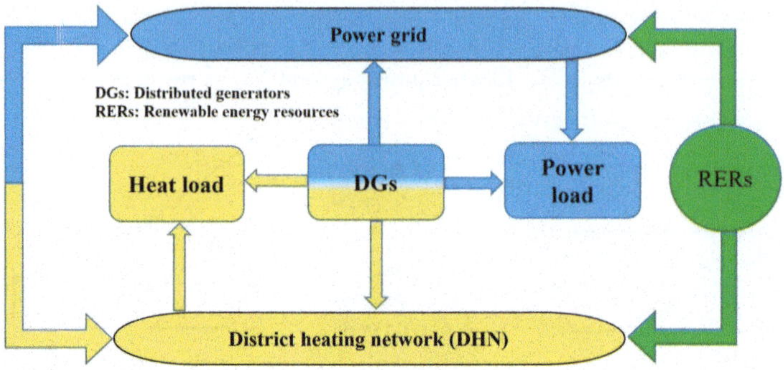

Fig. 1.5 A simple integrated structure of the electric power system and district heating network

inspire the integration of the power system with the district heating network (DHN). In other words, the heat-power integrated system and heat source electrification have become popular more than ever and are taken into account as the necessary step for grid modernization. A simple integrated structure of the electric power system and district heating network is depicted in Fig. 1.5.

Nowadays, current smart cities have been facing the challenge of environmental pollutions due to the utilization of a large number of fossil fuel-based systems. From the heating energy part of the modern MCENs' point of view, water and space heating systems that function basically depending on the fossil fuels are targeted to be replaced with electrical ones such as heat pumps and electric boilers in the future modern MCENs. Therefore, mushrooming of district heating networks (DHNs) along with the electrification of heating devices have structured the coupled multi-carrier energy distribution systems that harness different energy resources in the urban area.

1.5 Water Networks

With the development of convenience and socialization of human society, energy has become the foundation of human life in modern networks. This trend has sounded the alarm for the increasing requirements of water and energy more than ever in the modernization process of the MCENs. Regarding the appropriate response to the energy crisis, a growing interest is devoted to the different types of DERs due to their significant advantages. DERs can be operated close to the load centers to provide for the energy demand with lower power losses and improve the energy services to the power grid. The variety of DERs in type and size has created a flexible condition for the electric power system to adopt the appropriate ones considering different factors such as geographical conditions, and operation costs. Among the various types of DERs, the RERs are becoming even more attractive due to their positive effects in terms of economic and environmental aspects. However, the flexible and reliable operation of the power grid in the presence of high levels of

RERs has become one of the challengeable issues that future modern MCENs are facing. One of the effective ways to overcome such challenges is to use the capability of various energy structures and create effective interactions between them. In other words, converting different carriers of energy to each other can significantly increase the flexibility of the MCENs as well as facilitate the reliable energy supply. In addition to the natural gas and the DHN networks, the incorporation between the water distribution networks and the electric power system is also considered in the development plans of the future modern MCENs. Indeed, the developed plans regarding the future MCENs are not limited to the projects whose main players are the PV panels and wind turbines. This is while several other sources such as underwater currents and sea waves are taken into account for increasing the synergies among the water and electrical networks [21]. The water supply systems are seamlessly distributed throughout the territory, and the continuous water supply in this system directly depends on the availability of the electricity network for operating the water-electric-based devices in the water distribution networks. This integration not only is necessary for the water networks but also it has provided a great opportunity for recovering the electrical energy simultaneously with the water-electric hybrid systems. One of these systems is the electrolyzer that uses electrical energy for splitting the water molecule into hydrogen and oxygen. The produced hydrogen can be stored in the hydrogen storage for later use. Here, the fuel cell is another key device that uses the oxygen and discharged hydrogen from the storage as the entrance materials for generating electrical energy and water in the incorporated system. This evolution in the water-electric hybrid systems, as well as a mutation in the interactions between water and electrical networks, inspires this undeniable fact that we need to have a comprehensive and unified view of the hybrid networks to create effective interoperability between them in optimal modeling of the future modern MCENs. Therefore, the grid modernization process will seriously contain the effective interactions between the water and electricity distribution networks through the optimal operation of the water-electric hybrid systems.

1.6 Multi-Carrier Energy Networks (MCENs)

Traditionally, fossil fuels are considered as the dominant primary resource in energy networks [22]. As the energy mix is changed by leveraging the synergies among various energy carriers, the different energy structures, such as the electric power system, natural gas grid, and the DHN, that were traditionally operated independently in the past have become increasingly dependent at present [23]. In this regard, the progress in the cutting-edge technologies in energy generation, energy storage systems, or multi-carrier energy conversion devices has led to the appearing of other energy vectors, such as methanol, syngas, and hydrogen, that are expected to play a special role in the interactions of the future grids. This is a reflection of our strives in moving toward a low-carbon and more sustainable future modern energy system. In addition to the advanced technologies and hybrid systems, a much greater variety

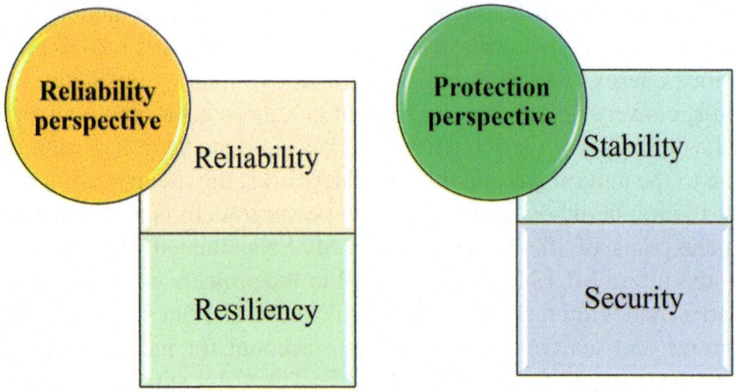

Fig. 1.6 Some key factors in the evaluation of future modern MCENs

of primary energy sources such as biomass, water, and solar is expected to be used in the new structure of the energy networks. All of this signifies the need for coordinated operation of several energy networks interdependently. In this book, we refer to this type of integral operation of the multi-energy networks as the MCENs.

Typically, in recent years, several studies have been focused on improving the key factors of energy networks such as reliability, resiliency, stability, and security without completely considering their effects on each other. However, investigations in the previous sections state this undeniable fact that we need to focus on the integrated version of them to achieve the near realistic results. When we scrutinize the different features of MCENs such as electricity, natural gas, the DHN, and water grids separately, we use certain factors and indexes in the system evaluation. However, our criteria will certainly require a significant overhaul for analyzing each of the aforementioned networks in the framework of the MCENs. Therefore, a great need is felt for assessing the key factors of the system in the scope of the future modern MCENs. In this assessment, given the different types of the DERs along with the various hybrid energy converting systems that are targeted to be in lieu of the traditional energy systems, the lack of complete and exact evaluations as well as the advanced technologies are a major obstacle in the evolvement of goals of the future modern MCENs. This issue has motivated us to fully investigate the status of the future modern MCENs from the viewpoints of some effective factors including reliability resiliency, stability, and security, which are depicted in Fig. 1.6. This investigation can give us a useful overview of the future MCEN structure, enabling us to adopt appropriate strategies in realizing modern society.

1.6.1 Reliability and Resiliency

The electric power system is going toward the realization of the green and sustainable society through equipping with high levels of RERs, involving various advanced hybrid systems, and integrating with different energy networks [24]. In this field,

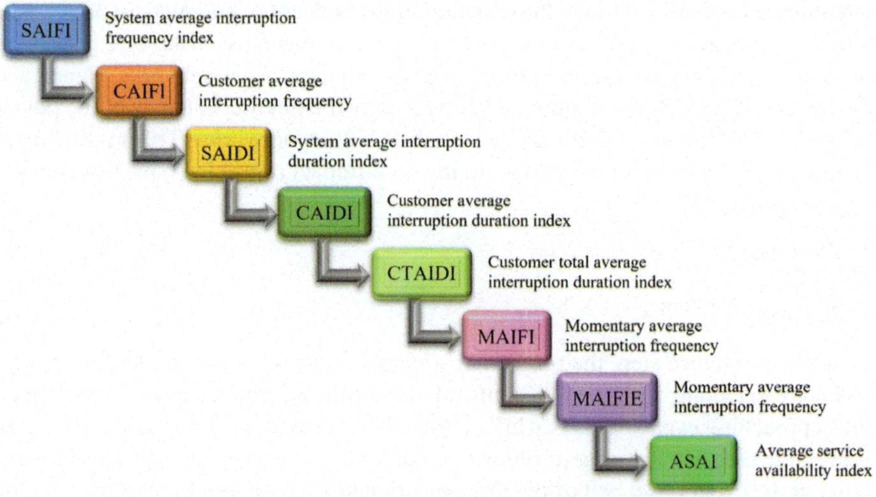

Fig. 1.7 Some key indexes for evaluating the reliability of the energy networks

the direction of all strives is going to identify all the existing challenges, analyze them professionally, and propose practical solutions for solving the existing obstacles. One of the critical challenges is the reliability of the system after integrating several networks. The reliability of the energy networks is one of the key indexes that should be provided for ensuring the continuous energy supply. Indeed, reliability is defined as "the ability of the system to satisfy the customer demand with acceptable quality" [25, p. 1]. There are many indexes for analyzing the reliability of the system, among which an important one is the Energy Not Supplied (ENS). The ENS is the difference between the amount of total energy consumption and the amount of energy generation in the system. The optimal scheduling of the system seeks to keep the amount of this index zero every time. Indeed, zeroing the amount of this index each time ensures continuous energy supply in the network. In addition to this index, some other important indexes are also introduced for evaluating the reliability of a certain system, which are demonstrated in Fig. 1.7. More information regarding these indexes such as their descriptions and formulations can be fully accessed in Ref. [26].

In the systems with a high or full share of the RERs, satisfying the reliability of the system has become one of the challengeable issues. The high level of the intermittences in the system treats the reliable energy supply in the absence of the controllable devices and advanced energy management schemes. Therefore, operating the controllable devices along with the clean energy production units can improve the reliability of the system. Also, applying advanced controlling and management technologies is expected for avoiding the blackouts that may occur due to different faults in the system.

In recent years, extreme weather events, such as tsunamis and hurricanes, and dangerous earthquakes have increasingly affected energy networks worldwide [27]. The catastrophic consequences of such problematic rare events have attracted global attention, which has led to the introduction of the concept of energy networks' resilience. The undeniable effects of such weather events can have an adverse impact on different systems in the electric power systems and natural gas networks. Hence, the resilience of an MCEN is significantly affected by the interdependence of the power grid, the DHN, natural gas, and water distribution networks. The multi-carrier energy networks' temporal response to the high-impact rare event typically consists of three steps [28]:

- Avoidance.
- Survival.
- Recovery.

In the avoidance step, the predictive system uses the potential of the forecasting technologies along with the other information in the system database to anticipate the approaching events. This is while it proactively provides the operations-oriented preventive measures aiming to remove, moderate, or postpone the system vulnerability at the event onset. All of the three mentioned steps should be seriously scrutinized for equipping the system with appropriate controlling devices for improving the resiliency of the hybrid network. In this field, recent studies have emphasized the importance of incorporated operation of the MCENs in improving the resiliency and reliability of the system [29]. Therefore, the coordinated operation of the future modern MCENs not only is essential for improving the two mentioned vital factors but also this structure needs to integrate with capable technologies to effectively control, manage, and coordinate the different hybrid components to achieve such goals.

1.6.2 Security and Stability

Regarding the MCENs, two other key factors that should be provided for the future modern MCENs are security and stability. In the traditional radial systems, the power only is supplied from the conventional power plants and the direction of power flow is from these power production centers to the consumers. However, this rule is not true for the modern systems that are engaged with numerous DERs. In these structures, we have a bi-directional power flow in the system that indicates the need for a system for equipping with suitable cyber-physical systems, relays, and controlling systems. Therefore, to realize the security in the modern hybrid networks, the advanced bi-directional relays along with the cyber-physical systems and other protection systems should be considered to enhance the ability of the system for detecting the different types of the fault while having quick interactions with the control center to take appropriate precautionary measures to prevent any load shedding in the system.

Although the integration of the large number of RERs in the future modern MCENs has substantial advantages in terms of economic and environmental aspects, their fluctuations have negative impacts on the stability of these networks. When the system is targeted to be equipped with high or full share of clean energy systems, the level of the uncertainties will certainly be high. In this circumstance, we may see the great ramps every time the control center has faced with a challenge of dynamic energy balance. In this respect, one of the main motivations for incorporating various types of energy networks is to enhance the stability of the system. Because the future modern MCENs are intended to be equipped with various types of DERs, this issue can give multiple options for the system to adopt the optimal strategy to keep the stability of the system at each status of the energy networks.

1.7 Smart MCENs

Currently, the demand for different types of energy as a popular tool is ever-growing and especially for electrical energy even faster [30]. Given the technical and societal developments, energy generation from the RERs for fulfilling the energy load is inevitable. As discussed in the previous sections, one of the main challenges regarding the RERs is that the volatility of their outputs is unpredictable. In this regard, smart grids are proposed as a potential solution for maintaining the stability and reliability of the energy supply through appropriate integration of the different energy sources in the MCENs. One of the effective strategies here is that smart grids are developed in a way to focus on the local energy generation and consumption considering the creation of multiple paths to meet the energy load. Because of employing the decentralized energy production protocol in the smart grids, the MCENs are targeted to be smart more than ever and structured based on the smart grid protocols. This type of structure enables the hybrid system to be more flexible and provide more options for future possible developments. In today's world, we are witnessing the increasing advancement in various technologies and the emergence of new versions of them to fulfill every day's need of the society. All of these technologies will require a proper platform for implementation. The smart platform will allow the MCENs to be involved with the capable technologies in all parts of the system to make life easier than before for the people. In addition to the aforementioned advantages, another significant motivation for moving from the traditional structure of the energy networks to the smart version of them is that deploying the smart grids' technologies and devices substantially improves the operational key factors, that is, the system security, reliability, efficiency, and resilience [31].

1.8 Why Modernization?

Recently, we can more see the "Modern" word in recent studies regarding the power grid, but this may raise an important question in our minds as to why modernization of the energy network, or what the grid modernization is for. To respond to the aforementioned questions along with the similar other ambiguities, the US

Department of Energy's (DOE) Grid Modernization Laboratory Consortium completed a study on how the distribution system planning issue is addressed by the utility regulators regarding the grid modernization and higher levels of DERs in December 2017 [13]. The Grid Modernization initiative strives across the DOE to solve the existing barriers in the way of modernizing future energy networks. The set of these efforts focuses on developing the conceptual requirements, relevant tools, and innovative technologies aiming to better predict, analyze, measure, control, and protect the system for providing the institutional conditions that enable the system for widespread adoption of various techniques and tools for rapid development. For this aim, several groups are formed by the DOE to pursue the special research trends in the context of RERs' integration, advanced energy storage systems, test procedures and the related standards, and a number of other significant issues in the grid modernization process. In this respect, the grid modernization aspects are not limited to the issues that are introduced in this book or are described in the previous works. All of the topics related to grid modernization are developing and will be certainly developed day by day. Now, we introduce the current motivations for the grid modernization or in some cases forecast future interests for this work. However, we certainly cannot speak for sure about the future and the expectations of that time. Thereby, in order to effectively respond to this question as to why we need grid modernization, we refer to our database and use the results of the conducted researches in this field and the relevant predictions made. In this respect, a clear vision of concepts, statements, definitions, objectives, motivations, and goals are important for appropriately designing the grid modernization schemes. Until recently, research institutions have provided several definitions for grid modernization. One of the most effective and comprehensive of them is related to the California Public Utilities Commission (CPUC) that has defined the grid modernization in its recent decision: "A modern grid allows for the integration of DERs while maintaining and improving safety and reliability. A modern grid facilitates the efficient integration of DERs into all stages of electric power system planning and operations to fully utilize the capabilities that the resources offer, without undue cost or delay, allowing markets and customers to more fully realize the value of the resources, to the extent cost-effective to ratepayers, while ensuring equitable access to the benefits of DERs. A modern grid achieves safety and reliability of the grid through technology innovation to the extent that it is cost-effective to ratepayers relative to other legacy investments of a less modern character" [13, p. 4.1]. Given this definition, grid modernization efforts have been structured based on the several key factors illustrated in Fig. 1.8.

1.9 Modernization of the MCENs

Nowadays, the necessity for integrated energy networks is felt more than ever. The justifications and advantages of the effective integration of the electric power system with the DHN, natural gas grids, and water distribution system have been discussed in the previous sections. Considering the advantages of this coupled

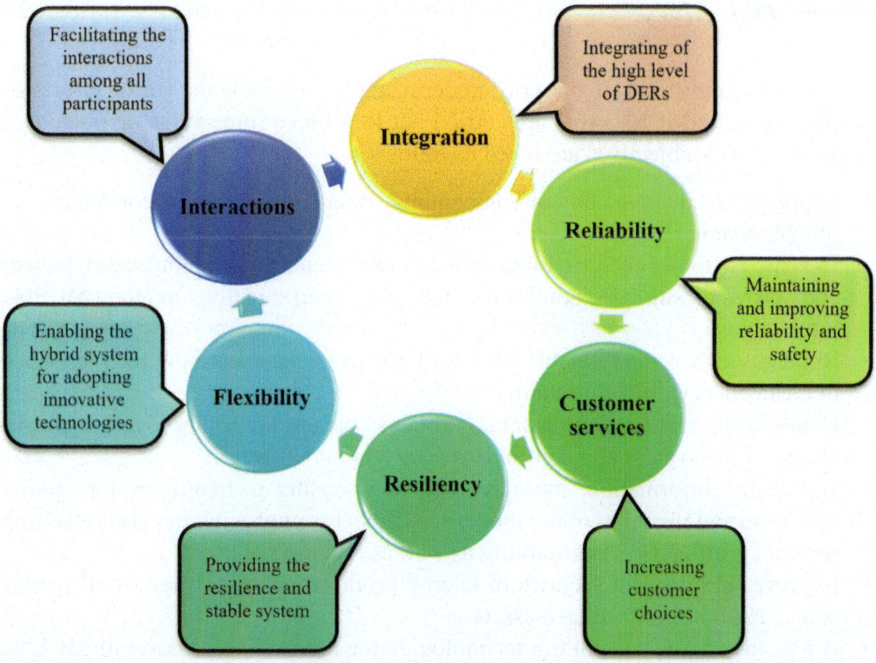

Fig. 1.8 Several key factors in the grid modernization process

infrastructure and modern grids reminds us that the grid modernization activities are not limited to some efforts for modernizing the power grid. This issue motivates us to expand the researches in the field of grid modernization to the MCENs. In other words, we strive to develop the concepts, models, goals, structures, technologies, and other basic requirements to broaden our activities by moving from the grid modernization to the MCEN modernization. This transition and mutation will probably multiply the activity area and increase our responsibilities to:

- Consider all the expectations and factors needed to increase people's well-being.
- Intend the restrictions of several energy networks together instead of a special network.
- Propose and model the hybrid schemes instead of one vector energy plans.
- Develop the current technologies tailored to MCENs' protocols.
- Increase the system reliability, resiliency, stability, security, and efficiency by adopting cutting-edge technologies engaged with the information technologies and advanced platforms.
- Provide adequate privacy for all users in the modern area.
- Facilitate the presence of all participants with different styles, interests, ideas, expectations, needs, beliefs, and attitudes in the modern MCENs' interactions.
- Prepare a suitable platform for the emergence of intelligent devices, hybrid controllable systems, user-friendly equipment, etc. throughout the future modern grids.

All of this compels us to have more clarity in the concepts regarding the MCENs' modernization such as objectives, challenges, opportunities, and principles.

1.9.1 Objectives

In this book, the mean of the grid modernization process is the set of efforts for modernizing the MCENs that are carried out to achieve some valuable objectives. Some of the key objectives are listed as follows:

- Supply cost-effective and environmentally friendly energy to all consumers in a reliable manner.
- Increase customer choices, improve their convenience by adopting smart devices, and provide suitable conditions for their participation in the MCENs' interactions.
- Build sustainable and flexible MCENs for effectively integrating the high levels of clean energy production units.
- Demonstrate and develop novel control systems aiming to improve the observability of the system and control it at a very high resolution.
- Utilize the information, measurement, and sensing technologies for optimal operation and planning of the modern MCENs for improving the grid reliability, resiliency, efficiency, and stability as well as reducing outages.
- Improve the physical security of energy production, the privacy of all participants, and other grid-related assets.
- Adopt the expert controlling technologies for monitoring the overall MCENs' interactions.
- Develop the system platform for adaptation to the emerging cycles such as food-energy-water nexus.
- Implement automated maintenance and operation systems as well as the self-healing paradigm.
- Develop the comprehensive architecture with several control centers aiming to simultaneously monitor the status of the electric power system, natural gas network, the DHN, and water distribution system to adopt the optimal strategies.
- Develop a holistic reconfigurable system capable of dealing with any changes in modern MCENs.

Certainly, the objectives of grid modernization are not limited to these, but they are being updated day by day as the relevant researches complete.

1.9.2 Challenges

In today's world, we all know that in the path of any great project, there are certainly challenges that can not only make us greedy for the ultimate goal but also make us more mature than before in this complex and difficult way. The MCENs' modernization issue is one of the greatest and complex plans in the energy world that faces with remarkable challenges at every step of it. In the face of these challenges, we can either become cold about the progress of the project or turn each of them into an incentive to continue. Recent substantial advances in the context of all technologies motivate us to think about the second option among the two aforementioned ways to continue. For this aim, we need to be aware of all these challenges to propose practical solutions for overcoming them. Some of these challenges are as follows:

- How to manage and control this great transition and evolve operations, planning, and marketing through subsequent infrastructure investment strategies and informed institutional decision-making that is both future-looking and prudent.
- How to provide the complete system to be sufficient, effectively address the future modern grid requirements, and rapidly foster and efficiently recover the large disruptions.
- How to handle the complexity of the large-scale hybrid system, the computational burden of the expert computing and control systems, and different kinds of computational language related to the various intelligent systems.
- How to develop a holistic platform for adopting emerging technologies with different requirements and a variety of applications.
- How to simultaneously monitor the electric power system, the natural gas grid, the DHN, and water distribution systems with different devices, hybrid systems, several control centers, and considerable restrictions in an optimal manner.
- How to integrate the different energy networks with each other considering the several players, market structure owners with different opinions and expectations, energy consumers with different consumption patterns, governments with different political policies, and regions with different geographical conditions.
- How to persuade the consumers, producers, and stakeholders with different requirements to follow the modern MCEN protocols and participate in the modern grid's interactions, energy management programs, and smartization schemes.

1.9.3 Opportunities

In addition to the challenges of modern grids, this great process creates valuable opportunities for human society through evolving and transforming the energy world. Indeed, MCENs' modernization provides opportunities to improve all the key factors of our national hybrid energy infrastructure in the face of evolving environmental threats and increasingly complex systems. Some of the special opportunities regarding the MCENs' modernization are summarized in Fig. 1.9.

1.9.4 Principles

Clearly, any plan for successful progress and implementation requires specific principles that are designed based on the objectives and rules of the relevant system. These principles guide all participants and network players to achieve specific goals and unite them to adhere to the approved rules. For future modern grids, also, some principles should be defined to effectively highlight the roadmap boundaries. These

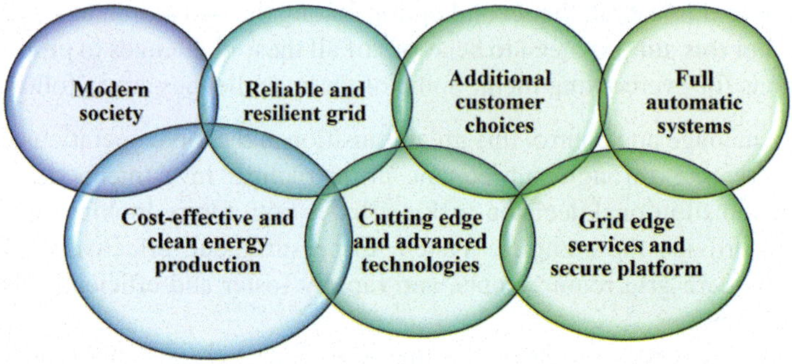

Fig. 1.9 Several valuable opportunities for the future modern MCENs

principles are designed based on the future modern MCENs' requirements, objectives, barriers, systems, technologies, active and passive players, and different structures. Some of these effective principles are listed below:

- Modern MCENs should provide the coordinated operation of the power system, gas grid, the DHN, and water network in a sustainable and reliable manner.
- Modern MCENs should maintain and improve the system reliability, efficiency, stability, and safety in the presence of unpredictable changes in any location of the integrated hybrid system.
- Modern MCENs should adopt the innovative technologies to provide the full clean energy production without concerning about their high level of intermittences by implementing capable technologies for establishing a dynamic energy balance.
- Modern MCENs should carry some special features including being auditable, scalable, observable, reliable, sustainable, efficient, adaptable, flexible, extensible, and reconfigurable in practice, not just in descriptions on different papers.
- Modern MCENs should provide an appropriate condition for the participation of all energy shareholders, producers, customers, and consumers in the energy market interactions.
- Modern MCENs should establish efficient interoperability among the hybrid energy network partners and leverage the synergies in the MCENs.
- Modern MCENs should meet the energy demands of all consumers every time, without load shedding, economically and environmentally.
- Modern MCENs should make people's lives comfortable, human society cleaner, and energy systems more integrated than ever before.

1.9.5 Framework

In the modern MCENs' area, all smart devices and innovative technologies need to be compatible with deployment in a unique structure. This is while data analytics models and system assessment schemes should be posed based on multi-variate and multi-modal methodologies capable of incorporating disparate and disperse data sets under the tent of a unified framework aiming to successfully identify incipient defeats. Therefore, the MCENs' modernization framework should be able to identify technology resource classifications, generalize their specific applications, and clearly specify methodologies for the widespread operation of the hybrid system. Given these items along with the other aspects of the MCENs' modernization presented in the previous sections, we define the MCENs' modernization framework as: *an integrated infrastructure of control, economic, and environmental mechanisms that allows the hybrid system to easily adopt multidisciplinary innovative technologies for maintaining system reliability and stability, improving system efficiency and safety, increasing system flexibility and privacy, enhancing the grid interoperability, providing grid edge services, adopting various types of clean energy production units, facilitating for effective interactions, and enabling the system for possible developments across the entire MCENs' infrastructure.* The scope of the future modern MCENs does not certainly limit to this framework and it will be and is expected to be developed in parallel with technology growth in the future.

1.10 Various Structures for MCENs

The structure of the future modern MCENs will certainly be not monotonous similar to the traditional power grid. It is expected to be a combination of the different structures based on their features and conditions. Here, three special structures are introduced for the future modern MCENs in which each of them has different characteristics and plays a key role in developing the future modern MCENs' plans.

1.10.1 Microgrids

Generally, microgrids are introduced to streamline the control, management, operation, and effective incorporation of the wide range of various types of DERs to supply multi-carrier energy loads in the proximity of the customer side in the distribution network [32]. Microgrids can be operated in the islanded and grid-connected modes. These two modes of operation allow the microgrids to be more flexible than before in continuous energy supply as well as maintaining the stability of the system. Microgrids as an alternative energy-producing system and resilience source in lieu of traditional energy generation units are trusted to meet energy demand with higher reliability and efficiency, improve resiliency and power quality, substantially decrease the power losses, and produce a high level of clean energy by employing state-of-the-art technologies for integrating the numerous RERs. On the other hand, in today's world, continuous and uninterrupted energy supply is a necessary issue for the sustainable development and well-being of the nation [33]. In this regard, smart electrification is recognized as one of the main

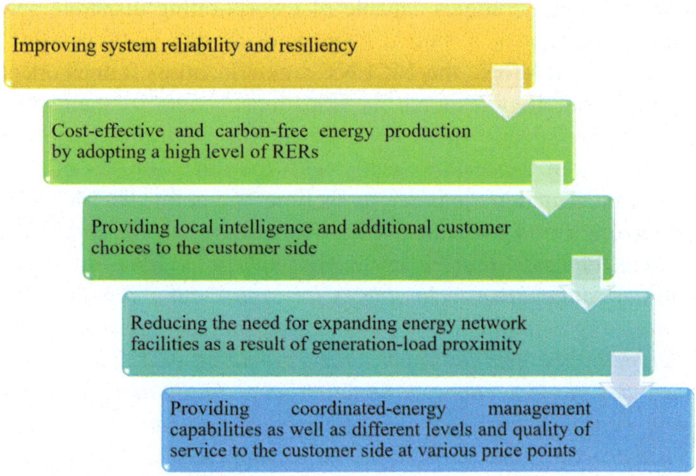

Fig. 1.10 Some key benefits of the microgrids for the future modern MCENs

steps for grid modernization and plays a critical role in developing a system to be advanced in terms of economic, operation, and reliable energy delivery [34]. For this aim, a decentralized energy supply system is essential, which provides a suitable platform for mushrooming of the several types of DERs throughout the energy networks. By proliferating of the decentralized energy resources around the world, microgrids are considered as an appropriate means for local energy generation and supply in the MCENs. In other words, microgrids are taken into account as the cluster of DERs that facilitates the optimal integration of the DERs and provides significant benefits for the hybrid system. Some of these benefits are demonstrated in Fig. 1.10 but are not limited to these items.

The substantial advantages of the microgrids have made them an effective option for developing the future modern MCENs. However, besides these beneficial characteristics of microgrids, their high investment costs have caused significant obstacles in evolving the capable technologies, which need thorough inspections for ensuring a reasonable time of return on investment. Nevertheless, they are still the main choice for the development of the modern structures for the multi-carrier energy systems. To effectively manage the energy demands in the future modern MCENs, the operation of microgrids as the attractive structures is almost mandatory for implementing a more flexible, informative, configurable, and physical hybrid system.

1.10.2 Prosumers

One of the important elements of modern grids is prosumers. A prosumer is a consumer who can produce and share the excessive energy to the energy network and other participants [35]. The prosumer term is derived from the combination of production and consumption indicating that each energy consumer can be a producer at

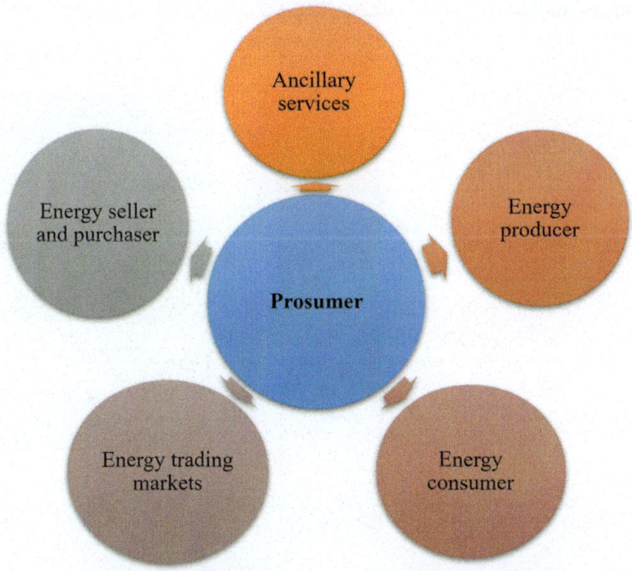

Fig. 1.11 The relationships of each prosumer in the future modern MCENs

certain times. Prosumers are recognized as one of the key stakeholders of the future modern MCENs, who have a vital role in implementing various energy management schemes. Generally, optimization techniques, as well as communication and information technologies, are taken into account as two key elements that are involved in the energy interactions among the prosumers. The role of the prosumer in the energy network's interactions is illustrated in Fig. 1.11.

According to Fig. 1.11, each prosumer plays a critical role in the energy interactions of the future modern MCENs. It can have effective negotiations with the local control center to participate in the energy market interactions with the aim of better monitoring the energy production and consumption process as well as improving the status of the hybrid network to be reliable and stable in the presence of the variety of events and natural disasters. Therefore, an effective potential of the prosumers can be engaged in the infrastructure of the future modern MCENs to allow the hybrid system to be more flexible than before and provide more options for the energy control centers for adopting optimal strategies in the system with a high or full penetration of the RERs.

1.10.3 Multi-Carrier Energy Hubs

In the field of multi-carrier energy systems, the energy hubs have been recognized as a new paradigm for future modern MCENs. The energy hub plays a strategic role in energy generating, converting, receiving, and storing as a coupled structure with a variety of components such as boilers, multi-carrier energy storages, combined

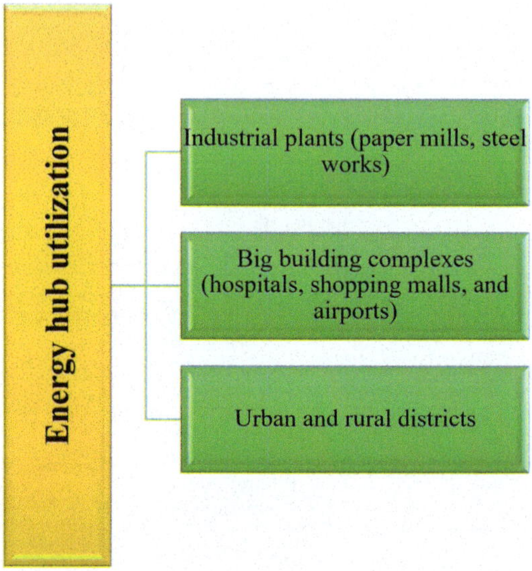

Fig. 1.12 Some examples of utilizing energy hubs in the real world

heat and power units (CHPs), power-electronic components, transformers, etc. [36]. Each hub is empowered to the capability of receiving and generating various types of energy carrier from the hybrid networks such as electricity, natural gas, the DHN, as well as water networks and storing them in the corresponding storages, converting them to each other by the related device, supplying them to the relevant consumers, and delivering them to the targeted other grids. Indeed, the key idea for energy hub is related to its vital role in linking the different energy structures with each other using the multi-carrier energy systems in an optimal manner. Some examples of utilizing energy hubs in the real world are illustrated in Fig. 1.12, which are not limited to these cases.

The redundancy in connections among the energy hubs' inputs and outputs empowers them to increase the flexibility of the MCENs. Increasing the flexibility of the system proposes new ways for energy supply, which can effectively improve the system condition in reliably delivering various vectors of energy to the customer part. Moreover, the system with a high amount of flexibility can be easily scheduled for different optimization goals in the presence of the hybrid components. These factors are only some considerable benefits of the energy hubs that incentivize their extension of utilization in the wide area of the future modern MCENs.

1.11 Energy Markets

Over the last decades, the energy industry is overhauled by transiting from the conventional centralized system to the decentralized competitive one in many developed countries around the world [37]. In 1996, a legal framework was enacted by

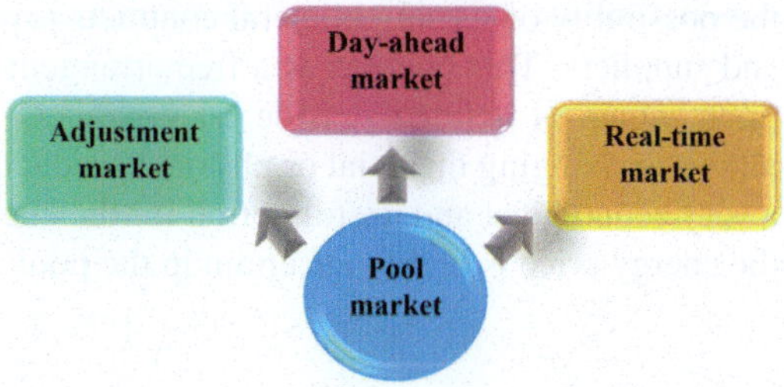

Fig. 1.13 Pool energy market organization

the Federal Energy Regulatory Commission (FERC) for promoting open access to transmission grids with the aim of increasing the competition in the US wholesale electricity markets [38]. Indeed, the energy markets are developed by proposing different competition frameworks to enhance the energy networks' efficiency as well as guaranteeing a satisfactory quality of the energy supply considering economic and environmental issues [39]. On the other hand, energy markets can regulate the effective incentives for capital forming and motivating different market players to participate in the market interactions for empowering it to be flexible in the various statuses of the grid.

Regarding energy markets, two typical energy-sharing arenas are developed aiming to facilitate energy commerce among the energy consumers and producers, which are called futures and pool markets. The energy is exchanged on a short-term basis in the pool market. Typically, this marketplace consists of three markets as shown in Fig. 1.13.

In the pool market, the bulk of energy transactions is generally covered in the day-ahead (DA) and adjustment markets within a day. In this regard, the organization of the adjustment markets is similar to the DA markets whereas they are cleared close to energy delivery in real-time (RT) and shorter energy-sharing horizon can be covered by them. In the pool market, in other words, the energy exchanging is mostly negotiated in the DA market. This is while the effective role of the adjustment markets is employed for making the required adjustments on the energy cleared in the DA market. In this respect, to ensure a dynamic energy balance in the system, the last-minute energy adjustments are considered in the RT balancing market. In the DA and adjustment markets, the energy blocks along with their corresponding minimum selling prices are generally submitted by the energy producers to the energy market for every hour of the market horizon. At the same time and on another side of the market, the energy blocks along with their corresponding maximum purchasing prices are submitted by the consumers and retailers to the market. After receiving the bids from consumers and offers of the producers, the market operator collects them and clears the energy market using a market-clearing procedure.

In the pool market, a futures market is an "auction market in which participants buy and sell physical or financial products for delivery on a specified future date" [38, p. 9]. In this market, the market structure allows energy exchanging on an intermediate- or long-term horizon through sales and procurements of standard products, which are called derivative or derivatives products [38]. In the futures market, the possibility of signing bilateral contracts is provided between energy consumers and suppliers. This contract is a free arrangement between a consumer and a supplier, which is defined outside an organized marketplace. Because this market allows exchanging financial or physical products in the future at today's prices, it is a very economical and useful option for the energy producers and consumers when the energy price is highly uncertain in the pool.

1.11.1 Reserve and Regulation Markets

Energy markets are developed in a way to be multi-commodity markets. These markets include at least four main products, which are depicted in Fig. 1.14. Among the four products illustrated in this figure, energy is taken into account as the main product of the energy market.

Although the energy is intended as the main product of the energy markets, the reserve is another key product for ensuring continuous energy supply by guaranteeing the availability of enough backup generation in case of sudden demand changes, drastic fluctuations of the stochastic producers, and equipment failures. The reserve market is targeted to be cleared either immediately following the DA market by the independent system operator (ISO) or jointly with this market. Indeed, an auction algorithm is used for clearing this market in which its complexity varies from market to market.

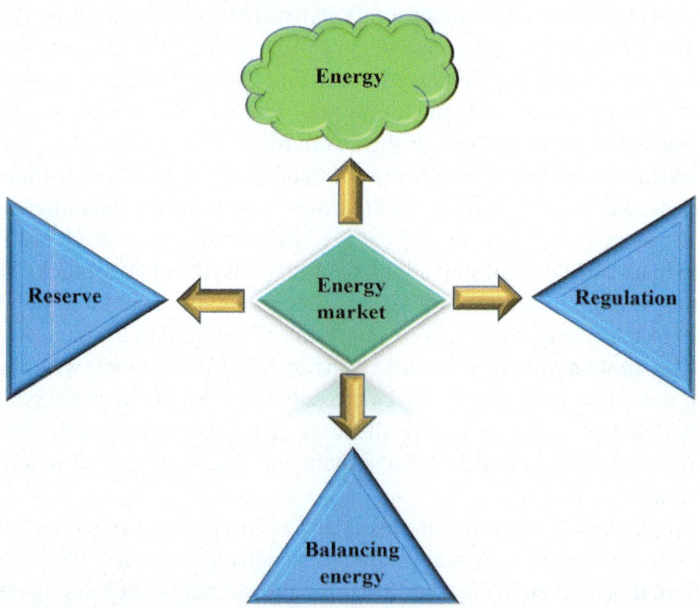

Fig. 1.14 Four main products of the energy market

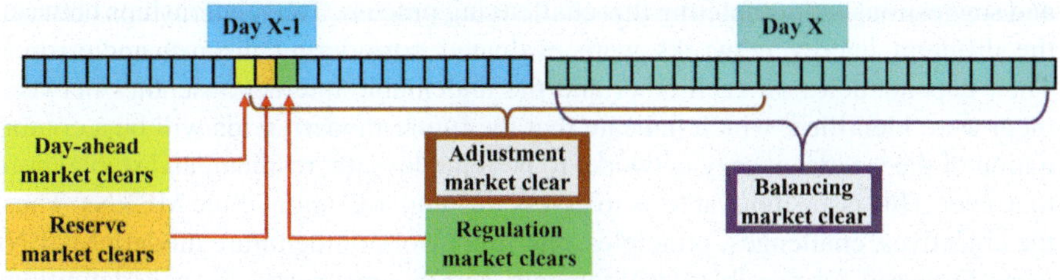

Fig. 1.15 Clearing sequence for the different daily energy markets

One of the energy markets that is structured to be cleared several hours prior to power delivery is the regulation market. One of the important responsibilities of this market is to allocate the load following bands between energy generation units that are interested and have the required capability to provide this service. Energy bands are devoted based on an auction, which follows an increasing price rule until attaining the sufficient regulating energy [38]. In the energy market, regulation and reserve markets are generally cleared once the DA market is closed. The clearing sequence for the different daily energy markets is shown in Fig. 1.15 [38].

1.11.2 Look-Ahead Market

The look-ahead energy market is an hour-ahead energy market [40], which considers the overview status of the system on the next day to effectively adjust the controllable devices in the current day. Indeed, this market allows the decision-maker to forecast some key factors such as weather conditions using the related techniques to use this information for optimally scheduling the different controllable devices such as energy storage systems in a way the system has an appropriate condition at the end of the current day to properly face with the new condition of the next day. For example, the charging and discharging of the energy storage systems can be consciously scheduled in the current day by considering the information of the next day status of the system to make the storage system more ready than before to participate in system interactions of the next day. This market can enable the future modern MCENs to act more optimally than before by automatically adjusting the controllable devices considering the probable status of the hybrid system on the next day.

1.12 Summary

This chapter is targeted to give a comprehensive overview of the grid modernization and smart grids. The features of the smart grids, as well as their superiority over current traditional networks, were investigated in detail. This was done for exactly analyzing why we need the grid modernization and what are our solutions and suggestions for completing this challenging process. The relationships between the different energy networks were evaluated considering their characteristics. Their dependencies on each other and the undeniable interdependencies between them were identified, which indicate that the future modern grids will be a combination of the several energy networks to be reliable, safe, resilient, and stable more than ever. This issue motivated us to assess the modernization of the MCENs. Thus, the objectives, challenges, principles, and framework of the future modern MCENs were discussed for possible developments, and in some cases, the new definitions were provided especially for the framework of this hybrid network. The different structures for the future modern MCENs were presented by investigating their effective roles in the new infrastructure of the system. At the end of the chapter, various energy markets were fully described and their effectiveness in the future modern grids' interactions was also analyzed in detail.

1.13 Questions

After careful reading and evaluating this chapter, we are expected to be able to answer the following questions.

1. Why do the energy networks need to transition from the traditional structure to the modern ones?
2. What are the significant features of smart grids that distinguish them from traditional networks?
3. What are the challenges facing smartization of current networks and to what extent can they be solved?
4. What are the main characteristics of the electric power system, natural gas grids, the DHN, and water distribution networks?
5. What are the impacts of the energy networks' key factors on the modernizing of the future MCENs?
6. Modernization should come to solve which problems and challenges of the current networks?
7. What are the key factors in the MCENs' modernization process that we need to consider in our researches and evaluations?
8. Which items are expected to be realized after MCENs' modernization and how are they supposed to be realized?
9. By considering what criteria, the MCENs' modernization objectives are designed, and what are those objectives?
10. How can we identify the challenges facing the MCENs' modernization, and what are those challenges?
11. Modernizing the MCENs creates what opportunities for the human society?
12. What is the role of principles in the progress of a certain project and what are these principles for the future modern MCENs?
13. Why do we need to define a clear framework for each plan and what is the framework of the future modern MCENs?

14. What effect does the use of different grid structures have on the process of modernizing MCENs?
15. Why microgrids are taken into account as the main candidate for developing the future modern MCENs?
16. Why the electrical and natural gas networks should be coupled in the future modern MCENs?
17. What are the limitations of the optimal integration of the electric power system and natural gas grids in modern grids?
18. What are the requirements for reliable and secure integration of the electricity and gas networks?
19. How can one prosumer play an effective role in the future modern MCENs' interactions?
20. How does the development of energy markets help to modernize energy networks and what are the different types of these markets?

1.14 Suggestions

In this section, some suggestions are provided related to this chapter that can give a good overview of some substantial issues in the future modern MCENs.

1. Employing the cutting-edge technologies in co- and tri-energy generation will be inevitable for reliable, secure, and stable operation of the future modern MCENs.
2. Developing the characteristics of the smart grids will be an introduction to implementing future modern MCENs.
3. Developing the new energy market structures will be very effective in creating a fair competitive environment with different participants.
4. Providing appropriate conditions for the presence of strong contender technologies in establishing a dynamic energy balance will be useful in the presence of the high or full share of the RERs.
5. Developing the technologies of the mediator systems will be an effective step in the optimal integration of the different energy networks in the future modern MCENs.

1.15 Future Trends and Discussion Topics

Here, future trends and some discussion topics are proposed for the interested readers regarding this chapter that can be useful for developing the MCENs' modernization process.

1. Analyzing the effects of applying the cutting-edge technologies in co- and tri-energy generation of the future modern MCENs.
2. Exactly evaluating the challenges of smart grids.
3. Investigating the new opportunities for the modernization of the future MCENs.

4. Focusing on the challenges in the context of the optimal integration of the electrical, natural gas, the DHN, and water distribution systems.
5. Assessing the required information and communication platform for handling the communications in the modern MCENs.
6. Scrutinizing the possibilities for developing the framework and principles of the future modern MCENs.
7. Feasibility study of the presence of new grid structures, energy markets, and participants in the future modern MCENs' infrastructure.

References

1. C. Gu et al., System dynamics modelling of urbanization under energy constraints in China. Sci. Rep. **10**(1), 1–16 (2020)
2. K. Rahbar, C.C. Chai, R. Zhang, Energy cooperation optimization in microgrids with renewable energy integration. IEEE Trans. Smart Grid **9**(2), 1482–1493 (2016)
3. G. Wong-Parodi, T. Krishnamurti, A. Davis, D. Schwartz, B. Fischhoff, A decision science approach for integrating social science in climate and energy solutions. Nat. Clim. Chang. **6**(6), 563–569 (2016)
4. S.M. Naik, M. Fernandes, G. Pushpakumar, R. Pathak, Smart grid communication protocol test automation along with protection test automation, in *2019 3rd International Conference on Trends in Electronics and Informatics (ICOEI)*, (IEEE, 2019), pp. 1313–1317
5. E. Kabalci, Y. Kabalci, *Smart Grids and Their Communication Systems* (Springer, 2019) https://doi.org/10.1007/978-981-13-1768-2
6. E. Gies, Can wind and solar fuel Africa's future. Nature **539**(7627), 20–22 (2016)
7. M. Daneshvar, B. Mohammadi-ivatloo, S. Asadi, S. Galvani, Short term optimal hydro-thermal scheduling of the transmission system equipped with pumped storage in the competitive environment. Majlesi J. Elect. Eng. **14**(1), 77–84 (2020)
8. M. Daneshvar, B. Mohammadi-Ivatloo, K. Zare, S. Asadi, Two-stage robust stochastic scheduling model for Transactive energy based renewable microgrids. IEEE Trans. Industr. Inform. **16**(11), 1–1 (2020)
9. M. Daneshvar, E. Babaei, Exchange market algorithm for multiple DG placement and sizing in a radial distribution system. J. Energ. Manag. Technol. **2**(1), 54–65 (2018)
10. M. Daneshvar, B. Mohammadi-Ivatloo, K. Zare, Two-stage optimal robust scheduling of hybrid energy system considering the demand response programs. J. Clean. Prod. **248**, 119267 (2020)
11. M. Daneshvar, B.M. Ivatloo, M. Abapour, S. Asadi, R. Khanjani, Distributionally robust chance constrained transactive energy framework for coupled electrical and gas microgrids. IEEE Trans. Ind. Electron, **68**(1), 1–1 (2021)
12. P. Olivella-Rosell et al., Centralised and distributed optimization for aggregated flexibility services provision. IEEE Trans. Smart Grid, **11**(4), 3257–3269 (2020)
13. A.L. Cooke, J.S. Homer, L. Schwartz, *Distribution System Planning-State Examples by Topic* (Pacific Northwest National Lab. (PNNL), Richland, 2018)
14. R. Ambrosio, Transactive energy systems. IEEE Electrific. Mag. **4**(4), 4–7 (2016)
15. Edison Tech Center, The history of electrification: the birth of our power grid (2013)
16. Comparison–Centralized,decentralizedanddistributedsystems(12July,2020).Available:https://www.geeksforgeeks.org/comparison-centralized-decentralized-and-distributed-systems/
17. E. Vaccariello, P. Leone, F.G. Canavero, I.S. Stievano, Topological modelling of gas networks for co-simulation applications in multi-energy systems. Math. Comput. Simul. **183**, 244–253 (2021)

18. F. Fallahi, P. Maghouli, Integrated unit commitment and natural gas network operational planning under renewable generation uncertainty. Int. J. Electr. Power Energy Syst. **117**, 105647 (2020)
19. O. Ellabban, H. Abu-Rub, F. Blaabjerg, Renewable energy resources: current status, future prospects and their enabling technology. Renew. Sust. Energ. Rev. **39**, 748–764 (2014)
20. J. Salpakari, J. Mikkola, P.D. Lund, Improved flexibility with large-scale variable renewable power in cities through optimal demand side management and power-to-heat conversion. Energy Convers. Manag. **126**, 649–661 (2016)
21. A. Dannier, A. Del Pizzo, M. Giugni, N. Fontana, G. Marini, D. Proto, Efficiency evaluation of a micro-generation system for energy recovery in water distribution networks, in *2015 International Conference on Clean Electrical Power (ICCEP)*, (IEEE, 2015), pp. 689–694
22. S. Samsatli, N.J. Samsatli, A multi-objective MILP model for the design and operation of future integrated multi-vector energy networks capturing detailed spatio-temporal dependencies. Appl. Energy **220**, 893–920 (2018)
23. Y. Chen, W. Wei, F. Liu, Q. Wu, S. Mei, Analyzing and validating the economic efficiency of managing a cluster of energy hubs in multi-carrier energy systems. Appl. Energy **230**, 403–416 (2018)
24. A. Bernstein et al., Renewables need a grand-challenge strategy. Nat. News **538**(7623), 30 (2016)
25. C.S.R. Atla, K. Balaraman, A. Patil, Estimating capacity credit of wind energy sources using probabilistic methods, in *2013 15th International Conference on Advanced Computing Technologies (ICACT)*, (IEEE, 2013), pp. 1–6
26. R. Billington, R.N. Allan, *Reliability Evaluation of Power Systems* (Springer US, 1984) https://doi.org/10.1007/978-1-4899-1860-4
27. M.H. Amirioun, F. Aminifar, M. Shahidehpour, Resilience-promoting proactive scheduling against hurricanes in multiple energy carrier microgrids. IEEE Trans. Power Syst. **34**(3), 2160–2168 (2018)
28. M.H. Amirioun, F. Aminifar, H. Lesani, Towards proactive scheduling of microgrids against extreme floods. IEEE Trans. Smart Grid **9**(4), 3900–3902 (2017)
29. S.D. Manshadi, M.E. Khodayar, Resilient operation of multiple energy carrier microgrids. IEEE Trans. Smart Grid **6**(5), 2283–2292 (2015)
30. A.S. Musleh, G. Yao, S. Muyeen, Blockchain applications in smart grid–review and frameworks. IEEE Access **7**, 86746–86757 (2019)
31. C. Timberg, U. Simovic, The future of the utility field area network. Nat. Gas Electr. **35**(12), 8–15 (2019)
32. V. Amir, M. Azimian, Dynamic multi-carrier microgrid deployment under uncertainty. Appl. Energy **260**, 114293 (2020)
33. S. Mishra, K. Anderson, B. Miller, K. Boyer, A. Warren, Microgrid resilience: a holistic approach for assessing threats, identifying vulnerabilities, and designing corresponding mitigation strategies. Appl. Energy **264**, 114726 (2020)
34. L. Gitelman, D. Sandler, T. Gavrilova, M. Kozhevnikov, Complex systems management competency for technology modernization. Int. J. Des. Nat. Ecodyn. **12**, 525–537 (2017)
35. R. Zafar, A. Mahmood, S. Razzaq, W. Ali, U. Naeem, K. Shehzad, Prosumer based energy management and sharing in smart grid. Renew. Sust. Energ. Rev. **82**, 1675–1684 (2018)
36. A. Shahmohammadi, M. Moradi-Dalvand, H. Ghasemi, M. Ghazizadeh, Optimal design of multicarrier energy systems considering reliability constraints. IEEE Trans. Power Deliv. **30**(2), 878–886 (2014)
37. M. Daneshvar, B. Mohammadi-Ivatloo, M. Abapour, S. Asadi, Energy exchange control in multiple microgrids with transactive energy management. J. Mod. Power Syst. Clean Energy, **8**(4), 1–8 (2020)
38. A.J. Conejo, M. Carrión, J.M. Morales, *Decision Making Under Uncertainty in Electricity Markets* (Springer, 2010) https://doi.org/10.1007/978-1-4419-7421-1
39. C. Etukudor, B. Couraud, V. Robu, W.-G. Früh, D. Flynn, C. Okereke, Automated negotiation for peer-to-peer electricity trading in local energy markets. Energies **13**(4), 920 (2020)
40. W.B. Powell, D.R. Jiang, Method for bidding battery storage into hour-ahead energy markets, ed: Google Patents, 2018

Chapter 2
Modernizing the Energy from Customer-Side

Abstract Modern grids are the incorporation between the current power grid, information, and communication technologies in which different networking techniques are effectively adopted for monitoring key factors as well as exchanging information between various sectors to improve the overall condition of the energy networks. Among the diverse networking technologies in the smart grids, the first one is the home area network that focuses on managing the customer-side by developing broad synergies among the energy management systems. Indeed, grid-modernization efforts are not limited to a special part of energy networks, and all of these schemes are started by organizing the energy from the customer-side as the base layer of this process. Therefore, this chapter contains a comprehensive analysis regarding the smart homes, various types of high-performance buildings, hybrid systems, and their roles in modernizing the multi-carrier energy networks. The application of home energy management systems is discussed considering the different intelligent appliances in smart homes. Moreover, the structure of various types of buildings, as well as their design strategies and hybrid equipment, are described aiming to propose the appropriate ones for the future modern grids.

Keywords Supply-side energy management · Demand-side energy management · Modern homes · Modern buildings · Green buildings · Hybrid systems

2.1 Introduction

Nowadays, energy is taken into account as a strategic need and a critical foundation of human society and modern life. This inalienable and undeniable commodity has made it a key asset in everyday life so that we are witnessing a significant increment in various types of energy carriers day by day. In recent decades, heavy industrialization, as well as the rapid expansion of the world population, has led to a substantial increment in global energy demand leading to drastic changes in the global energy landscape [1]. This issue has critically threatened the availability of

© Springer Nature Switzerland AG 2021

M. Daneshvar et al., *Grid Modernization – Future Energy Network Infrastructure*, Power Systems, https://doi.org/10.1007/978-3-030-64099-6_2

multi-carrier energies, and energy security has become more challenging. Due to the necessity of energy existence for sustainable development and human well-being, improving the reliability and availability of the multi-carrier energies is of paramount importance. In this respect, various segments of the multi-carrier energy networks (MCENs) have moved toward fundamental changes to improve the structure of energy production, transmission, and distribution to further strengthen the security and reliability of continuous energy supply for different consumers. All these efforts are underway and are being developed day by day to empower the MCENs to provide the consumers with energy demand at any time [2]. In other words, the consumer-side is the final destination of the energy path and the results of all these efforts lead to this key part of the energy network. Therefore, this issue further induces the importance of the customer-side and its key role in the process of modernizing MCENs. Although modernizing the customer-side is essential for future modern energy networks, there are currently several gaps regarding the concepts, models, designs, and implementation plans that require immense actions. This prompted us to fill in some of the considerable gaps by examining the process of modernizing the consumer-side of energy. Therefore, this chapter focuses on evaluating the various dimensions of modernizing the customer sector.

Before we start to evaluate the modernization of the customer-side, we need to identify the scope of the customer-side in the first step. The customer-side analyzed here is the distribution part of the MCENs that includes residential consumers, smart homes, different levels of buildings, and commercial and industrial consumers. Figure 2.1 illustrates the scope of the customer-side in the future modern MCENs.

In the customer-side, the title of the largest energy consumer may be given to commercial and industrial consumers with certainty. Due to this, these consumers play a key role in implementing the grid-modernization initiatives at the customer-side of the future modern networks. Because one of the main goals for modernizing the future energy networks is to effectively deploy the full or high level of the renewable energy resources (RERs), the commercial and industrial consumers are

Fig. 2.1 The scope of the customer-side in the future modern MCENs

targeted to be equipped with a large number of proper types of the RERs for more clean energy production in the customer-side. This scheme not only enables the large consumers to meet a high portion of their energy demand through the cost-effective energy production units but also gives the opportunity of participating in the different energy markets for energy trading aiming to maximize their economic benefits. This is while modernizing the customer-side provides a fully automatic procedure for large consumers to elevate the quality of energy interactions. This is realized by using intelligent systems to create dynamic negotiations with the local control centers for establishing both the energy and information flows. On the other hand, this scheme is scheduled to increase the contribution rate of these consumers in a way to enable them for a prominent presence in the different types of demand-side energy management programs. The variety of these programs in the modern system empowers the consumers to be more flexible than before in participating in the energy network interactions. On the customer-side, by moving downwards in terms of energy consumption, we can reach the modern building sector. The grid modernization is aimed to convert the buildings from a typical consumer to an active prosumer. This means that all types of buildings can be equipped with the multi-carrier energy generation systems and storage devices to be a consumer and a producer. The advantages of modernization are further enhanced by providing appropriate conditions for the buildings to be involved in the market interactions in the role of small microgrid and energy hub. The last layer of the customer-side is smart homes. Smart homes are the group of residential consumers that have always faced challenges in actively engaging in energy interactions. Most of these challenges have been related to their small size for participating in energy interactions. Because the energy consumption of the smart homes is smaller than the typical energy range in the energy markets, the aggregator had to be considered for participating in the energy market interactions as an intermediary agency representing them. However, modern energy networks enable smart homes to effectively act in the energy networks by controlling their energy interactions through the home energy management systems (HEMSs) and advanced control centers. In the customer-side, the energy players are not limited to commercial or industrial consumers and smart homes, and one of the prominent components of the customer-side is the transportation system. Many players, such as electric vehicles, buses, and city trains, play a key role in the transportation system, and analyzing the random behavior of each of them has a significant impact on improving the energy cycle. In this respect, the stochastic behaviors of the transportation system have entered high levels of uncertainties in the modern hybrid energy networks that have led to a substantial increase in the complexity and optimal handling of the system. Due to this, future modern MCENs are targeted to be integrated with innovative technologies and expert systems to prevent the customer-side against a large number of congestions in the system. To achieve this goal, we need to examine each part of the customer-side in detail to suggest practical solutions for modernizing the customer-side by carefully assessing the existing needs, challenges, and obstacles. In fact, this goal is the mission of this chapter, to assist the modernization process of the customer-side.

2.2 Energy Customer-Side

Until recently, when we heard the word "customer-side," energy consumers and energy customers concepts came to our mind. However, like many other concepts, when we are talking about modern energy networks, we should reconsider most of the previous definitions, one of which is the customer-side. In the traditional networks, the consumers and customers at the customer-side were involved only in the energy purchasing schemes from the supply-side [3]. This issue has created significant limitations for all types of consumers in improving their economic benefits and consumption patterns. In other words, the chance of participating in the energy market interactions has been taken away from the consumers in conventional energy networks. This is while the future modern MCENs is targeted to be designed and developed in a way to provide the opportunities of being consumer and producer for all sectors, including buildings, transportation, and industrial. From now on, we call the combination of commercial and industrial parts, buildings and apartments, and smart homes as the residential and commercial sectors, and we will use this term hereinafter instead of them. In the modern MCENs, different structures in the various layers of the customer-side will be equipped with the logical capacity of the clean energy production units for energy generation, hybrid energy storage systems for ensuring the energy supply in the presence of the RERs, and controllable energy consumption devices. Indeed, residential and commercial buildings can play a role as both consumers and producers in the MCENs. On the other hand, transportation systems such as electric vehicles can charge energy sometimes (consumer role) and discharge it at other times (producer role). All of these concepts insinuate this undeniable fact that we can no longer see the customer sector only in terms of consumption. Therefore, we need to revise the word customer-side and replace it with the new word that fits with the future modern MCENs. This new word is the prosumer-side that will then be used instead of the customer-side. The process of the conversion of consumers in the traditional energy networks to the prosumers in the future modern MCENs is depicted in Fig. 2.2.

After going from the concept of the customer-side in conventional energy networks to the concept of prosumer-side in the future modern MCENs, we intend to discuss the notable challenges ahead of the modernization of the prosumer-side. As mentioned before, future modern MCENs are aimed to use intelligent systems and innovative technologies to provide significant benefits for the users. This means that almost all appliances around us should be controllable and will have a smart performance feature. For example, in a modern network, all systems in a smart home from the lighting system to the washing machine and refrigerator are expected to be automated. This is while all this equipment will be managed by a central control system that will be connected to the local control system and the main network through the same system. However, forming such a process first requires providing the right basic conditions. For example, to establish a modern city and make optimal use of all its programs and benefits, we will need all the residential and commercial buildings in that city to be equipped with smart equipment. However, the question here is whether the building occupants/owners, for example, agree to these

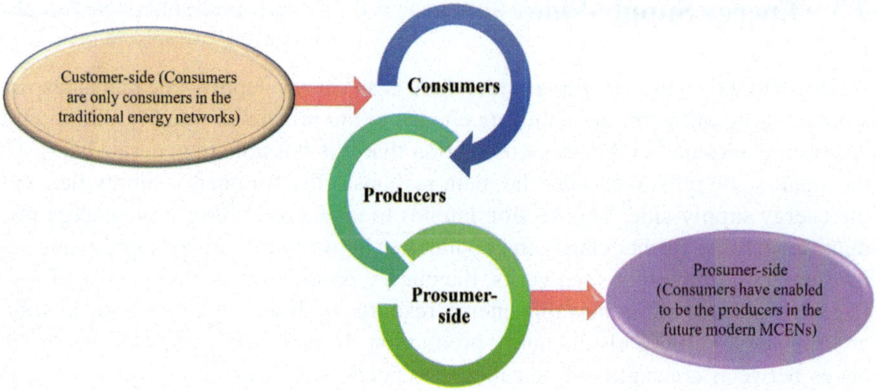

Fig. 2.2 The process of conversion of customer-side in the traditional energy networks to the prosumer-side in the future modern MCENs

intelligent systems and will have the necessary cooperation to equip their own buildings with such smart devices? This is where the challenges of modernizing the prosumer-side begin, as we see many similar questions at this stage. Some of the key challenges in this step are summarized as follows:

- Is the intelligent version of all the equipment available to use in the modernizing process of the different parts of the prosumer-side?
- Do all equipment have adequate potential to become smart and connected to a modern MCENs infrastructure?
- What solution does the modern system have for evaluating and controlling the energy consumption of those devices that are not intelligent?
- Given that the focus is on the modern hybrid networks, is there the technology needed to convert one-dimensional energy devices to multi-dimensional energy systems?
- Do all people with different tastes and interests agree and be satisfied with the transition from a traditional lifestyle to a modern one?
- What are the modernization plans to justify the majestic effects of the implementation of modern grids to convince those who oppose modernization?
- Does the current structure of the prosumer-side have enough potential to become a modern structure?
- What are the available solutions to overcome the physical and technological barriers of the modernizing of the prosumer-side?
- How will different governments react to network modernization schemes? Could there be a possibility of political obstacles to grid modernization schemes?
- What strategies have been proposed to persuade different consumers to cooperate in the MCENs modernization projects?

The aforementioned challenges are only some of the substantial challenges ahead of modernizing the prosumer-side. However, clarifying these challenges and recognizing them make it possible to provide more practical solutions and develop existing technologies to meet the needs of the future modern MCENs.

2.3 Energy Supply-Side

Adjacent to the energy prosumer-side, there is an energy supply-side that is responsible for generating and transmitting energy to the prosumer-side [4]. Although it has been mentioned in the previous section that it is possible to generate energy in the small-scale prosumer-side, the main responsibility for energy supply lies with the energy supply-side. This section intends to take a brief look at the energy production sector to further clarify the boundaries between the energy supply-side and prosumer-side. In today's networks, the energy production sector consists of a set of renewable and non-renewable energy resources. Modern networks, meanwhile, are targeted for 100% clean energy production. In fact, one of the obvious differences between conventional and modern energy networks is the use of modern features and conditions for maximum use of the capacity of the environment to produce clean energy [5]. In other words, the future modern MCENs are scheduled to operate different types of RERs to enable the system for clean energy production at all times of the day. The modern supply-side uses the potential of the cutting-edge technologies in co- and tri-energy production to extend the production of clean energy from electrical energy to other energy carriers. In this regard, a group of efforts has been made for effectively managing the supply-side that seeks environmentally friendly and cost-effective ways for energy generation and distribution in the modern structure of the energy network. Some important considerations in supply-side management are demonstrated in Fig. 2.3.

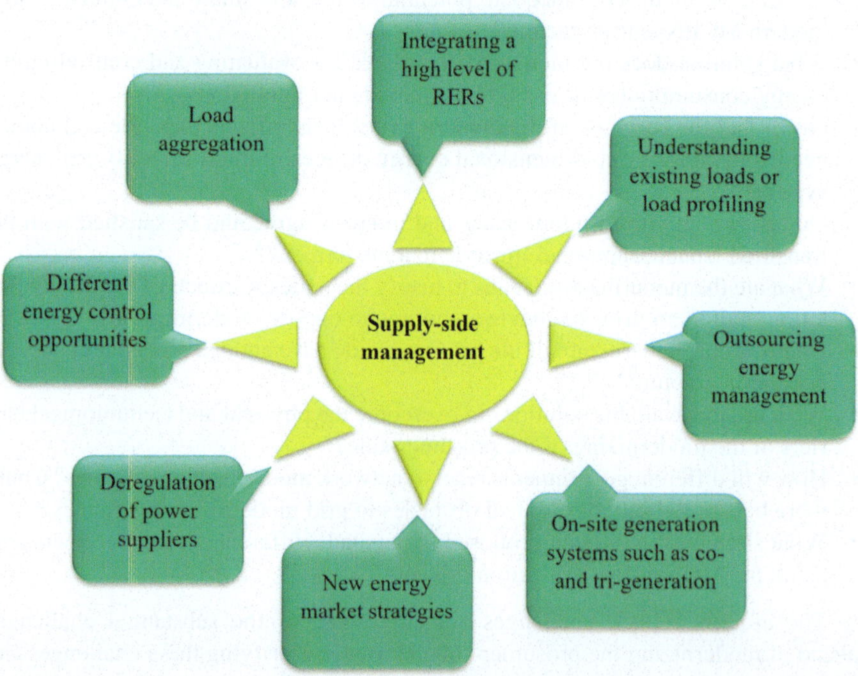

Fig. 2.3 Several important considerations in supply-side management

Supply-side management is one of the prerequisites for modernizing future MCENs, which is influenced by several important factors, some of which are shown in Fig. 2.3. In the grid-modernization process, similar efforts are increasing day by day, especially by more focusing on the operation of various types of RERs.

2.4 Smart Homes

Recently, the word "smart" has been recognized as an umbrella term for controllable devices and innovative technologies throughout the energy networks [6]. Smartness evokes various features in our minds, some of the most important of which were described in Section I of Chap. 1 on smart cities. The ability to monitor the information acquired from the surrounding environment and respond accurately and properly is only one of the key attributes of the smart systems [6]. The main objective of modernization is to increase the well-being of people, which has led to the emerging of new concepts such as "smart homes." A smart home is a residence that is targeted to equip only with intelligent appliances to provide tailored services for occupants. Indeed, a smart home is aimed to be designed in a way to improve health, comfort, and productivity for its residents by monitoring and controlling intelligent devices aiming to promote independent living and enhance the quality of life.

In recent years, the substantial developments in the technologies of the automation systems have increased the consumer adoption of residential automation systems and products ranging from automatic lightings and smart thermostats to the keyless entry systems in residential buildings. One of the vital automation devices in smart homes is the home energy management system (HEMS). The effective role of the HEMS motivated us to discuss the application of this system in making the smart home structures suitable for successfully joining the future modern MCENs' infrastructure.

2.4.1 Home Energy Management Systems (HEMSs)

The HEMS is developed to implement the automation control process in homes with smart appliances. The HEMS is based more on the approaches, optimization, and algorithms tripods, which enable the controlling system for optimal scheduling of the residential appliances such as heating, ventilating, and air-conditioning (HVAC) system and lighting. This is done for different objectives such as energy cost minimization and peak shaving while maintaining and improving the level of convenience and comfortability of the building occupants. The HEMS is enabled by the sophisticated technologies to act on behalf of the homeowner in providing satisfactory conditions in the smart home [7]. For this aim, in addition to performing some forecasts such as energy consumption, the HEMS receives some prominent information such as price signal to use them in satisfying the homeowner objectives at best possible conditions. Until before, the HEMSs were connected only with the electric-based systems and had interactions only with the electricity network. However, this type of HEMS is not appropriate with the future modern MCENs with multi-carrier energy systems. In the future modern MCENs, the existence of several dimensions of energy has caused us to not be satisfied only with electrical energy in network interactions. Thus, in the path of modernizing the prosumer-side, like most energy systems, we need to change the structure and functions of the HEMSs so that it can meet the goals of future modern MCENs in the prosumer-side. The new HEMSs will be responsible for converting ordinary houses to modern homes in order to meet the desired conditions in accordance with the related wishes of the homeowner and act considering the MCENs' limitations and other required conditions. This prompted us to introduce an improved model of the HEMSs in this book, which we will henceforth call it the modern home multi-carrier energy management system (HM^2EMS). The design of the HM^2EMS should be such that it cannot only simultaneously integrate energy interactions with multi-carrier energy systems but also be able to simultaneously establish interoperability with an electric power system, natural gas grid, district heating network (DHN), and water distribution system. However, the HM^2EMS must be able to upgrade and adapt to emerging technologies at any time. This means that a new smart appliance can be easily either added or removed into or from the system. On the other hand, this flexibility should not be limited to this level and the new system should be able to receive various signals from different hybrid networks at the same time and be able to monitor and respond to them at the minimum time. The HM^2EMS is expected to be very applicable for modernizing the prosumer-side due to its special control and management features. The suggested structure for the HM^2EMS is shown in Fig. 2.4.

The proposed structure for the HM^2EMS in Fig. 2.4 is the complete version of the home energy management system in the integrated electric power system, natural gas network, district heating network (DHN), and water distribution system, which is the first of its kind in the world. According to Fig. 2.4, all controllable appliances in the smart home are connected to the HM^2EMS system and users can enter their desirable sets into the system for considering in the smart home process. After receiving the user preferences and some required information from the local control center such as energy price signals, the HM^2EMS system is started to optimally schedule all controllable appliances by optimizing the goals and users' required conditions. This optimization will be conducted in accordance with the local control center by simultaneously considering the status of the electric power system, natural gas networks, the DHN, and the water distribution system.

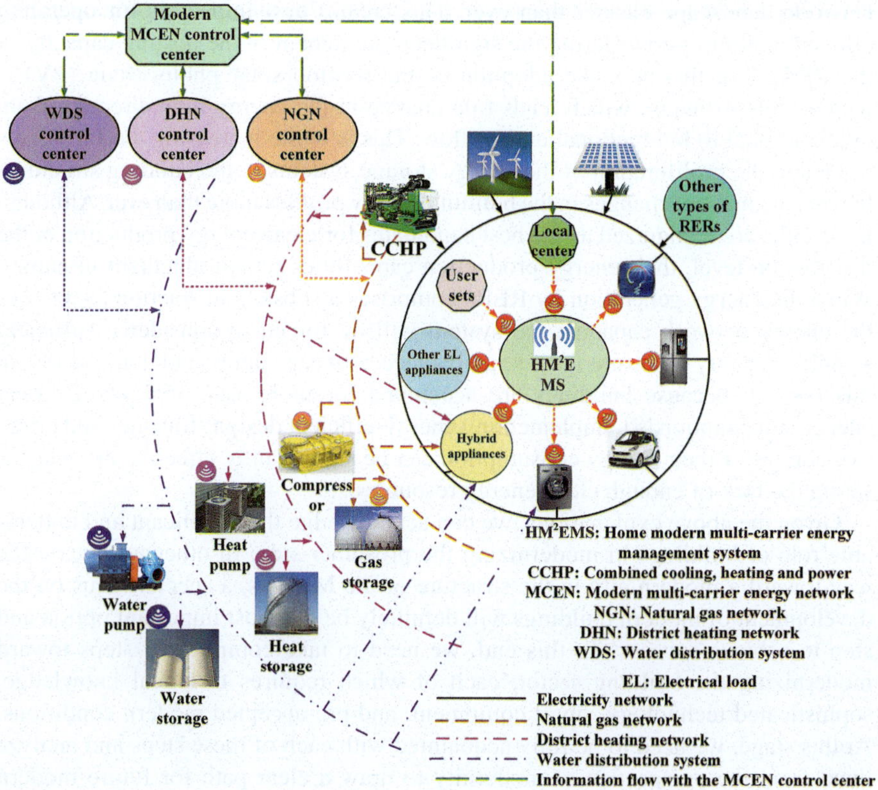

Fig. 2.4 The schematic of the HM^{2}EMS in the modern structure of the MCENs. HM^{2}EMS; home modern multi-carrier energy management system; CCHP, combined cooling, heating and power; MCEN, modern multi-carrier energy network; NGN, natural gas network; DHN, district heating network; WDS, water distribution system; EL, electrical load

2.5 Buildings

Approximately more than 30% of the world's energy is consumed by buildings, and they are also responsible for emitting about a third of the world's greenhouse gas emissions [8]. This statistic for the United States in the energy consumption sector is about 40% and the greenhouse gas emissions are about 38% [8]. In the United States, there are almost 100 million single-family homes that account for 36% of the electrical energy demand and substantially influence the peak load, especially during summertime when the HVAC systems' use is high [7]. Nowadays, traditional building energy profiles are changing due to an increase in the use of energy-efficient building designs and materials, leading to a significant reduction in the buildings' energy load. The need for alleviating environmental problems is shifting the focus more on designing sustainable buildings for making the energy networks landscape cleaner than ever. This creates new impetuses for operating efficient buildings considering the significant increment in the hosting capacity of the RERs. For this aim, the adoption of the rooftop solar photovoltaic (PV) is increasingly growing, which leads to a change in the energy flow direction from unidirectional to bidirectional energy flow. This is while increasing the PV panels has led to the proliferation of the energy storage systems in the building structures that results in more changes in the building energy profiles more than ever. Although these units are recognized as the best candidates for clean energy production at the distribution level, their energy production capabilities remain fluctuant in nature. When the energy generation by RERs comprises a substantial fraction (>20%) of the energy network capacity, the system will be forced to consider a sufficient standby capacity to ensure the energy balance between supply and demand [8]. In this respect, because buildings are accounted for more than 75% of electrical energy consumption [8], implementing energy-efficient design strategies and effective control of their energy consumption can be significantly effective in compensating the lack of enough clean energy resources.

Given the above explanations, we can acknowledge the significant and undeniable role of buildings in modernizing the prosumer-side. In other words, on the way toward a modern city in the structure of the MCENs, a special focus on the development of modern buildings will definitely be the most important option and step to achieve success. To this end, we need to take complicated steps toward modernizing the building sector, each of which requires technical knowledge, sophisticated technology, smart equipment, and pre-accepted modern conditions. At this stage, we need to be fully acquainted with each of these steps and analyze them from the perspective of modernity to draw a clear path for future modern buildings.

Several prominent steps need to be taken to achieve the modern structure of buildings. Most of these steps in practice are only specialized in the field of civil and architecture engineering, so these topics are out of the scope of this book. This is while we examine only some key parts of this complex process that have great impacts on modernizing future buildings. As seen in Fig. 2.5, the first step in reaching the modern buildings is to determine the building function and consequently decide the best geographical location for it. This inspires us the key fact regarding the importance degree of the geographical location for constructing it. The geographical location of the buildings in modern cities is important since it determines its access to public areas, hospitals, schools, etc. Then we have to think about what goals are important to us in this building and whether these goals are in line with the goals of modern buildings. In this way, in the next stage, the building plan is designed, taking into account its related goals and modernization aspects. After designing the relevant plan and determining the framework of the building, we need to examine the costs in more detail and the feasibility of realizing the goals of the project in the modern city. If the plan is in line with the standards required for a modern city, the use of capable technologies for constructing a building in accordance with the mentioned standards and their implementation in the next step will be on the agenda. One of the key steps in modern building

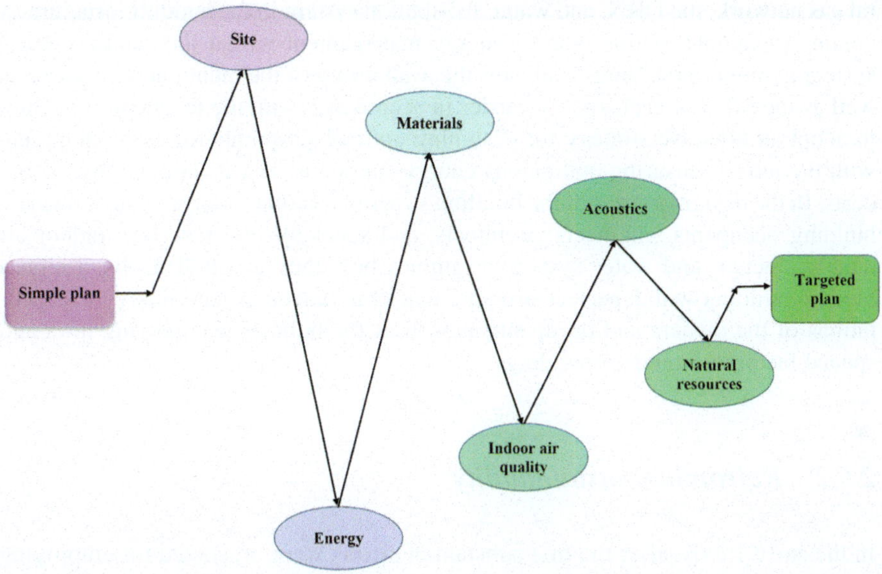

Fig. 2.5 Some key factors required for transmitting from the simple design of building to a desired one

construction is that its architectural design fits within the scope of modern cities. In this step, there are several key design factors that are effective, such as lighting, building view, design tailored to the type of building use, the appropriateness of the design to the designs of adjacent buildings, and so on. In this process, equipping the building with the smart appliances and intelligent control systems as well as examining the success of a building energy management system are the two final steps to achieving a modern building in the MCENs' landscape.

2.5.1 Structure

Generally, a set of different professions and interdisciplinary sciences such as civil, electrical, and architecture engineering play a key role in designing the structure of a building. Figure 2.5 illustrates such factors that need to be completely analyzed in order to polish a simple building design to make it the desired design [9]. For the future modern buildings, the integrated design will be unavoidable for effectively intending the interrelation of depicted factors in Fig. 2.5 with one another. Indeed, the integrated design for modern buildings refers to a set of considerations that must be intended in order to meet the goals of future modern MCENs. Energy management in modern buildings will no longer be limited to electrical energy as before, but a comprehensive energy management system will be considered in order to conduct extensive interactions simultaneously with the electric power system, natural gas network, the DHN, and water distribution system in the building structure. A common example of this type of energy management is that the central control system in modern buildings evaluates the availability of the mentioned networks as well as the price of each energy carrier in accordance with the local control center to adopt an effective strategy for designing optimal set-points for each equipment with the aim of ensuring continuous energy supply as well as minimizing energy costs. In the design process of the building structure, a collaborative team of owners, building occupants, engineers, architects, and specialists in materials, indoor air quality, energy, and water systems examines how they can holistically design a modern building with a perfect structure that simultaneously can satisfy the expectations of the owners and occupants as well as the building engineering laws and special factors for modern buildings.

2.5.2 *Reframing Sustainability*

In the early 1990s, after the first substantial efforts were organized for employing the sustainability architecture on the built environment in the short quarter-century, the world has witnessed a significant growth in the momentum and strength of the sustainable construction movement [9]. Sustainable construction is a set of widespread efforts that seeks to balance the requirements in the society, economy, and environment by employing innovative technologies and integrated flexible systems to provide win-win solutions. In order to create clear features for sustainable future modern buildings, we need to be more transparent about sustainable building design goals in the process of modernizing the prosumer-side. Thus, sustainable design for modern buildings is targeted to:

- Avoid all strategies and efforts that lead to a waste of critical resources such as water and energy.
- Make building resilient enough to natural hazards and unforeseen events.
- Create the potential of deploying environmentally friendly and cost-effective facilities considering the sufficient level of occupants' well-being.
- Provide modern built environments that can instill a sense of comfort and safety in the residents.
- Ensure the self-healing feature for buildings against the different disturbances and other harmful events.

Regarding sustainability, determining principles is essential to provide well-being conditions for the region's population considering their very survival. In this regard, even though buildings can be designed as sustainable buildings from the outset, one of the key challenges is how all existing buildings should be able to meet the standards of a sustainable building. Before answering this important question, it is important to get familiar with the general principles of sustainable building design. While the descriptions regarding the sustainable building design are ever-changing, six fundamental principles are defined for the sustainable buildings by the National Institute of Building Sciences, which are as follows [10, 11].

1. *Optimize site potential*: Selecting the appropriate site is the first step for creating a sustainable building. In this respect, incorporating modernization principles, as well as physical security issues, into the location and landscaping of a building has substantial impacts on critical items such as local ecosystems and energy use. In short, integrating the sustainable principles in the site design is necessary for successfully reaching the modern version of the buildings, whether in retrofitting the existing buildings or designing new buildings.
2. *Optimize energy use*: With ever-increasing energy demand as well as subsequent growing concerns regarding the environmental pollutions and economic issues in the presence of the fossil fuel-based systems and continuous energy supply concern in the renewable-based grids, improving the energy management cycles in the buildings is essential for increasing the energy efficiency and independence.
3. *Protect and conserve water*: A sustainable building must be able to eliminate the bad habits of water consumption in buildings by implementing autonomous and practical mechanisms. This will save significant water consumption especially when the world is facing a freshwater crisis.
4. *Optimize building space and material use*: As the consumption of natural resources is expected to increase along with the world population growth, a sustainable building should be designed by focusing on effectively using the materials by applying the integrated solutions and reusing the materials in a sustainable way.
5. *Enhance indoor environmental quality*: A sustainable building is expected to improve the moisture and ventilation control, avoid using unhealthy materials, optimize acoustic performance, and maximize daylighting, aiming to ensure the residents' comfort, health, and productivity.
6. *Optimize operational and maintenance practices*: A sustainable building should meet the preliminary design protocols regarding the optimal operating and maintenance of a building to reduce the costs and consumption of energy, water, and other sources, prevent potential system faults and disfunction, and improve the productivity.

Considering these principles in designing new buildings is essential for constructing a sustainable building. However, retrofitting existing buildings based on these principles is recommended by the sustainable design advocates rather than building a new one [9]; this is because retrofitting existing buildings can be more cost-effective than developing new ones. Indeed, considering the sustainable design attributes in the building construction whether for the new buildings or for existing ones that are targeted for renovations not only can reduce the operation costs and negative environmental impacts but also can make the buildings more resilient than ever before.

Regarding sustainable buildings, we usually face some considerable terms such as sustainable construction, green, and high performance that can be often used interchangeably. Among them, the economic, social, and ecological issues are most thoroughly addressed by the term sustainable construction in the context of the

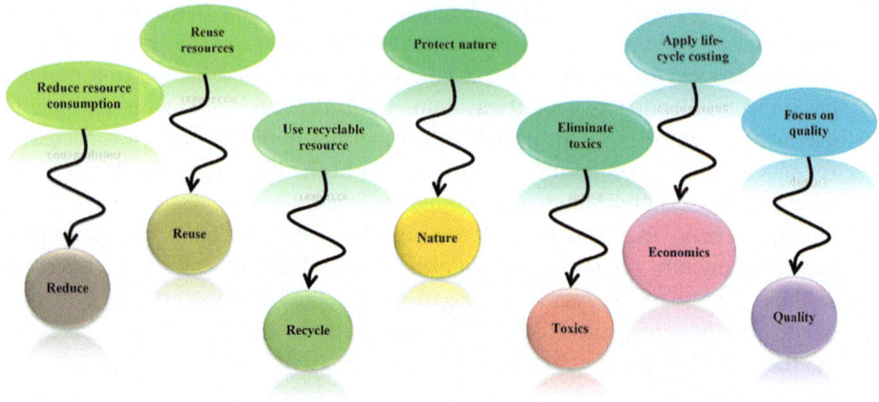

Fig. 2.6 Seven principles for sustainable construction articulated by an international construction research networking organization

community. An international construction research networking organization defined sustainable construction in 1994 as: "creating and operating a healthy built environment based on resource efficiency and ecological design" [12]. This organization developed seven principles for sustainable construction that ideally articulate decision-making for each step of the sustainable design and cover the building's entire life cycle [9]. These principles are demonstrated in Fig. 2.6. In this figure, each of these principles is summarized in a keyword that is shown inside the duplicate circles.

2.5.3 Electric Systems

An electric system in the buildings has a specific structure that is fully described in different references. Here, we are going to briefly express only some distinguishing features that the modern buildings are expected to completely include them aiming to make them suitable for the modern infrastructure of the future MCENs. In other words, modern buildings should be equipped with advanced intelligent systems to have appropriate interactions and connections with the modern city. Some of the important features are as follows but not limited to them. The modern building should:

- Equip with smart appliances that can simply connect with the control center system and quickly react after receiving optimal set-points from this center.
- Use wireless systems and advanced communication protocols to minimize energy losses at the lowest level.

- Include a smart panel to easily receive the occupants' preferences and consider them in the shortest possible time in an optimal manner.
- Have expert systems for establishing energy interactions with the other buildings aiming for uniting with them to achieve a specific goal in the necessary times under the supervision of the local central control system in the modern city.
- Empower with the sophisticated central intelligent system that has the ability to manage the receiving and transmitting of a large number of information signals with different frequencies with the aim of enabling the residences to easily participate in the energy network interactions.

2.5.4 Heating and Cooling Systems

Heating and cooling systems in a building play a key role in providing comfort for the occupants during a year. There are several fundamental factors that affect the performance of the HVAC system that should be considered in their design for providing comfort for the occupants. These factors are constantly changing with the development of related technologies. Although all of these changes are made with the aim of evermore increasing the well-being of the people, their development trends are moving more toward complicating existing heating and cooling systems. From the point of view of modern buildings, in addition to the features mentioned for electric systems, heating and cooling systems must also have some special features as follows.

- The integration of the separated cooling and heating systems into the unique hybrid system is preferred for the modern buildings to:
  - Avoid occupying extra space in the building.
  - Avoid complicating the operation of the energy management system in the presence of separate equipment for each of the cooling and heating energy.
  - Avoid excess plumbing.
  - Prevent the possibility of various disturbances getting increased due to using separate heating and cooling systems.
  - Avoid inefficient use of separate systems throughout the year (each of the cooling and heating systems remains unused at certain times of the year).
  - Avoid high thermal energy losses.
- The hybrid version of the heating and cooling systems should be developed by considering the ability of simultaneously working with multi-carrier energies in the coordinated operation of the natural gas network, the DHN, and the water distribution system. For this aim, they are connected to the central control system for optimal operation and monitoring of the smart devices.

2.5.5 Local Climate and Shading

Generally, outdoor weather conditions have direct impacts on energy consumption in the buildings. Building envelope plays a major role in affecting the climate changes on the internal conditions of the buildings. The climate factors such as ambient temperature and humidity, wind speed, and solar radiation can influence the indoor temperature of the building [13]. This indicates that the building's energy consumption and the indoor environment directly depend on the robustness of the building envelope against the surrounding climate. Therefore, local climate characteristics should be considered as the critical items in the architects' lists in the early design stage of the building envelope. For this aim, various climate zones are defined by the American Society of Heating, Refrigerating and Air-Conditioning Engineers (ASHRAE), and the required physical properties are determined for building envelope in each of the climate zones [13]. However, the future modern buildings should have special specifications in facing with different climate conditions. Some of the most important specifications are as follows:

- The central control system in the building should be updated in each time period by receiving climate information signals from the local control center connected to the meteorological control panel. The forecasted information signals, which are generated very near to the real time, give the near-realistic information regarding the meteorological events. This valuable information can enable the control center in the building to organize all smart appliances to properly deal with any meteorological events considering the restrictions of the MCENs.
- Modern building envelopes should be designed in a way that are responsive and adaptable to local weather conditions. Sensors can be implemented in envelope structure to receive and send proper information from and to the central control system. This is done with the aim of taking corrective actions in a timely manner if there are any imperfections in the envelope of the modern building.

Changes in the local climate can affect the building's lighting needs. In this regard, shading is a key issue in protecting the inside of the building from glare and reducing the cooling load. Recently, smart external shading devices are targeted to be used in the modern buildings to control the lighting of the inside based on the occupants' wishes that are already entered into the boiling control center system. This solution is suitable for rooms that have access outside the building through windows, while rooms that do not have access to the outside of the building in any way will be deprived of natural light. On the other hand, even in daylight hours, these rooms must use a variety of lamps for lighting, which will greatly increase the energy consumption in the building. For this type of room or some hallways that are not accessible outside the building according to the design conditions of the building, modern buildings need to provide appropriate solutions. Due to recent advances in physics, especially in the field of reflective science and various lenses, it is recommended to use reflection mechanisms for developing modern buildings. For this aim, in different parts of the building envelope, window-shaped valves are embedded so that each of them is equipped with

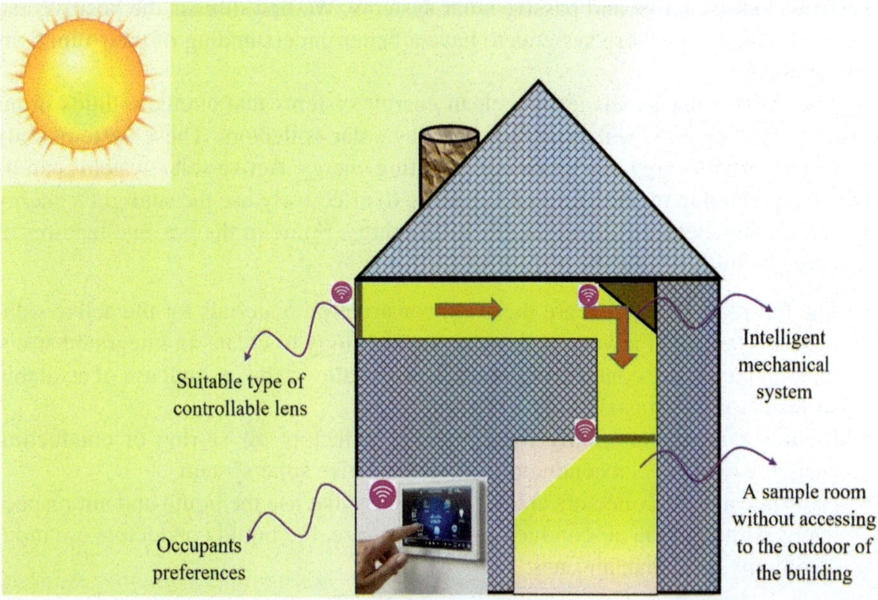

Fig. 2.7 The schematic of the proposed solution for delivering natural light to the rooms that do not have access to the outdoor of the modern building

suitable types of lenses depending on the angle they are placed. Each of these valves is connected to the relevant rooms inside the building through appropriate channels so that in each screw of this path, suitable lenses for reflecting light have been re-embedded. In this solution, each of the lenses in the connection path of the valve installed in the building envelope to the rooms or corridors without windows are mounted on the super-sensitive and intelligent mechanical systems so that they can be controlled by the central control system of the building. Based on the amount of light intensity that residents set to the building's control system through the touch panel, this system adjusts the angle of the lenses in such a way that this light is reflected from the outside of the building to inside the related room using the reflection technique. This is one of the solutions that can allow the use of natural light in all rooms of modern buildings with the least amount of energy consumption based on the reflection mechanism. The schematic of this process is illustrated in Fig. 2.7.

2.5.6 Active and Passive Solar System

One of the main goals of the future modern buildings is to effectively use the natural resources in producing clean energy. Proper design and construction of buildings can enable them to choose various types of clean energy production technologies for cost-effective and carbon-free energy generation. There are two types of solar systems, that is, active and passive solar systems. We first discuss the positive and negative features of these systems to have a better understanding of their functions and applications.

The active solar panels are the clean energy systems that circulate fluids or air using pumps or fans, which are activated by solar collectors. These systems only rely on external energy resources in generating energy. Active solar systems can be widely operated in modern buildings aiming to effectively use the sunlight's energy for generating clean electrical and thermal energy. Some of the positive features of this type of solar system are as follows [14]:

- The flat-plate PV panels are the main construction materials for the active solar panels. The use of advanced designing technology to create an integrated structure (all panels are connected with each other) allows the optimal use of available surfaces for their installation in the buildings.
- Liquid or air is commonly used in solar collectors for storing or conducting energy, which plays a conductor role in the active solar system.
- Hydronic and air collectors are the conductors that use the liquid and air, respectively. Although the air conductors do not freeze, the liquid conductors are more commonly used than air ones.

On the other side, like any other product, the active solar systems have some drawbacks that need to be considered in selecting the appropriate kind of PV panels. Some of these disadvantages are as follows:

- The required equipment for this kind of PV panels is expensive, which leads to an increment in the total investment cost of the building.
- Maintenance costs are high due to the usage of expensive and extra equipment.
- There is the possibility of toxins releasing in the air from the fluids that are used for storing heat in the PV panel.

Generally, passive solar energy systems are structured for optimal usage of the solar energy for generating the cooling and heating energy in the buildings. The structure of these systems mostly depends on the construction and design of the buildings. This is while the operation of passive solar panels is conducted without reliance on external devices. In these systems, the passive collectors are used for converting rays into the sunlight. These systems are used for transferring the heat energy from warmer to cooler surface and work based on the thermodynamics laws. The amount of thermal mass in the walls and overall orientation are two important factors that affect the success of the passive solar system, which is independent of external devices. Like the active solar panels, the passive version of them also has some advantages as follows [14]:

- The entire setup cost is cheap due to a lack of external equipment.
- These systems have a cost-effective and environmentally friendly energy generation process.

In addition to the aforementioned advantages for the passive solar panels, there are some disadvantages for them, which are as follows:

- The efficiency of passive solar systems depends on the weather, which can lead to overheating conditions for the buildings located in the hot areas.
- A suitable type of windows should be chosen for achieving satisfactory results in generating energy.

2.5.7 Thermal Envelope for Thermal Comfort and Indoor Air Quality

As mentioned before, the local climate in the outdoor of the building can affect the thermal conditions inside the building and impact the energy performance in the buildings. The building envelope is a physical barrier that separates the outdoor and indoor environment of a building [15]. This has a key role in maintaining and controlling the quality of indoor air irrespective of transient outdoor air. Building envelope contains different materials and components such as [15]:

- *Foundation*: This forms the lower portion of the building structure that has an important mission to maintain the strength of the building and transfer its gravity loads to the earth. Shallow and deep foundations are recognized as the two typical foundations for the buildings.
- *Roof*: This is one of the critical parts of the building envelope that can influence the indoor conditions due to its direct exposure to climate, especially solar radiation.
- *Fenestration*: This part of the building envelope refers to openings, including doors and windows. The fenestration not only plays an effective role in providing illumination levels and thermal comfort for occupants but also they contribute to aesthetics aspects of building design.
- *Walls*: They are taken into account as a significant fraction of a building envelope, which has a substantial role in providing both acoustic and thermal comfort within a building, considering the avoidance of negative effects on the aesthetics of the building.
- *Thermal mass*: It is a set of high heat capacity materials that can absorb heating energy, store, and release it later. The thermal mass contains building components such as floors, ceilings, furniture, partitions, and walls of a building that can store thermal energy.
- *Thermal insulation*: This is a combination of materials or a special type of material that can significantly reduce the heat flow losses by radiation, convection, and conduction when it is properly applied in the building structure. The different physical forms of the available thermal insulation in buildings are illustrated in Fig. 2.8 [15].
- *External shading devices*: These devices are used for automatically controlling the daylighting using special sensors that are connected to the control center of the building. The external shading devices play a key role in efficiently designing the building shading envelope; one significant applicability of these devices is blocking the summer sun while permitting the winter sun.

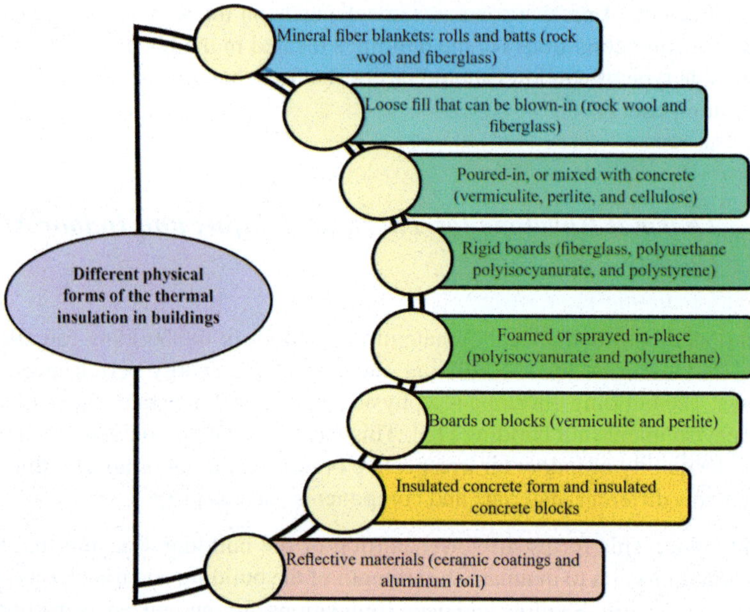

Fig. 2.8 The different physical forms of the available thermal insulation in buildings

One of the important parts of the building envelope is the thermal envelope, which is known as a heat flow control layer. In the buildings, the thermal envelope can be generally found in various building components, such as in a ceiling. The future modern buildings are targeted to be equipped with the different types of more sensitive sensors and infrared cameras for detecting any anomalies such as interstitial condensation in the building and informing occupants through the building control center and employing a self-healing process in some possible cases. The application of the building envelope, especially thermal envelope, has a key effect on building indoor air quality. Indeed, developing an appropriate thermal envelope for the modern buildings not only can successfully separate the outdoor conditions with buildings' indoor environment and guarantee indoor air quality using the HVAC and other smart devices controlled by the buildings' control center but also it can provide the acceptable range of thermal comfort for the occupants by optimally managing the application of the smart hybrid cooling and heating systems and other related devices in the building.

2.5.8 *Architectural Acoustics*

Architectural acoustics focus on designing the building considering the sound issues from various viewpoints and are typically subdivided into building acoustics and room acoustics [16]. In this respect, optimal conditions for speech communication and music perception are taken into account as the main goals in classical room acoustics. This is while the scope of room acoustics is not limited to the aforementioned items and they also focus on the other spaces such as classrooms, offices, workplaces, and foyers in which noise control aspects are also considered in the optimum acoustic conditions [16]. Regarding the sound with different intensities transmitted between buildings' rooms, not only the airborne sound insulation of the rooms should be taken into account but also structure-borne excitation aspects, for example, air condition or ventilation systems and water installation, need to be intended for determining the overall acoustic performance in the buildings [16]. Indeed, predicting the acoustic performance can be carried out using the different theoretical models such as statistical energy analysis, geometrical acoustics, and numerical or analytical wave models [16]. In the future modern buildings, the acoustic performance is expected to be analyzed using the building control center. In other words, the advanced theoretical models are defined for the central control system to automatically detect any acoustic faults in the building and take action in proportion to each fault. Despite the fact that the architecture of the modern buildings should be designed from the acoustic standpoint based on the latest standards in the building construction, the building central control center is intended to evaluate the different types of sounds with various intensities in each period of time. In this assessment, the control system compares the intensities of sounds with their standard level from different viewpoints such as health issues, and reports the required alerts for the occupants. This capability allows the occupants to effectively control their interactions in the home environment, create the right mood for relaxation with the desired music, and be aware of the state of sound transmission in the building environment.

2.6 Green Building

The term "green building" is developed based on the methodologies and principles of sustainable construction, which consider the characteristics and quality of the actual structure [9]. Until now, different definitions are provided for the green building. For example, in Ref. [12], green buildings are defined as "healthy facilities designed and built in a resource-efficient manner, using ecologically-based principles." Green buildings are targeted to be truly sustainable, integrate with a high level of RERs for adequate clean energy production, and structure in accordance with the conditions of modern grids. The future modern buildings are also targeted to be constructed as the green buildings, which not only should include the aforementioned features but also need to be included with some other special features. Indeed, the term "green building" should not only refer to this point that green buildings should appear with zero emissions in the modern city but also the extended concept is required for the modern buildings from a green point of view. Some of these features are as follows:

- If the building is not well designed in terms of appearance, it will somehow pollute the appearance of the city. Therefore, beauty in a design should be considered for modern buildings in accordance with modern city standards.
- All potential of the modern building should be considered for equipping suitable types of RERs aiming to meet all energy demands of the building through clean energy production units. This means that modern buildings are designed as an environmentally friendly structure with zero emission.
- The architectural design of the modern buildings should be conducted in a way to avoid any acoustic pollutions to the outdoor of the building.

To sum up, modern buildings must be green in terms of greenhouse gas emissions, acoustic pollution, and good looks. For green buildings, different terms are used in recent literature that refers to the same features for the buildings. Two common versions of them are net-zero energy buildings and healthy buildings. The green, net-zero energy, and healthy buildings have the same definition and scope of features for the buildings. Perhaps each of them has some differences in emphasizing some of the specific features that have led to them being used in the right place, depending on the topic of discussion. For example, the green building seems to be more used when we want to intensely emphasize on using clean energy production units, net-zero energy buildings are more used when the emphasis is more on the use of a set of equipment that ultimately leads to zero emissions in the building, and healthy buildings when the focus is more on using a set of equipment that does not endanger human health.

2.6.1 Design Strategies

2.6.1.1 Active and Passive Design Strategies

Recently, the green building movement is rapidly growing by extending and developing the indices for building design and construction as well as improving the quality of materials. This movement seeks to develop a model for other parts of economic endeavor for providing a consensus-based market-driven technique aiming to increase the effectiveness of the model. In the green buildings, maximum clean energy production through the RERs is targeted and has a great impact on the whole performance of the building; also, the energy consumption control is essential for improving the building performance. In this respect, designing appropriate strategies for green buildings has remarkable impacts on improving their performance in the presence of a high level of clean energy production units. For this aim, the combination of active and passive strategies is considered for the green building performance to make their performance in different sectors better than before. The passive design strategies are structured by focusing on the usage of the natural ability of the external environment, such as daylight, for improving some effective indexes inside the building, such as energy consumption. This is while the active design strategies are created using the potential of the controllable devices in the

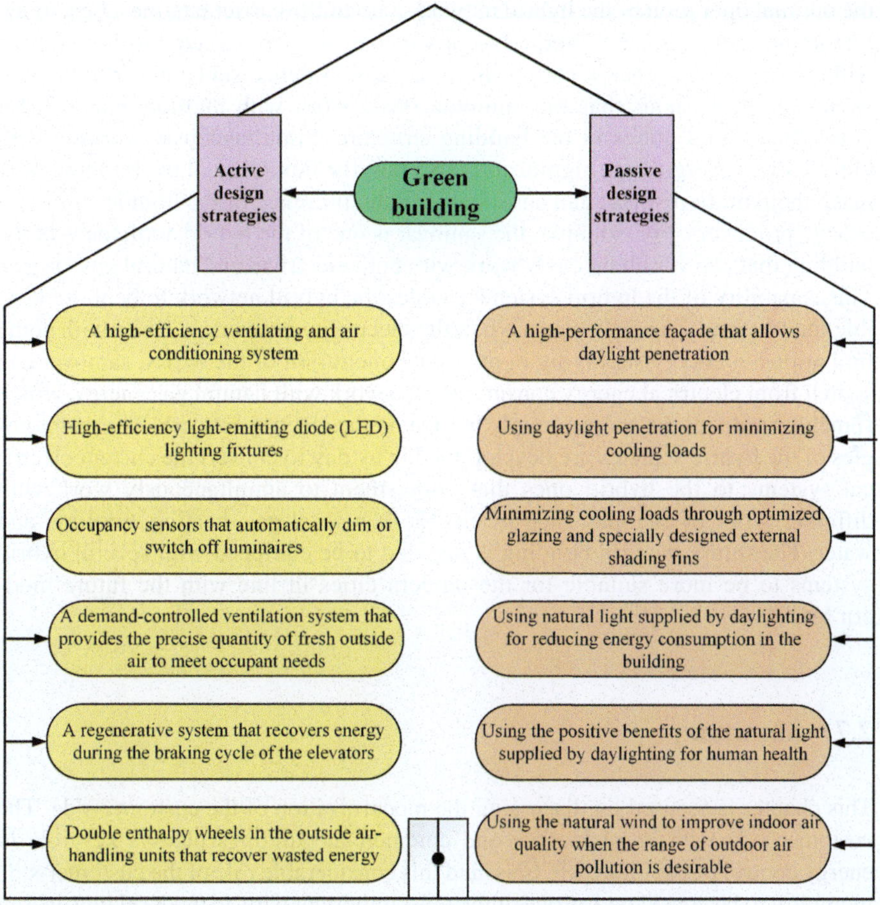

Fig. 2.9 Active and passive design strategies for green buildings

building to manage all possible parts inside the building, such as air quality, for improving the building performance. The detailed information about the intended parts in the building by the active and passive design strategies are illustrated in Fig. 2.9 [9].

2.6.1.2 Hybrid Systems

In the future modern MCENs, hybrid systems will play a vital role in establishing purposeful interactions between the electric power systems, natural gas grid, the DHN, and water distribution system, especially at the prosumer-side. In this regard, buildings have substantial impacts on greatly enhancing these interactions due to their higher-level contribution in the prosumer-side activities. Therefore, using the different types of hybrid systems in the buildings will be necessary for the optimal operation of the hybrid network with multi-carrier energies. One of the important features of the hybrid systems is that they are targeted to be working with more than one energy carrier. Indeed, these systems can be bi- or tri-dimensional energy systems that can simultaneously work with multiple energies for doing the special duties in the building structure. Simultaneously working with multi-carrier energies can significantly increase the flexibility of the buildings and make them more resilient and sustainable in the presence of the high level of stochastic producers. For example, the samovar is one of the hybrid appliances in the building that can simultaneously work with both electrical and natural gas energy. This capability of the hybrid systems enables the hybrid network to be more flexible than before. This is because when the electricity network is faced with some limitations in energy supplying in the high penetration of the RERs, samovar can switch from electrical energy consumption to work with natural gas energy, which somehow reduces the energy supply burden on the electric network. The technologies of the hybrid systems are developing day by day to convert the current electrical systems to the hybrid ones that allow them to simultaneously work with different types of energies such as electrical, natural gas, heating, cooling, and water. The future modern building is targeted to be equipped with several hybrid systems to be more suitable for the modern cities in line with the future modern MCENs.

2.7 Summary

This chapter is focused on discussing the modernization of the customer-side. The customer-side with its residential and commercial buildings plays a key role in energy control of the future MCENs, and this considerable role of the customer-side requires it to be considered as one of the most important parts in modernizing energy networks. This issue motivated us to evaluate the modernization aspects in the customer-side, which lead to a switch from the customer-side to the prosumer-side in the future modern MCENs. In the first step of this chapter, the boundary between energy supply-side and customer-side was clearly identified by providing relevant details and players for both parts of the hybrid networks. The smart home's role in modernizing the prosumer-side was discussed by assessing the application of the smart hybrid systems and other appliances. The novel HM^{2}EMS was proposed for energy managing and controlling in the modern homes of the future modern MCENs. Afterward, the different types of buildings were investigated along with evaluating the different steps in the building construction from the grid modernization point of view. The key results summarized in that the future modern buildings should not only include the previous green buildings characteristics along with the new special features in designing, constructing, and equipping, but also they should be developed with full renewable-based systems as well as a high level of smart hybrid systems for improving the building performance in all sectors especially in terms of sustainability, resiliency, and health issues. Moreover, some effective ways were proposed for the future modern buildings with the aim of facilitating the achieving of the mentioned goals in line with the grid-modernization objectives.

2.8 Questions

After careful reading and evaluating this chapter, we are expected to be able to answer the following questions:

1. Why do we need to modernize the customer-side?
2. What are the boundaries between the energy supply-side and the customer-side?
3. What are the challenges ahead of modernizing the prosumer-side with high penetration of the RERs?
4. Why do we need to switch from the customer-side in the conventional energy networks to the prosumer-side in the future modern MCENs?
5. What are the distinguishing features of the supply-side in the modern MCENs in comparison with the traditional energy grids?
6. Which considerations should be intended in the modernization process of the supply-side?
7. What are the distinguishing features of the smart homes that are required to make them suitable for the modern structure of the future MCENs?
8. What is the application of the HEMSs in smart homes and why their structure should be changed for the modern hybrid networks?
9. Why the HM^2EMS is proposed for the future modern homes and what important tasks this system is supposed to perform on the future modern MCENs?
10. What are the differences between HEMSs and HM^2EMS in smart homes?
11. Why the role of buildings is key in modernizing the prosumer-side?
12. Which changes are needed for the current buildings to make them suitable for the modern structure of the hybrid networks?
13. What are the key steps for modernizing the buildings and why they are necessary for grid modernization?
14. Which factors are required for transmitting from the simple design of building to a desired one in the modern cities?
15. Why do we need sustainable buildings in modern cities?
16. Sustainable design of the modern buildings should be conducted for meeting which considerable goals?
17. What are the principles for sustainable buildings and why they should be intended for modern buildings?
18. How can we define sustainable construction for modern buildings?
19. What are the principles for the sustainable construction of the modern buildings in the future MCENs?

20. The modern buildings should include which features from the electric systems point of view?
21. What are the special features that the heating and cooling systems should have in modern buildings?
22. What are the significant specifications for the modern buildings in suitably facing with the different climate conditions and shading issues?
23. What is the innovative way for those rooms that do not have access to the outdoor of the modern building to receive natural daylight during a day?
24. What are the advantages and disadvantages of active solar systems for modern buildings?
25. What are the advantages and disadvantages of passive solar systems for modern buildings?
26. Why effectively designing the thermal envelope is essential for thermal comfort and indoor air quality in the buildings?
27. What are the components that are used to form the building envelope?
28. What are the different physical forms of the available thermal insulation that can be used in the buildings?
29. Which factors should be considered in designing the architectural acoustics for the future modern buildings?
30. What are the green buildings and why the future modern buildings need to be green?
31. What are the features of the green buildings that are essential for building construction?
32. What are the differences between green, net-zero energy, and healthy buildings?
33. What are the design strategies for green buildings?
34. What are the differences between active and passive green building strategies?
35. What are the hybrid systems and why their presence in the future modern buildings is necessary?

2.9 Suggestions

In this section, some suggestions are provided related to this chapter that can give a good overview of some substantial issues in the future modern MCENs.

1. Switching from the customer-side in the traditional energy grids to the prosumer-side in the future modern MCENs will create the opportunity of the presence in the energy market interactions for selling or purchasing energy for all participants of the prosumer-side.
2. Integrating the supply-side of the future modern MCENs with the full level of the RERs will make the hybrid network cleaner than ever before, but the innovative technologies and practical solutions will be needed for dynamic balancing between the energy supply and demand.

3. Employing the HM^2EMS in the prosumer-side of the future modern networks will allow the effective presence of the hybrid systems in the buildings and other parts of the residential and commercial structure for improving the flexibility, reliability, and resiliency of the future modern MCENs.
4. Developing green and sustainable buildings will be essential for modernizing the prosumer-side incorporated with a high level of clean energy production units.
5. Applying the innovative technologies as well as appropriately using the various sciences such as reflection theory in physics can give practical solutions for facilitating the implementation of the modern MCENs.

2.10 Future Trends and Discussion Topics

Here, future trends and some discussion topics are proposed for the interested readers regarding this chapter that can be useful for developing the MCENs' modernization process.

1. Analyzing the key challenges ahead of modernizing the prosumer-side for providing practical solutions.
2. Improving the supply-side management systems in line with the objectives of the future modern MCENs.
3. Developing the hardware and software for the HM^2EMS to make the hybrid systems plan implementable in the structure of the modern networks.
4. Improving all steps of building construction based on the modern structure of future MCENs.
5. Designing the effective building envelope, especially thermal envelope, for realizing the modern building structure.
6. Developing hybrid systems technologies for increasing the interactions between multi-carrier energy networks.

References

1. S. Kosai, H. Unesaki, Short-term vs long-term reliance: Development of a novel approach for diversity of fuels for electricity in energy security. Appl. Energy **262**, 114520 (2020)
2. M. Daneshvar, B.M. Ivatloo, M. Abapour, S. Asadi, R. Khanjani, Distributionally robust chance constrained transactive energy framework for coupled electrical and gas microgrids. IEEE Trans. Ind. Electron. **68**(1), 347–357 (2020)
3. M. Daneshvar, B. Mohammadi-Ivatloo, K. Zare, S. Asadi, A. Anvari-Moghaddam, A novel operational model for interconnected microgrids participation in transactive energy market: A hybrid IGDT/stochastic approach. IEEE Trans. Industr. Inform., 1 (2020). https://doi.org/10.1109/TII.2020.3012446
4. M. Daneshvar, B. Mohammadi-Ivatloo, S. Asadi, M. Abapour, A. Anvari-Moghaddam, A transactive energy management framework for regional network of microgrids, in *2019 International Conference on Smart Energy Systems and Technologies (SEST)*, (IEEE, 2019) Porto, Portugal, pp. 1–6

5. M. Daneshvar, B. Mohammadi-Ivatloo, S. Asadi, S. Galvani, Short term optimal hydro-thermal scheduling of the transmission system equipped with pumped storage in the competitive environment. Majlesi J. Electr. Eng. **14**(1), 77–84 (2020)
6. D. Marikyan, S. Papagiannidis, E. Alamanos, A systematic review of the smart home literature: A user perspective. Technol. Forecast. Soc. Chang. **138**, 139–154 (2019)
7. A. Pratt, D. Krishnamurthy, M. Ruth, H. Wu, M. Lunacek, P. Vaynshenk, Transactive home energy management systems: The impact of their proliferation on the electric grid. IEEE Electrif. Mag. **4**(4), 8–14 (2016)
8. S. Katipamula, J. Haack, G. Hernandez, B. Akyol, J. Hagerman, VOLTTRON: An open-source software platform of the future. IEEE Electrif. Mag. **4**(4), 15–22 (2016)
9. C.J. Kibert, *Sustainable Construction: Green Building Design and Delivery* (John Wiley & Sons, Hoboken, 2016)
10. C.F. Albrecht, J.C. Carlo, L.D. Iulo, P.D. Buckland, International transdisciplinary approach to sustainability research related to place: Sustainable, affordable homes and ecosystem services in the US and Brazil, in *Universities and Sustainable Communities: Meeting the Goals of the Agenda 2030*, (Springer, Cham, 2020), pp. 187–201
11. WBDG Sustainable Committee, Sustainable building (28 April 2020). Available: https://www.wbdg.org/design-objectives/sustainable
12. C.J. Kibert, Establishing principles and a model for sustainable construction, in *Proceedings of the First International Conference on Sustainable Construction*, Tampa, Florida, 6–9 Nov 1994
13. D.-S. Lee, J.-H. Jo, S.-H. Koo, B.-Y. Lee, Development of climate indices using local weather data for shading design. Sustainability **7**(2), 1884–1899 (2015)
14. YellowLite, The difference between active and passive solar systems. Available: https://www.yellowlite.com/blog/post/the-difference-between-active-and-passive-solar-systems/
15. S.B. Sadineni, S. Madala, R.F. Boehm, Passive building energy savings: A review of building envelope components. Renew. Sust. Energ. Rev. **15**(8), 3617–3631 (2011)
16. M. Vorländer, D. Schröder, S. Pelzer, F. Wefers, Virtual reality for architectural acoustics. J. Build. Perform. Simul. **8**(1), 15–25 (2015)

Chapter 3
Technical and Theoretical Analysis of the Future Energy Network Modernization from Various Aspects

Abstract Generally, system availability for an aging electric power system infrastructure diminishes in the absence of modernization and maintenance activities. The degradation of power system infrastructure has made the prioritization of these schemes a more complex problem when it is coupled with the limited resources and growing backlog of needs as two other problematic items. By increasing the number of capable technologies as well as penetration of the stochastic producers in the energy production process, the current structure of the energy networks and existing techniques utilized for solving this problem lacks a topological aspect, holistic approach, and sufficient potential. Hence, the great need is felt for adopting innovative technologies and intelligent systems in the context of modern networks. In order to effectively fill these key gaps, this chapter is aimed at analyzing the modernization of the future multi-carrier energy networks from various aspects, including energy generation, storage, and management systems. In this respect, technical and theoretical requirements are discussed from the different viewpoints for the grid modernization due to the growing trend of energy consumption and the presence of various participants with different goals and strategies in the hybrid network.

Keywords Centralized energy generation systems · Decentralized energy generation systems · Modern energy production · Renewable energy resources · Energy storage systems · Multi-carrier energy conversion · Demand-side energy management · Demand response · Food-energy-water nexus · *i-Energy*

3.1 Introduction

As people's living standards and awareness have increased following the emergence of a variety of energy-consuming systems and significant advances in different technologies, energy grid modernization has become a more popular and complex term. Indeed, energy grid modernization is intended to respond appropriately to expectations for substantially increasing the well-being level in human

© Springer Nature Switzerland AG 2021

M. Daneshvar et al., *Grid Modernization – Future Energy Network Infrastructure*, Power Systems, https://doi.org/10.1007/978-3-030-64099-6_3

society over recent years [1]. The word "modern" can be considered as an umbrella term for sophisticated technologies, expert systems, information and communication paradigms, etc. [2]. This issue indicates that modernization is not only limited to the special part of the energy networks but also it should be investigated for various energy sectors that have significant effects on the energy network interactions. Indeed, in the process of modernizing the future multi-carrier energy networks (MCENs), we need to fully analyze different segments of the MCENs from various viewpoints in terms of technical and theoretical assessments. Some of the more considerable parts in the MCENs are the energy generation sector, energy storage systems (ESSs), demand-side energy management techniques, and integration of the distributed energy resources (DERs), which, evaluating them from the modernization point of view, are necessary for successfully implementing the modern structure for the future MCENs. Generally, the energy production sector can be categorized into two parts: centralized energy generation and a decentralized one. Until recent decades, the energy generation was completely dependent on the centralized energy production mechanisms. In these mechanisms, the large power plants, such as thermal and natural gas units located near the load centers, are responsible for meeting the energy demand of the consumers during a day. Although their controllable feature in energy generation has been provided a significant benefit for the energy network, they are the reason for creating some considerable challenges in the system. Higher investment, construction, operation, and maintenance costs, contributing to a high amount of greenhouse gas emissions, and a large amount of energy losses are some prominent disadvantages of these units that have made them an unacceptable option for developing the energy networks. Due to this, the research trends in this field were changed to find alternative solutions for generating energy by focusing on some considerations such as reliability, resiliency, efficiency, stability, and economic and environmental issues. The results of these efforts led to proposing decentralized energy production mechanisms, and proper DER technologies are developed and implemented in the energy production process of the current energy networks. Among the different types of DERs, the renewable energy resources (RERs) have attracted notable attention due to their economic and environment friendly features in energy production [2]. Their dramatic benefits have led to the introduction of various technologies for producing clean energy in the system such as wind, solar, hydro, tidal and wave, geothermal, and biomass energies. In spite of the significant benefits of the RERs for the energy networks, they have brought special challenges that need to be seriously analyzed for reliably operating them at high levels in the system. In other words, the stochastic behaviors of the RERs in energy generation have created difficulties in reliable operation and scheduling of the energy systems [3]. Hence, the studies in the field of clean energy production units are extended by proposing practical solutions that are supposed to effectively support the renewable-based systems. One of these solutions is to store energy when the amount of energy production by the RERs is greater than energy consumption, using the different technologies of the storage systems, and use the stored energy in times when the amount of their outputs is less than energy demand. The ESSs are proposed for alleviating the negative impacts of the RER fluctuations, which leads to the introduction of the different technologies for storing energy with special features. Herein, the optimal integration of a large number of RERs in the system is another key suggestion for renewable-based energy networks, which enables the system for appropriate usage of the advantages of the different types of RERs in the clean energy production process during a day. Although this solution is recognized as an effective way to increase the reliability of the energy supply in the systems with a high share of RERs, the implementation of the related plan requires innovative technologies in the hybrid networks [4]. In addition, to consider the energy generation sector for addressing the RERs' challenges, several studies have also focused on the prosumer-side aiming to propose the energy management schemes for renewable-based grids. In this regard, the novel demand-side energy management schemes are proposed to not only manage the energy consumption in the prosumer-sector but also improve the flexibility of the system by changing the consumption patterns of the consumers. Demand response programs are common examples of these schemes that are provided in various classifications.

All of the mentioned efforts have been carried out to effectively increase the penetration of low carbon energy production technologies in hybrid networks. This is while the future modern MCENs are targeted to be fully equipped with the RERs for 100% clean energy production. In response to this need, a comprehensive investigation is required for all routes whose final destination is to equip the system with a high level of clean energy resources. For this aim, this chapter is aimed at analyzing how the current systems and strategies can be changed to be suitable for the modern structure of future MCENs. Due to this, we need to scrutinize all the effective parts of the system in detail to examine their potential for assisting the realization of the grid modernization goals.

3.2 Energy Generation in the Future Modern MCENs

Since the time of operation of the first central energy generation unit by Edison on Pearl Street in lower Manhattan, New York [5], the energy generation has undergone many changes, and these changes are still ongoing. This evolution in energy production is quickly going toward the low carbon technologies so that it is expected to become fully RERs in the near future. Due to this trend, the future modern MCENs are targeted to equip with 100% RERs to meet the energy demand through cost-effective and carbon-free energy production units. Although the goal of the energy system is to be fully equipped with RERs for the future energy networks, the existence of conventional energy production units in the current structure of the network has made it necessary to examine how the current energy production units will be operated in future energy networks. Generally, energy production can be categorized into two groups: centralized and decentralized energy generation systems. Both energy generation mechanisms are fully discussed in the following subsections.

3.2.1 Centralized Energy Production Units

Centralized energy production units refer to huge power plants throughout the energy networks. Because these units have a high energy production capacity as well as the expensive construction and/or operation, their number is usually small for each region. The amount of energy generation by conventional power plants can be controlled that enables the power grid for reliable scheduling in the system. In addition to these valuable characteristics that enable load following service, the traditional centralized energy generation systems have some remarkable disadvantages and are as follows:

- Nearly 70% of the traditional centralized energy generation systems are fossil fuel-based units, which not only impose the high fuel consumption cost for the system but also release a considerable amount of greenhouse gas emissions in the atmosphere. This is while continuously consuming non-renewable energy resources has resulted in day by day decreasing them all over the world.
- Constructing the conventional power plants near the load centers has created significant health problems for humans and other living organisms due to releasing the high volume of environmental pollutants.
- Because of the limited number of traditional power plants that operate in the system due to their large size and high costs, the energy production should be transmitted from the power plants to all load centers in the region. Therefore, a considerable amount of the generated power is lost in the transmission lines. In addition to the wasted power in the lines with more costs, the foundation of the transmission lines and their maintenance costs have made the centralized energy production mechanisms as one of the costly plans in the energy networks.
- In the centralized mechanism, the occurrence of any faults in the system, such as in the power plants or transmission lines, will result in system disruption and cause a vast blackout in most cases. If the faulted unit is one of the key and largest power production units in the system, exiting such systems from the energy generation process can lead to serious difficulty in recovering the system and may lead to a blackout in the overall system. Such events in the centralized systems impose irreparable costs on the system.
- Given the large production capacity of the centralized power units, the cost of transmitting and distributing energy from these units to the end-users is also high.

The aforementioned disadvantages of centralized energy generation mechanisms have made them not to be the sole candidate for developing energy networks. The construction of new centralized energy production systems or the development of their current units is no longer preferred to respond to the increasing growth of energy consumption. Therefore, many efforts have been made to propose new energy generation mechanisms for appropriately addressing the aforementioned challenges.

Here, three types of centralized energy production units are briefly described in the following subsections to clearly highlight their characteristics for evaluating how they can effectively be joined to the modern structure of the future modern MCENs. This is while the variety of the centralized energy generation units is not limited to the aforementioned power plants and there are other types of such units in the power system that have similar energy generation processes with the discussed units such as combined-cycle power plants and so on.

3.2.1.1 Thermal Power Plants

Thermal power plants are a type of traditional systems that use heat energy for electricity generation. The core of these units is the generator, and that its shaft is coupled to the shaft of a steam turbine. The energy production process in such power plants starts by burning fossil fuels and generating heat to convert water into steam. The produced steam is reheated to the point where it turns into dry steam. When the steam is generated at the desired temperature, it is directed to the steam turbines. As a result of the collision of dry steam with the turbine blades, the turbine's shaft rotates and leads to the rotation of the generator rotor. Since the speed of the generator rotor determines the amount of electrical energy generated in the system, it is possible to adjust the speed of rotation of the turbine shaft, and thus the speed of the generator rotor, by adjusting the amount of steam entering to the steam turbine. This is done automatically in the thermal power plants for stabilizing the frequency in the system by balancing energy between supply and demand. The fuel used in this type of power plant is usually gasoline and mazut, which cause significant amounts of environmental pollutants to enter the atmosphere. Thermal power plants have a high amount of ramp up and ramp down, which slows down their rapid response to changes in the energy balance, especially in the presence of a high level of RERs. On the other hand, the amount of minimum up and down times for them is also high so that makes these units stay on or off for a certain period of time before changing the on/off status. This limitation imposes additional costs for the system when the grid needs low energy generation at certain times while these units should stay on due to their minimum up time limitations that lead to higher energy production cost.

3.2.1.2 Nuclear Power Plants

In nuclear power plants, electrical energy is generated using the process of nuclear fission. The only difference between thermal and nuclear power plants in generating electrical energy is related to how to provide heating energy for converting water to steam. Indeed, in the thermal power plants, this energy is obtained by burning fossil fuels. This is while the energy is provided from the nuclear fission process in the nuclear power plants. In these units, the heating energy production directly depends on uranium. The process of energy generation in nuclear power plants can be fully accessed in Ref. [6]. Some of the significant advantages of these units are as follows:

- Energy production by nuclear power plants has lower environmental problems.
- Their technology is mature and easily available.
- One nuclear power plant alone can generate a lot of electricity.

In addition to the above advantages, there are some drawbacks to them that need to be considered in operating them. Environmental problems caused by nuclear waste and limited uranium resources are some of the problems associated with this power plant.

3.2.1.3 Natural Gas Power Plants

The natural gas power plants can also be named gas-fired power stations or gas-fired power plants. Compressors, combustion chambers, and gas turbines are the three main components of a gas power plant. In these units, the electricity generation process is started by burning natural gas for producing heating energy. The generated heating energy is combined with the air that is compressed by the compressor to form a compressed hot air. The heated air is then directed to the gas turbine to rotate the turbine blades, causing the electric generator shaft to rotate. The natural gas power plants use natural gas to generate heating energy, which creates lower greenhouse gas emissions in comparison with gasoline and mazut used in the thermal power plants. The amount of ramp up and down (the minimum up and down time) parameters for these units is also higher (lower) than the thermal power plants, which enables them to quickly react to any changes in the energy networks. Moreover, the interval between turning on these power plants and generating energy is only a few minutes. However, the mentioned time may be a few hours for thermal power plants and a few days for nuclear power plants.

In the systems with a high or full share of stochastic producers, the operation of the regulated capacity of the fast-response and controllable devices is necessary to increase the system reliability and timely react to the changes in the RER outputs for continuous energy supply. As mentioned before, modern MCENs are targeted to operate with 100% RERs for fully clean energy production. This does not mean that we should turn off all centralized power plants that cost a lot of money for their construction and operation. In other words, we can see the concept of 100% RERs for the modern grids from this viewpoint that constructing new units or developing current traditional power plants is not expected in the plans for the future modern MCENs. However, now we need to effectively schedule for the current power plants in a way to efficiently use their potential in future power grids. In the energy networks, some large factories in the industrial level are usually incorporated in forming the baseload in the energy network. Due to the fact that this type of consumer is considered as indispensable loads, the energy network is responsible for their continuous energy supply and if this energy is not provided at any times to them, it will have to pay a lot of penalties for these consumers. Despite these, loads can be reliably met using the full RER system by employing the innovative ways that we will discuss in Chap. 5, but the energy network can also use the potential of those centralized power plants with minimum greenhouse gas emissions in supplying continuous energy to the indispensable demands. For this aim, existing nuclear power plants can be operated to cover a large portion of the baseload in modern energy networks. Although these units are taken into account as the carbon-free energy production units, a major concern about them from the environmental viewpoint is the production of radioactive wastes such as spent reactor fuel, uranium mill tailings, and other radioactive wastes. This is while if there are capable technologies to reliably minimize nuclear wastes, these units can be considered for limited presence in the future modern networks.

Although the power generation capacity of gas power plants is normally much lower than the nuclear and thermal power plants, they are the best choice among other centralized energy generation units for modern networks due to advantages such as fast-response, lower start-up and downtimes, as well as lower environmental issues. In other words, the natural gas power plants are the suitable units for future modern MCENs with 100% RERs as the backup systems that can quickly react to any changes in the hybrid network for dynamic energy balancing. This means that they can be intended as the first priority units among the conventional power plants that can rapidly start energy generation as the backup system. Indeed, these units are expected to operate only for limited hours to support a renewable-based system in balancing energy supply and demand. The produced CO_2 emissions by gas power plants in these limited hours can also be converted to the carbon and oxygen using the conversion technologies at first and then can be used for creating new combinations such as CH_4 (methane gas) using the energy conversion technologies such as Power-to-Gas (P2G) or Power-to-Hydrogen (P2H) in the modern grid. Thereby, applying such strategies can effectively reduce the negative effects of the gas power plants in the modern structure of the MCENs. On the other hand, thermal power plants are typically operated in the 100-megawatt ranges and create harmful effects on the environment. Therefore, these units can be considered as the last option for the future modern grids for establishing energy balance in the system. Indeed, thermal power plants are expected to be used only in the hours that the energy network is in the emergency conditions and the possibility of widespread blackouts in the network due to the large fluctuations in the energy production of the stochastic producers is very high if the huge backup units like thermal power plants do not enter the energy generation process. Therefore, the main mission of the thermal power plants in the future modern MCENs can be system support and prevention of global blackout in the hybrid network on sensitive moments. The priority of the centralized energy generation units in supporting future modern MCENs with a full share of the RERs is depicted in Fig. 3.1.

3.2.2 Decentralized Energy Production Units (DEPUs)

In recent decades, due to disadvantages associated with the centralized energy generation units, the research trends in the energy sector have shifted sharply toward finding alternative solutions [7]. As a result, decentralized energy production units

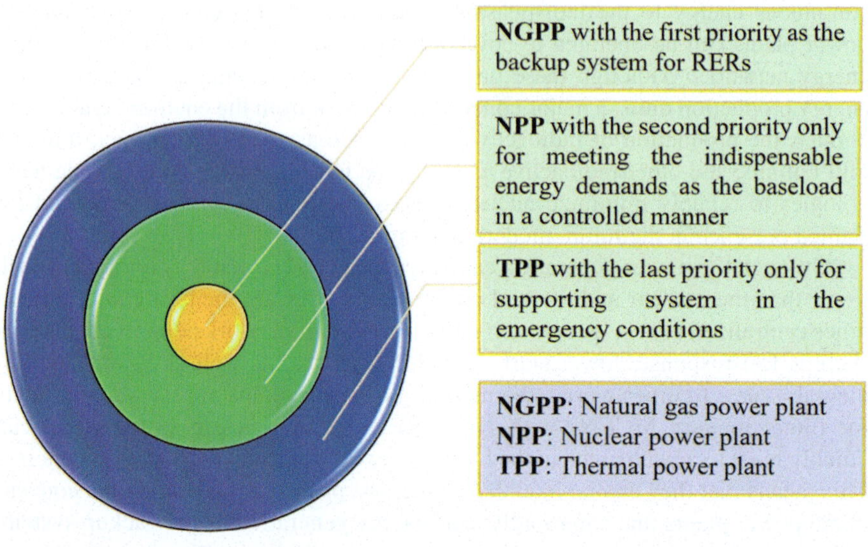

Fig. 3.1 The priority of the presence of centralized energy production units in the future modern MCENs

were introduced and distributed throughout the energy networks for decentralized energy generation [8]. Due to this, decentralized energy production units were known as the DERs and were introduced with the aim of eliminating the disadvantages related to centralized energy generation systems and solving some of the key challenges in the energy production sector of the hybrid network. The decentralized energy generation units are divided into two categories: renewable and non-renewable energy resources. Some of the important advantages of these units are as follows but are not limited to the following items.

- The DERs cause fewer environmental problems in comparison with the centralized energy generation units. The non-renewable types of DERs typically burn natural gas for producing energy like the natural gas power plants, which release a lower amount of greenhouse gas emissions in comparison with burning the gasoline and mazut as the fuel. This is while the renewable version of the DERs does not release any greenhouse gas emissions to the environment.
- The required capacity of the DERs can be installed in remote areas that cannot be connected to the energy network for supplying the energy demand. This issue not only reduces the cost of developing the transmission systems for transferring energy to the remote area with lower energy consumption but also significantly reduces the energy losses in the transmission lines.
- The DERs can be easily distributed throughout the energy network without requiring special developments in the transportation parts.
- Because of existing DERs and distributing them throughout the network, if any of the DERs goes out of the energy network's circuit due to the faults in its system,

only a very small capacity of the network's production capacity is reduced and the network can easily supply the cut-off unit's energy without any interruptions.

- The variety of the DERs allows the energy network to maximum use of the environmental potential for clean energy production or to operate the gas-fired systems in different capacities.
- The operation of the numerous DERs increases the flexibility of the energy network in facing any changes in the system, enhances the reliability of the continuous energy supply, and improves the energy efficiency and stability of the grid.
- The multi-carrier DERs such as combined heat and power systems not only provide the opportunity of simultaneously generating electricity and heating energy but also enhance the total energy conversion efficiency in the system.

3.2.2.1 Non-Renewable Energy Resources

The first group of the DERs is the non-renewable energy resources that mainly use fossil fuels to generate energy. The common fuel for most of these units is natural gas, which produces a lower amount of greenhouse gas emissions in comparison with burning oil products. These units are the controllable systems that can be set for energy generation in accordance with the amount of energy load. This feature enables the system to reliably meet the energy load and improve the sustainability of the power grid. Although the non-renewable energy resources bring this valuable advantage to the system, they use fossil fuels for generating energy that not only dwindles the sources of natural gas, oil, and other non-renewables throughout the universe but also result in negative environmental issues. Some of the most important non-renewable-based systems are illustrated in Fig. 3.2. Each of them is described in detail in the following subsections.

3.2.2.1.1 Diesel Generator

A diesel generator (DG) consists of a combination of an electric generator with a diesel engine for electrical energy production. The DGs usually burn diesel fuel for generating energy, but some of them are designed to run on natural gas or liquid fuels. Some of the features of the DGs are as follows:

- They are typically operated in specific places, especially in remote areas, without connecting to the energy network.
- They are considered as a suitable backup option for some sensitive places such as the hospitals for emergency energy supply if the power grid fails for any reason.
- They can be effectively used for some important supports in energy networks such as peak shaving.

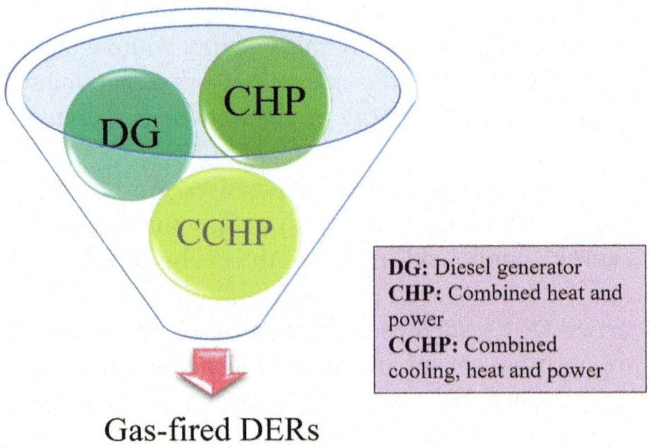

Fig. 3.2 Some common non-renewable energy resources in the energy network

3.2.2.1.2 Combined Heat and Power (CHP)

With the initial use of DGs in power networks, their significant benefits quickly became apparent, and the widespread use of these units at the network level was on the agenda. Given the growing popularity of these units in power grids, researches into them have strengthened their technology to increase energy efficiency. Studies on energy loss in these units led to valuable results that eventually concluded in the emergence of new ideas in the field of construction of these units. Given that a large amount of heating energy is generated by these units when electrical energy is produced, the researchers focused on the optimal usage of this generated heat energy. On the other hand, the need for heating energy along with electricity for the areas where these units were used was also felt more and more. Therefore, all these reasons had increased the motivations for improving the manufacturing technology of these units, so a new version of them, which had the ability to simultaneously generate electricity and heating energy, was unveiled for energy networks. Currently, we know these resources as the combined heat and power (CHP) systems. These units are taken into account as a part of the cogeneration systems and have satisfactory energy efficiency compared to the centralized energy generation units. Indeed, the CHP is a common word that refers to any energy production unit that can produce both electrical and heating energies at the same time. According to the second law of thermodynamics, in some of the conventional power stations, approximately 40% of the thermal energy used in the energy generation is wasted [9]. This is while a large proportion of this wasted heating energy is captured for meeting the heating energy demand of buildings or other places. A comparison between the separate generation of the electrical and heating energy and their cogeneration in the CHP units is demonstrated in Fig. 3.3 [9].

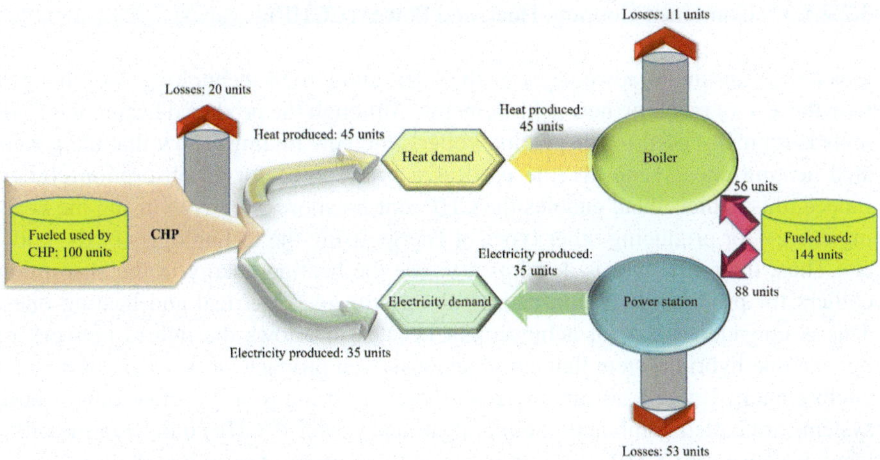

Fig. 3.3 A comparison between the separate generation of the electrical and heating energy and their cogeneration in the CHP units [9]

To sum up, CHP units can be preferred for simultaneously generating electricity and heating energy due to the following reasons:

- They can reduce greenhouse gas emissions.
- They have lower operation costs.
- There are significant incentives in government policies for them.
- They increase the security of energy supply in hybrid networks.
- They can efficiently use renewable fuels.
- They are one of the key systems that can be used for increasing the interoperability between the electric power system and district heating networks (DHNs) in the future MCENs.

Moreover, there are different types of CHP units that enable the hybrid network to operate a suitable type of them based on the various conditions in the network. Some of the important types of CHP units are as follows. The detailed descriptions for each of the CHP units can be fully found in Ref. [9].

- Spark-ignition gas engines (50 kW to 10 MW)
- Mini or small-scale CHP (<50 kW)
- Micro gas turbine CHP
- Individual dwelling CHP (<2 kW)
- Larger-scale CHP
  - Gas turbines
  - Dual-fuel reciprocating engines
- Fuel cells

3.2.2.1.3 Combined Cooling, Heat, and Power (CCHP)

Given the substantial advantages of the CHP units, their technology was changed with the aim of reaching new achievements. Although the produced heat in the CHP units is typically used for displacing boilers, the new findings show that the generated heating energy can also be used as a heat source for an absorption chiller. Indeed, this achievement enables the CHP unit for more effectively using the heating energy for producing other types of energy using the technologies of the chillers. Thus, their structure is developed to use the heating energy in the absorption chillers for producing cooling energy along with the electrical and heating ones. This is considered a great achievement because it allows the hybrid network to operate one hybrid system that simultaneously can produce electrical, heating, and cooling energy, which are mainly required in most areas at the same time. These systems are called combined cooling, heat, and power (CCHP) units. On the other hand, because the CCHP units can generate three types of energy at the same time, they are considered as one of the tri-generation systems in the hybrid networks. The CCHP unit consists of two main parts, the power generation unit (PGU) and the boiler. The PGU is responsible for generating energy and has a gas turbine to support the system for heating energy production. This is while the main mission of generating thermal energy (heating and cooling energy) is the responsibility of the boiler system [10]. The structure of the CCHP unit is illustrated in Fig. 3.4.

The technologies of the non-renewable energy resources started from DGs and developed day by day to the CHP and CCHP systems. It is also expected to be developed to the multi-generation systems in the future.

Given the features of the different types of non-renewable energy resources, their capability can be scrutinized for effective presence in the future modern MCENs. Due to this fact that the future modern MCENs are targeted to be structured based

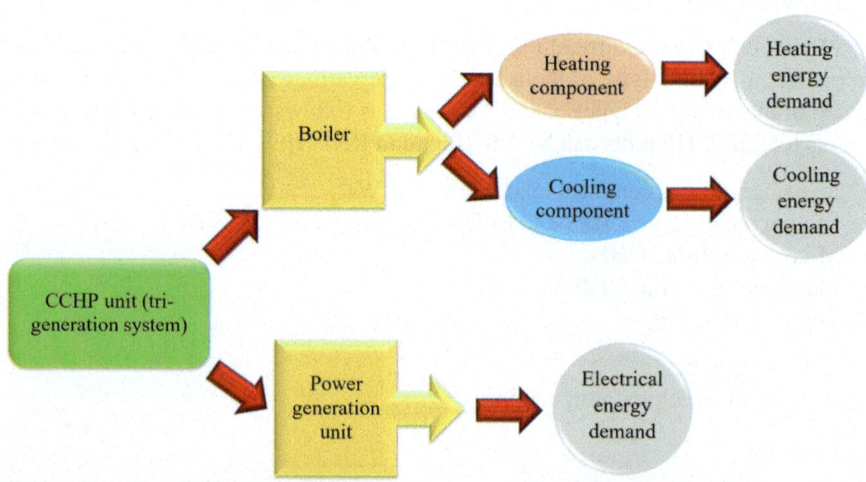

Fig. 3.4 The structure of the CCHP unit

on the different dimensions of energy, the multi-energy generation systems will be more applicable to create the widespread interactions between the electric power system, natural gas network, the DHN, and water distribution system. It is obvious that all of the non-renewable energy resources burn fossil fuels for energy production. In order to present such units in the future modern MCEN infrastructure, the gas-fired version of them is a more appropriate choice due to releasing a lower amount of greenhouse gas emissions in comparison with other fuels. This is while the CO_2 produced by these units should be decomposed into elements such as carbon and oxygen during various chemical reactions, and new environmentally friendly compounds need to be produced using different technologies. Indeed, the new compounds should not only sound reasonable from the environmental aspects but also their production process should be justifiable in terms of economic viewpoints. By observing these conditions, the gas-fired hybrid systems can be effectively operated in the future modern MCENs. Among these three different types, the gas-fired CCHP unit is more deserving than others due to its ability to simultaneously generate electrical, heating, and cooling energies in the hybrid network.

3.2.2.2 Renewable Energy Resources (RERs)

The environmental problems and high fuel costs are two general concerns associated with non-renewable energy resources that have prompted the current research in the field of energy to find alternative solutions [11]. The result of these efforts has led to new achievements to use the environmental potential for energy production. In this theory, the energy of the natural drivers in the environment such as wind speed and solar radiation is targeted to be used for energy generation. Because these energy sources are unlimited, they are recognized as renewable-based energy resources. The RERs allow the energy networks to generate clean energy using the appropriate devices such as wind turbines without requiring any fuels and releasing any greenhouse gas emissions to the atmosphere [12] and with minimum cost [13]. Figure 3.5 depicts some important and common types of RERs that are used all over the world for clean energy production.

Moreover, some of the key advantages of RERs are listed below.

- RERs will not run out and their sources are unlimited.
- RERs provide more economic benefits with zero operation and fuel costs and lower maintenance costs.
- RERs provide substantial environmental benefits without requiring any fuel for energy production.
- RERs significantly reduce the dependency of the energy network on the international non-renewable energy resources and increases energy security.
- RERs require less personnel due to low maintenance needs.

Given the International Energy Agency (IEA) reports regarding the amounts of the renewable electricity capacity additions [14] along with the renewable electricity generation by region and scenario [15], the operation of RERs has always been

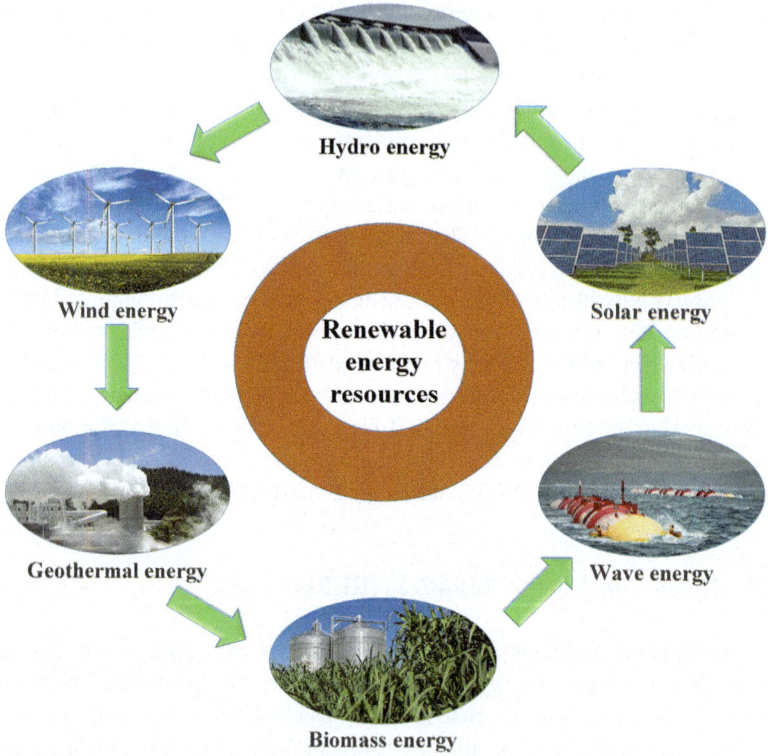

Fig. 3.5 Some common types of RERs in the energy network

on the rise, and forecasts show that this trend will continue until the goal of 100% RERs is reached for different regions. Among different countries that are scheduled for high adoption of RERs, China has the most investment in RERs to meet the high portion of the energy demand using clean energy production [14]. The forecasts for renewable electricity generation are carried out by IEA for different regions until 2040 [15]. These forecasts show that the amount of clean energy generation in the world in 2040 will exceed 25,000 TWh. Countries in Asia will have the largest portion of the forecasted clean energy production in 2040 [15]. All these forecasts indicate this inevitable fact that the world is going to substantially increase the operation of RERs and this trend is expected to continue until reaching the main goal of 100% RERs.

In addition to the valuable advantages of the RERs, they also have considerable disadvantages that can cause significant challenges to the energy network. Some of the remarkable disadvantages are as follows:

- Though RERs bring more economic benefits for the energy networks, their investment costs are pretty expensive.
- The energy production by the RERs directly depends on the climate conditions and is very uncertain, which has made the optimal scheduling of the system more unreliable.

- The intermittences in RERs' outputs require other equipment and technologies to make optimal use of the clean energy produced, which is mainly complex and expensive technology.
- Providing a sufficient level of reliability and resiliency for the energy networks is becoming more challenging than ever before in the presence of RERs.
- It is not possible to exploit them anywhere due to the geographic limitations and variety of weather.

Some of the important types of RERs are described in the following subsections, aiming to clearly identify their characteristics for effectively investigating their strong presence in the future modern MCEN structure.

3.2.2.2.1 Wind Energy

Wind energy is one of the remarkable natural capabilities of our universe. The effective usage of wind energy can result in generation of a large amount of clean energy in minimum time. Due to the use of wind power for clean energy production, wind turbines are developed in different capacities and structures. The wind speed is the vital element for generating energy in the wind turbines, and its intensity determines the level of the produced energy in the system. In the wind turbines, the turbine blades are designed in a way that can effectively absorb the wind power for rotation. These blades, whose number in each wind turbine is typically three, are connected to the shaft of the turbine that is coupled to the generator rotor. Therefore, by rotation of the wind turbines' blades, the rotor of the generator subsequently rotates to generate the electrical energy. A set of wind turbines in a certain region forms a wind farm. One wind farm can produce a huge amount of clean electricity when the wind speed is in the desired conditions in the related area. They can inject a high amount of clean energy into the energy network and in a case that their generated energy is more than demand, the surplus energy should be stored or it will be wasted. On the other hand, the amount of wind power generated from 1 hour to the next may change drastically, which can be even in the range of megawatts. In other words, the output power of wind turbines has very high ramps, which causes a lot of uncertainty in the energy network. Therefore, under these conditions, using appropriate methods to effectively model these random behaviors of wind turbines in energy production will be inevitable in optimizing the energy grid.

3.2.2.2.2 Solar Energy

Solar energy is one of the RERs that has remained as one of the most viable sources of energy [16]. This energy has enough potential to be sustained and can effectively support all activities regarding the following items [16]:

- Supplying heating energy to the lands and atmosphere
- Supporting the life of animals
- Supporting the life of plants

- Generating fossil fuels
- Warming the oceans
- Driving the water cycle
- Generating its wind

Solar energy is one of the infinite and powerful sources of energy that can strongly facilitate the transition process from the dark environment caused by the fossil fuel-based systems to the green and pollutant-free environment. In such a process, solar radiation is a key parameter that is essential for solar energy production and its intensification determines the amount of energy production in the solar-based systems. Some of the significant applications of solar energy, which have necessitated global attention, are as follows [16]:

- Solar heating
- Solar architecture
- Solar thermal energy
- Supplying energy to natural processes
- Exciting electrons in a photovoltaic cell
- Molten salt power plants

As mentioned before, the amount of energy production by the solar-based systems directly depends on the amount of solar radiation. During a day, solar radiation has the maximum amount at noon when the amount of energy consumption in the energy network is at the highest level. This means that solar panels, for example, have the maximum clean energy production at peak times when the energy network needs to receive more energy from the energy production sector for balancing energy. Indeed, solar panels are taken into account as the key RER units to support system especially in addressing some challenges such as peak shaving. In addition to the aforementioned benefits of the solar-based systems for the energy networks, similar to the other RERs, they also have substantial fluctuations in their outputs, which cannot be exactly predicted for supporting confident schedules in the system. This inspires the serious requirement of the RERs to the innovative solutions for optimal operation of the energy networks in the presence of a high level of them.

3.2.2.2.3 Hydro Energy

Generally, a set of efforts for harnessing falling water to generate electrical energy has led to the definition of a new type of RERs that is called hydro energy. In this regard, hydroelectricity or hydropower (an electrical form of the hydro energy) remains as the widely used energy resources, most established, and long-lasting RERs for electricity production in energy networks. In other words, they have a great potential for generating a large amount of clean energy at different scales and can be operated from very small (~kW) to very large (~GW) capacities [16].

However, the power generation depends crucially on rainfall and topography to provide enough water flow and fall. This is while their installations are not only limited to energy production but also they are used for other purposes such as pumped storage of water for subsequent system, the supply of water, and flood control. In order to effectively use this source of energy, large dams are usually built along rivers so that proper turbines and electric generators are installed in the right places inside the dam's foundation. The location of the mentioned equipment inside the dam foundation depends more on the type of dam, but, in general, these devices are installed in the lower parts of the dam to create a high height with the water level collected behind the dam. In fact, high altitudes cause the water to hit the turbine blades with a lot of pressure when released, and, as a result, the high speed of rotation of the turbine blades results in the production of more electrical energy. Although the hydropower plants can participate in the frequency control in the energy networks due to their high speed in energy production and large capacity, their potential for energy generation directly depends on the geographical conditions and climate changes. In other words, geographical constraints prevent the construction of these units in any place. On the other hand, climate changes have affected the pattern of rainfall in different regions and the key element in energy production by these units, that is, the amount of available water, is strongly affected.

3.2.2.2.4 Tidal and Wave Energies

The oceans all over the world offer a vast amount of clean energy resources [17]. The leverage of technologies is developing to harness the energy of seas day by day. The origin and characteristics of ocean energy are illustrated in Fig. 3.6 [18]. In this regard, tidal and wave energies are two common energy sources to generate clean energy in vast ranges. Tidal energy is typically generated by the tides of the ocean. The spin of the earth and the pull of gravity from the moon are two reasons for producing tides in the oceans. There are three ways for harnessing tidal energy, which are as follows [19, 20]:

- *Tidal barrages:* Tidal barrage plays the role of a dam in the energy production process. In this process, going high and dropping of tide have respectively led to filling up and down the water in the reservoir. In both states, the water in its path of movement hits the turbine blades, causing them to spin, resulting in energy production.
- *Tidal fences:* The structure of the tidal fences is smaller than a barrage. In this structure, a fence is formed using several vertical turbines between two landmasses, which allows for energy production through the collision of water with the turbine blades and spin for generating electricity by the electrical generator.
- *Tidal turbines:* These turbines are usually placed anywhere that the tidal flow is strong enough for spinning the turbines' blades for energy production.

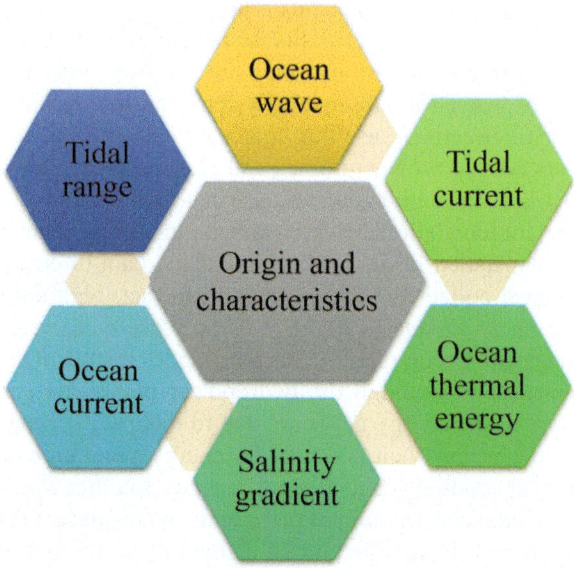

Fig. 3.6 Origin and characteristics of ocean energy

The clean energy produced by the waves of the ocean is called wave energy. The existing wind across the surface of seas or oceans forms the waves that can be used for generating wave energy. Three typical ways can be used for capturing the energy of waves and are as follows:

- *Surface devices:* These devices take the energy of the waves through their ups and downs and convert them into electrical energy.
- *Underwater devices:* These devices are long tubes stretched over a long distance or balloon-type objects in the ocean floor that are equipped to the suitable type of turbine and electricity generator. By hitting the ocean waves with the turbine's blades, the electrical energy is subsequently generated.
- *Reservoir:* In these devices, water is moved into a reservoir by the ocean waves on the coastline. When water returns to the ocean after hitting the coastline, it hits the tube, causing the turbine blades to spin and generate electrical energy.

Although the tidal and wave energies have created significant opportunities for clean energy production in the hybrid networks, they have also brought some challenges, which are as follows [21]:

- The number of appropriate locations for economically installing them is small.
- The construction, installation, and maintenance costs of the tidal and wave power plants are very high in comparison with other RERs.
- Their operation has negative impacts on the life of marine animals. For example, migration of fishes is very difficult in the presence of large tidal barrages.
- The activities related to the operation, manufacturing, and maintenance of the ocean power plants' devices can injure marine animals.

3.2.2.2.5 Geothermal Energy

Geothermal energy refers to the energy deep inside the earth. The word geothermal is the combination of the two Greek words: geo, meaning "earth," and thermal, meaning "heat" [22]. Geothermal energy is an almost infinite source of clean energy that can provide a large part of the world's energy. The core of the earth is very hot and its temperature is estimated at about 7600 °F [22]. In geothermal energy, pipes made of suitable materials are connected to the appropriate depth of the earth by drilling, so that at this depth, the temperature is sufficient to convert water into dry steam. Water is pumped from the surface of the ground and released through these inlet pipes to the depth of the ground, where it is converted to dry steam due to high heat and then transferred to the surface of the ground through the outlet pipes. At the surface of the earth, the ends of the outlet pipes are connected to the turbines so that the dry steam from them leaks into the turbine blades, causing them to rotate and generate electrical energy from the electric generators. Due to the increasing energy consumption and the need to use clean energy resources for energy production, considering their benefits in terms of economic and environmental aspects, geothermal energy can effectively provide a high portion of this energy in the future modern MCENs. Despite the undeniable benefits of the geothermal energy in clean energy production, they have some remarkable drawbacks as follows [22, 23]:

- The construction and operation of geothermal power plants is a very difficult and expensive process.
- They need innovative technologies for effectively using the energy in the depth of the earth for extracting energy with minimum losses.
- Geographical limitations negatively affect the development and construction of these power plants.

3.2.2.2.6 Biomass Energy

Biomass is the first-ever fuel that is used by mankind for energy generation. Although solar energy is recognized as the "mother" of all other types of RERs, for all multi-cellular organisms, the biomass is taken into account as one of the sources of energy [24]. The term biomass plays a role in the general term and it is like an umbrella term that includes plant biomass or phytomass and animal biomass or zoomass. Sun's energy is stored or fixed in the form of aquatic vegetation and terrestrial when it is intercepted by plants and converted into chemical energy by continuing the photosynthesis process.

When the animals use the vegetation as food (grazing), they convert the vegetation into animal biomass (zoomass) and excreta. In this process, the terrestrial animals' excreta can be considered as a source of energy. This process fits for dairy animals while the collection of the excreta and processing them for generating energy is not possible for the aquatic animals.

In the aforementioned subsections, we attempted to briefly describe the key points of the various technologies of the RERs and their detailed descriptions can be accessed in Ref. [25]. Given the features of the different types of RERs, their advantages and disadvantages are more highlighted aiming to provide a suitable overview of them to facilitate the decision process for adopting them in the hybrid networks. One of the common features of some of the RERs is that their energy production has uncertainty. Indeed, they show stochastic behaviors in producing energy during a day. If the future modern MCENs are aimed to equip with 100% RERs, the innovative solutions are necessary for reliable operation of the system in the presence of a high level of them. For this aim, the trend of the researches is going toward the finding of new practical solutions for mitigating the negative impacts of the RERs' fluctuations in the energy networks. In this regard, the ESSs were introduced that emphasize on using new devices that have the ability to store energy when the amount of RERs' outputs is greater than energy consumption and supplying it to the energy network when the energy consumption dominates its production. By increasing the penetration of RERs in the energy network, the exploitation of ESSs is subsequently increased, which has led to substantial developments in their technologies and operating the various types of them in the system. The following section discusses the applications of the different types of ESSs in the energy network and analyzes their potential for effectively participating in the future modern MCENs with a full share of RERs.

3.3 Energy Storage Systems in the Future Modern MCENs

As the operation of RERs and their popularity have increased in the energy networks, the ESSs' presence in the system has subsequently witnessed fundamental changes and significant overhaul. In parallel with proliferating the different forms of RERs, the intensity of these changes regarding energy storage technologies has grown in a way to introduce various types of them in the energy network. Alleviating the stochastic behaviors of the RERs perhaps can be considered as the main mission of the ESSs in the grid. In other words, they are known as one of the very effective systems for facilitating the operation of high level of clean energy production in the system. Thus, their presence will be inevitable in the future modern MCENs with 100% RERs. This is the main reason that motivated us to investigate how different types of ESSs can effectively participate in future hybrid network interactions. The following subsections discuss different forms of storage systems. Figure 3.7 shows a brief overview of the applications of the ESSs in the energy networks [26].

ESSs can be classified into different groups [26]. They can be ordered as an electrical, compound, mechanical, warm, crossover vitality stockpiling, and electrochemical framework [26]. This is while the ESSs frameworks can be additionally grouped based on their materials and their procedure of developments. The complete classification for the ESSs considering their composition materials and energy formations is reported in Ref. [27] but this list is evolving due to advanced technologies. In addition to the significant advantages of the ESSs for the energy networks, especially for the renewable-based systems, they have negative impacts when they are operated in the grid and are enumerated as follows [28]:

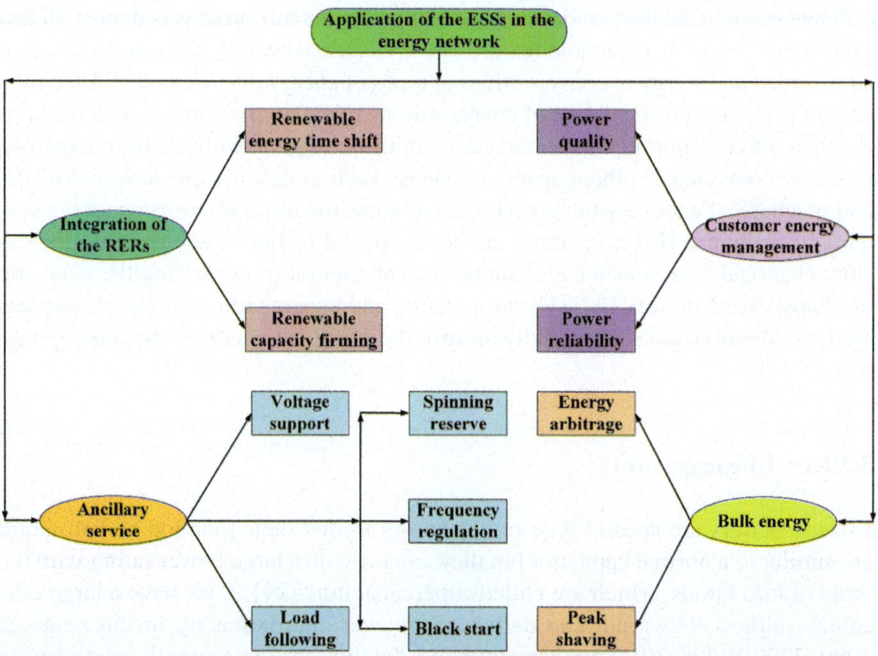

Fig. 3.7 The schematic of the applications of the ESSs in the energy networks [26]

- Their construction technology is very expensive and complex in most cases.
- A portion of the energy stored is usually lost in the energy conversion and reconversion processes of the ESSs.
- There are some substantial challenges in creating the required space and desirable infrastructure for the ESSs.
- Some types of ESSs, such as batteries, cause concerns regarding manufacturing emissions.

Additionally, challenges and barriers regarding the development of the ESSs highlight some remarkable issues and are as follows:

- ESSs should be flexibly placed in a way to provide suitable conditions for the widespread operation of RERs.
- ESSs should meet the requirements of the regulators and industry to prevent the grid interruptions in a reliable manner.

3.3.1 Electrical Storage Systems

Electrical energy is a special type of energy that can be easily converted to other types of energy. Indeed, electrical energy is a high-grade form of the energy that their conversion efficiency is higher than other energy carriers. High-grade energy is a certain type of energy that can be converted to other types of energy with more efficiency, while the low-grade energy is defined when this process is done with low efficiency. Due to this valuable feature of the electrical energy, it is usually selected as intermediate energy to convert different types of energy into each other. Electrical energy is the most critical type of energy among other energy carriers with the largest share of consumption and interactions in the energy networks. This energy can be easily converted to other carriers of energy such as gas, hydrogen, heat, kinetic, and mechanical energy, which can effectively use the different energy storage systems to maintain. However, there are some special technologies that can directly store electrical energy as the electricity that can yield it by separating the magnetic or charges field by flux [27]. Ultracapacitors and superconducting electromagnets are two common ways for directly storing the electrical energy in the energy networks [27].

3.3.1.1 Ultracapacitors

Ultracapacitors are special type of capacitors in that their function and structure are similar to a normal capacitor but they can provide a large power rating with the scale of kilo farads, which are called supercapacitors [29]. They have a large efficiency (almost 95%) and can deliver the power approximately in the range of 1000–2000 W/kg [30]. Ultracapacitors are famous for their significant ability in storing energy and are accepted as efficient devices for electric vehicle applications. Their special features in storing high-level power have led to valuable results including no need for maintenance, a long operation time, and temperature insensitivity. Ultracapacitors are typically used as an energy source during rapid acceleration required for hilling in electric vehicles and as energy storage during electric braking considering their fast charging and discharging profile [31]. Ultracapacitors are usually classified into three types, including hybrid capacitors, pseudocapacitors, and electric double-layer capacitors [30]. Among them, electric double-layer capacitors have a higher power density while they impose high cost and have high self-discharge with a lower range of energy (5–7 W h/kg) [30]. Due to this, ultracapacitors are intended for operating with fuel cells, batteries, and other electrical energy storage systems in electric vehicle applications aiming to achieve an extension of overall life, high energy densities, and a combination of high power [32].

3.3.1.2 Superconducting Magnetic Energy Storage (SMES)

Superconducting magnetic energy storage (SMES) systems act based on storing electrical energy in a magnetic form [27]. They can usually achieve power ranging from kW to MW. This is while they require a power converter for DC/AC energy supply and a refrigeration system for maintaining the low temperature in the system [32]. The SMESs are not only introduced for hybrid electric vehicle applications with batteries but also are known as one of the best candidates for utilization in the power quality and power systems applications [33]. Some of the most important features of the SMESs are as follows [27]:

- They possess a high energy storage efficiency (approximately 97%).
- They have a full energy discharge capability.
- Their life cycle is very long (100,000 cycles).
- They can quickly respond to any request in the system (quick response of milliseconds).
- Their initial cost is high (205–340 $/kW).

3.3.1.3 Battery Energy Storage Systems (BESSs)

One of the famous energy storage systems for electrical energy is battery energy storage systems (BESSs). The BESSs are a combination of one or more cells, which unite with each other for participating in the special chemical reactions for creating a flow of electrons. All BESSs consist of three main parts: anode and cathode as the catalysts and various types of electrolytes. BESSs can directly store electricity for later use and can be structured in different capacities. These features of the batteries have made them as one of the strong contenders for facilitating the high level of RERs in the future modern MCENs. Although they create special opportunities for renewable-based systems, their construction cost is very high, which has led to considering the economic aspects of their exploitation. On the other hand, their life is also significantly reduced after several charges and discharges. However, substantially developing their technologies in recent years has led to the development of different types of batteries in the energy networks and are as follows:

- Sodium sulfur (NAS) battery
- Advanced lead-acid battery
- Lithium-ion battery
- Electro-chemical battery
- Nickel-cadmium battery
- Lithium polymer battery

3.3.1.3.1 Sodium Sulfur (NAS) Battery

Sodium sulfur battery is known as one of the most promising and applicable types of the BESSs that is constructed from liquid sulfur (S) and sodium (Na). It is taken into account as a member of the molten-salt batteries and can be used in the large-scale energy with rating 1–34 MW. The NAS battery is charged and discharged through a two-way chemical process and carries the following advantages with it [34]:

- High energy density
- Low price
- High instantaneous power capacity
- Lower size and weight compared to nickel-cadmium battery
- Longevity and high efficiency
- Ability to quickly respond to changes
- Easy maintenance and monitoring due to the lack of valves, pumps, etc.

In addition to the aforementioned advantages for the NAS batteries, they have some remarkable disadvantages, which, considering them, can be helpful in effectively adopting the NAS battery in the future modern MCENs' infrastructure. Their disadvantages are as follows:

- Requires high temperature to function
- Reliability problem of the cell set
- Safety problems associated with sulfur flammability
- At high discharges, large amounts of undesirable solids (Na_2S_2) are formed

The NAS battery can be used for various objectives and possesses diverse applications in the energy networks; some of them are as follows:

- Peak shaving and load leveling
- Microgrid stabilization
- Power quality
- Wind power stabilization
- Electric vehicle applications
- An appropriate alternative option for developing the new substations in the system

To ensure the long-term performance of NAS batteries, various tests are performed on them during their manufacturing, aiming to guarantee their operation without any errors for a long time period. Some of these important tests are as follows:

- Module firing test
- Module crush test
- Module drop test
- Cell burning test

In addition to the aforementioned features of the NAS battery, it offers some special benefits for the utilities who are motivated to install them. These distinguishing attributes are as follows:

- It can be commercially available in a very short time period.
- Its purchasing cost is very economical than constructing new electric substations in the power grid.
- It possesses a favorable field experience in its previous operation for different applications.
- It has a suitable size for most of the applications (compactness).
- It allows the system to be more flexible by increasing or reducing its capacity (modularity).
- It can be easily transferred to anywhere in the energy networks (transportability).

3.3.2 Mechanical Energy Storage Systems

Mechanical energy storage systems are recognized as one of the proper options for taking the advantages of some natural forces to store inputted energy. They can be classified to work based on some special principles such as potential energy, kinetic energy, forced springs, and pressurized gas [28]. The most beneficial advantage of this type of energy storage systems is that they can be used for readily delivering the required energy whenever the mechanical works can be adopted. They are easily adaptable for converting and storing energy from tidal and wave sources as well as the water current. The mechanical energy storage systems mainly comprise of compressed-air energy storage (CAES), pumped storage systems, and flywheel.

3.3.2.1 Flywheel Systems

The main part of this storage is a massive rotating cylinder that the magnetically levitated bearings support on a stator [35]. When the amount of electricity production by the intermittent energy resources is greater than the energy consumption, the excess energy produced is used for working the motor/generator system. This system is connected to the flywheel's axes of rotation and can be operated in the role of both motor and generator. The motor/generator system consumes the surplus of energy generated in the system for spinning the axes of rotation and then the energy is stored in the kinetic form. When the amount of energy consumption dominates energy production in the system, the related clutch is dropped in a controlled manner to use the stored kinetic energy in flywheel for spinning the rotor of the generator to produce electrical energy. Flywheel energy storage systems can be classified as low speed and high speed, which can be employed for storing energy in electric locomotives to support their motion in the different sections of the rail lines [28].

3.3.3 Pumped Storage Systems

Pumped storage systems use the water movement between two reservoirs at different altitudes for storing and generating energy aiming to create energy balance in the system. In this regard, pumped hydro energy storage systems can be mainly used for building large-scale reservoirs with enough volume of water. Pumped storage systems are mostly targeted to firm the fluctuations of the RERs such as wave, solar, and wind power. In this storage, the water pump motor in the lower reservoir consumes the energy for pumping the water to the upper reservoir at a higher altitude when the outputs of the RERs are larger than the energy demand. However, at times of lower output and high energy demand, the stored water in the upper reservoir is released and is directed to the turbines placed in the lower reservoir through

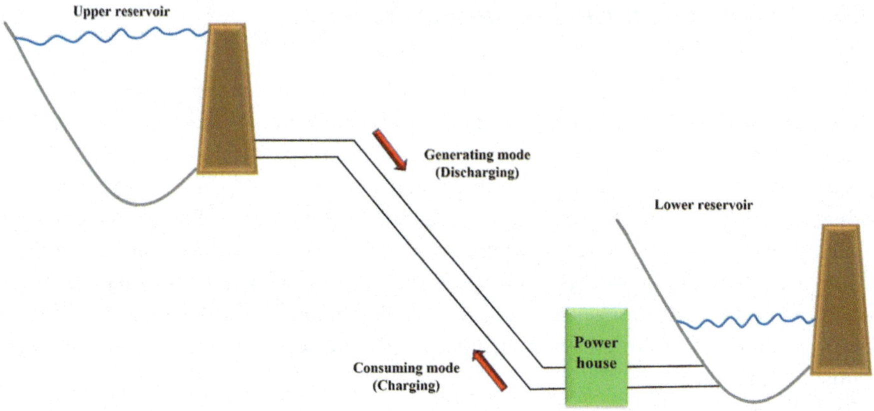

Fig. 3.8 The schematic of the pumped storage system

the corresponding channels. Afterward, water hits the turbine blades, causing them to rotate and generate energy by rotating the turbine shaft connected to the rotor of the electrical generator. The simple schematic of the pumped storage system is shown in Fig. 3.8.

Pumped storage systems can be categorized into three main types [28]:

- *Closed-loop:* This system consists of two reservoirs that are placed at different elevations, which completely rely on the water that was already transferred into an upper reservoir. In this system, none of the upper and lower reservoirs are connected to another body of water.
- *Semi-open:* This system consists of one modified or artificial reservoir and one modified lake, which can be a river impoundment with natural water flow.
- *Open-system:* This system is also recognized as a pump-back system, in which both of the reservoirs are accessed to the continuous flow of water.

The previous version of the pumped storage systems is equipped with the fixed-speed pump turbines that can participate in the frequency regulation only during its energy generation mode. The new advanced technologies allow for the development of variable pump turbines, which enable the storage system to get involved in the frequency controlling process in both generating and pumping modes. The substantial developments in the pumped-storage systems have led to the development of new technologies for storing and generating energy and some of the most important are as follows:

Seawater pumped storage This type of storage system is constructed near the sea in a way that the open sea is used as the lower reservoir while the upper reservoir is created at a higher altitude.

Compressed air pumped storage This storage system is innovatively designed so that its construction is independent of geographic requirements. Indeed, both upper and lower reservoirs in this storage are artificial that makes its construction possible at almost any location with scalable and flexible capacity. This storage consists of one pressure vessel as the upper reservoir and an open pond as the lower reservoir. The upper and lower reservoirs are connected with each other by suitable channels, and between them motor pump/generator system is placed for storing and generating energy. When the amount of energy generation in the system is greater than energy consumption, the water in the lower reservoir is pumped into the upper one and due to the isolation of this reservoir, it causes the water to compress the air inside the reservoir by rising inside it. When energy is needed in the system, the compressed air inside the upper reservoir pushes the water at the right pressure to the outside of the tank and to the lower reservoir, causing it to rotate and generate energy by hitting the turbine blades along the way.

Undersea pumped storage system In this system, a submerged pressure vessel is transferred to the floor of the sea so that it is connected to the sea through a pipe. The motor pump/generator system is installed at the surface of the vessel for storing and producing energy. In this system, a submerged pressure vessel plays the role of the lower reservoir while the sea itself is considered as a high reservoir. The extra energy produced in the network is consumed by the motor pump to transfer water from the tank into the sea and the energy is somehow stored in seawater. For energy production, as the tank's valve opens, seawater fills into an empty tank that collides with the turbine blades along the route, generating energy from the generator.

Although fast responsivity and large capacity are the key advantages of the pumped storage systems, the geographical conditions are also considered as a considerable drawback for this type of storage.

3.3.4 Compressed-Air Energy Storage (CAES)

Compressed-air energy storage (CAES) systems have great potential for creating a potent energy reserve. This storage can be classified into two models:

- Diabatic CAES
- Adiabatic CAES

In these systems, the effective potentials of underground salt caverns or abandoned caves are used for compressing air. In the diabatic CAES, the excess electricity is consumed for compressing air into the underground cavern for later use. When energy production is low in the system, the stored air is transferred toward the high and low-pressure turbines. In the path of the air leaving the cavern, it is heated using natural gas to prevent the turbines from icing up. However, in the adiabatic CAES, the heating energy generated in the comparison process is stored in the thermal energy and then after absorbing the heat energy of the air, it is compressed in the cavern. When more energy is required in the system, the stored heating energy in the thermal storage is reused to heat up the air again when it is released from the cavern.

In this process, it is no longer necessary to burn fossil fuels like the diabatic CAES for heating the exhausted air from the cavern before hitting the turbines' blades. Exemption of using fuels such as natural gas for heating the withdrawn air from the cavern has made the adiabatic CAES as one of the pure power-to-power storage systems. The efficiency of the diabatic CAES is around 42–54% while this index is up to 70% for the adiabatic storages [28]. The general advantages of the CAES are as follows:

- They have a high start-up speed.
- They have high storage capacity.
- They can effectively use existing potentials such as abandoned and salt caverns for storing energy.

3.3.5 Thermal Energy Storage Systems

Thermal energy storage (TES) systems are the appropriate devices for storing heating or cooling energy at the temperate conditions for further usage at different locations with various power or temperature. TESs are useful options for residential and industrial applications and are applicable in some processes such as hot water production, heating, and cooling processes, space cooling or heating, and electrical energy production. These systems can be categorized into three various classifications, including absorption and adsorption system, latent heat, and sensible heat [28].

- *Sensible heat system*: This system changes the temperature in the process. The capacity of the storage system is determined based on the mass of the medium used and the specific heat capacity [36]. The storage medium can be in various phases such as liquid with solid filler material (stone/molten-salt), solid (concrete, stone, the ground, or metal), or liquid (thermal-oil or molten-salt and water) [36].
- *Latent heat system*: Phase change materials are used in these systems as storage media and the latent heat cannot be sensed in contrast to sensible heat storage by the change of temperature [36]. The latent heat system uses the thermal conductivity as a key facility, and the enthalpy as well as the density are important in determining the volumetric storage capacity. The different states of the phase change materials such as bio-based (salt hydrates), organic, and inorganic are available to use in this type of storage system [28].
- *Absorption and adsorption system*: One of the indirect ways of storing heat is the thermo-chemical storage system [28]. In the contract to the sensible heat or latent, the heat is stored through a physicochemical process. For example, absorption and adsorption systems can consume or release heat in a similar process when they are in the charging and discharging modes [36]. One of the prominent advantages of the sorption energy storage systems is the high energy density due to the small volume of material.

In the energy networks, one of the common problems, especially in the summer days, is that the air conditioning systems dominate peak energy use. To address this challenge, one of the primary solutions was to use the DGs in the peak time for meeting the energy consumption of the air conditioning systems with the aim of controlling the peak energy demand. However, high investment and operation costs as well as using fossil fuels for generating energy, and subsequently creating environmental problems, are the main concerns related to this solution. On the other hand, relying on the RERs for supplying the energy needed by the air conditioning systems has imposed the investigators to annual operating cost risk and seasonal peak rate. Additionally, the RERs cannot provide controllable conditions in a reliable supply of energy in the system. All of these issues have led to the introduction of the new types of energy storage systems for storing thermal energy, which is called TES systems. The main mission of the TES systems is to store energy at various temperatures associated with heating and cooling energies. The operation of these storage systems can be most cost-effective in the following situations:

- A facility's maximum cooling energy demand is much larger than the average energy demand.
- The related tank for storing cooling energy is available.
- The availability of the electrical power system for supplying the required energy is not adequate in the relevant site.
- There is the desirable capacity of the backup cooling system in the relevant site.
- The amount of energy consumption is not well matched with the energy supply system.
- The amount of energy price is varying from hour to hour, so storing energy at times of lower energy prices and selling it at the higher ones will be economical.

In order to overcome the challenge of the energy load of the air conditioning systems that is mainly high at peak times, using the appropriate types of cooling energy storage systems is strongly proposed. There are different technologies for cooling energy storage systems, which are briefly described as follows.

- *Ice storage:* This system uses the ice for storing cooling energy for later use. Ice storage systems typically receive and accumulate the cooling energy capacity from the refrigeration plants. When the cooling energy is required in the system, they can release the stored energy at different rates and target hours. One of the useful advantages of the ice storage is that this system uses a smaller tank and offers the potential for the use of a low-temperature air system. This issue provides the desired amount of saving in space and the cost of the storage system. However, they need a more complex chiller system for generating cooling energy, which increases the system's operation costs.
- *Chilled-water storage*: In these systems, the sensible heat capacity of water is used for storing cooling energy. The operation temperature of the chilled-water storage systems is at the ranges that have made it suitable for working with the standard chiller systems. Storing the coldest water possible in these storages has

increased their capacity, which is due to absorbance of much heat from the chilled water. The chilled-water storage systems can work with the largest tanks, but they are compatible with the existing chiller systems.

- *Eutectic salts:* This type of storage system is recognized as phase-change materials. The combination of the water, inorganic salts, and other materials is used for creating a mixture that can be easily frozen at a favorable temperature. In the eutectic salt storage systems, the plastic containers are used for encapsulating the material that are accumulated in a storage tank [37]. These systems use existing chiller systems and work at the warmest temperatures, which reduces the efficiency of the storage system.

Some of the significant benefits of the cold thermal storage systems are given as follows:

- *Capital cost-saving*: Utility cash incentive programs along with the downsizing cooling equipment have provided a special opportunity for cost savings.
- *Energy cost savings*: It happens due to effectively reducing the time-dependent energy costs such as time-of-use energy charge and electric demand charges.
- *Energy savings*: Chillers usually operate at night when the ambient temperature is lower than other hours of the day, which increases their efficiency.
- *Improved HVAC operation*: The thermal load profile is decoupled from the operation of the facilities aiming to increase the system backup capacity, reliability, and flexibility.

In spite of the beneficiary features of the TES for the energy networks, a portion of the energy is lost in the thermal process of the TES. Moreover, some types of TESs require chiller systems with advanced technologies that substantially increase the total costs of the energy storage systems.

There are two common strategies for operating the TES, including full storage operating and partial storage. In the full storage operating strategy, the total amount of the peak cooling load is considered to be met only by operating the suitable capacity of the TES. This means that the chiller systems work only at off-peak hours and generate cooling energy for storing in the TES. However, at peak hours, the TES is responsible for meeting the cooling energy demand alone and the chiller systems are off at these hours. Because of using the full potential of the TES for meeting the cooling energy load at peak times without the participation of other systems like chillers, this strategy is named the full storage operating strategy. In this strategy, the effective potential of the chiller systems is not used during a day and they have to be off at peak hours. This is while the TES with high capacity is required for covering the total cooling energy consumption at peak hours, which increases the investment cost of the cooling system.

In the partial operating strategy, a suitable capacity of the chiller systems operates during all hours of the day and produces a constant amount of cooling energy for storing in the TES. This indicates that the chiller systems are planned to work at peak hours along with the TES. This issue not only allows the cooling system to effectively use the potential of the chiller system for generating cooling energy but also the system's need for the high capacity of the TES is eliminated, which results in a significant reduction in the energy cost of the system. Here one important point needs to be considered in the networks with the time-of-use energy price rate. In such networks, the energy price is at the maximum amount at peak times, and working chiller systems in these hours may not be economical for the system. Therefore, the exact evaluation is required for adopting an appropriate strategy considering the energy prices in the network as well as the investment cost of the different facilities.

The application of TES systems is not limited to the storage of the cooling energy but also effective storage of the heating energy is a key for the DHNs. The DHNs as the prominent part of the future modern MCENs is responsible for reliably supplying the heating energy in the system. In this regard, the heating energy storage systems are at the core of realizing the passing of this goal that can provide satisfactory conditions for proper participation of the DHNs in the future modern MCENs' interactions.

3.3.6 *Natural Gas Storage Systems*

The future modern MCENs is targeted to be a combination of several energy networks to effectively increase the correlation between energy networks. One of the keys of these networks is the natural gas grid that is expected to be connected with other energy networks like electric power systems, heating energy, water distribution system. In this network, gas energy plays a vital role in creating all interactions associated with the natural gas network. Due to rapidly developing energy competitive markets, we will witness the widespread presence of the natural gas networks in these markets in the near future. This will be due to the fact that the price of gas will fluctuate during the day so that in high gas consumption hours, the gas price will be high to control energy consumption, and in hours with lower energy demand, it will be the opposite. On the other hand, the natural gas grid is not an infinite source of energy and it faces different limitations sometimes in supplying a sufficient amount of natural gas in the system. In this regard, natural gas storage systems are introduced to increase the reliability of the gas supply in the energy networks. These devices can provide more economic benefits for the energy market participants by purchasing gas when its price is low and gas supply is at the desirable conditions in the system and selling it at times of higher gas prices. In addition to the economic benefits, natural gas storage systems can effectively increase the reliability of gas supply by charging and discharging gas energy in the required hours. Given the development of the hybrid systems' technology, they can support the hybrid systems by delivering an adequate amount of gas to them for converting to other carriers of energy and improving the energy supply conditions in the MCENs. Thus, they offer special opportunities for enhancing the flexibility of the MCENs. Due to the valuable features of the natural gas storage systems, the necessity of their presence in the future modern MCENs' interactions for improving the financial and technical aspects will be clearly undeniable.

3.3.7 Water Storage Systems

Water storage systems are one of the key components of the water distribution system that play a critical role in continuous energy supply in the system. Until recently, the storage of water only referred to the storage of both non-potable water for agriculture applications and potable water for human consumption. However, by the advent of new concepts in the context of integrating the energy networks, the different energy networks, such as water grids, have found a great chance to be involved in the interactions of the interconnected energy networks. Due to this, the term water storage is expanded by adding another goal, that is, the water storage with the aim of effective presence of water distribution systems in the MCENs' interactions. This is a key issue because water can be used as a fuel for different hybrid systems as well as can be intended as the main component of various chemical reactions aiming to generate special products, which can be the critical elements for achieving other carriers of energy. Hence, this interdependency condition between energy networks indicates the necessity of storing different energy carriers such as water for supporting the hybrid networks' reliability, resiliency, stability, and flexibility. This is an interoperability feature that we expect to witness an expanded and developed version of it in the future modern MCENs.

3.3.8 Hydrogen Storage Systems

Hydrogen is a clean energy carrier material, non-toxic, and highly abundant renewable fuel that is not found in pure state on this planet [38]. It can be generated from most of the primary energy sources such as from gasification of biomass and methanol, from reformation of fossil fuels, and from water by electrolysis and thermolysis [39]. After the combustion reaction, water vapor is the only release of the water as emission [40]. Meanwhile, hydrogen is known as one of the useful intermediate energy forms.

In the hydrogen storage, when the amount of the RERs' outputs is greater than energy consumption, the excess energy is used by the electrolyzer system that is responsible for the decomposition of water molecules into hydrogen and oxygen. Oxygen is released into the atmosphere, but hydrogen is compressed in a suitable pressure by the compressor and stored in the hydrogen storage system. The stored hydrogen can be used for different applications such as in direct hydrogen marketing, hydrogen refueling station, and as a primary fuel for the fuel cells. Fuel cell receives the hydrogen from the hydrogen storage systems and combines it with the oxygen for producing water and electrical energy. Then, the generated electricity is injected into the energy network and water can be used again for different objectives. The key role of the hydrogen energy storage system in the electricity generation process by hydrogen energy is illustrated in Fig. 3.9.

Hydrogen has some special features that highlight its advantages and disadvantages as a medium type of energy for storage. The amount of energy density for

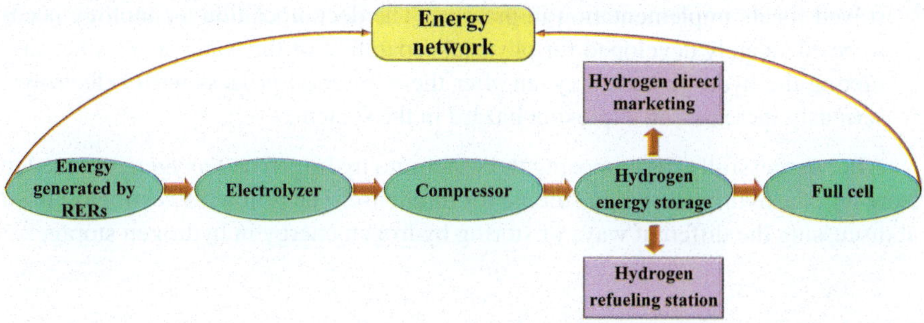

Fig. 3.9 The key role of the hydrogen energy storage system in the electricity generation process

hydrogen energy is high (116,300 kJ/kg) on a mass basis in comparison with liquid methane and other fuels [41]. This is while the amount of volumetric energy density index is very low for it, which creates the need for operating a large storage volume in the system. Employing the hydrogen storage system with a larger volume not only requires more space for operation but also imposes more investment costs on the system. On the other hand, hydrogen energy should be stored in the storage based on the high-pressure process. Storing it with high pressure increases the risk of explosion in the system and requires spending more costs to provide an acceptable degree of pressure in the storage system.

The distinguishing features of the hydrogen energy have made it one of the most promising alternatives to fossil-based fuels that are mainly a combination of carbon. It can be easily produced using the energy production of the different types of the RERs and when coupled with the fuel cell system, becomes a key element for producing energy and water in the net-zero emission process without any release of the greenhouse gas emissions. The first step for storing and generating energy using hydrogen energy is decomposing it by the electrolyzer system. There are two usual ways to decomposing the hydrogen energy and are as follows:

- *Direct thermal decomposition*: This method involves decomposing water in one step by thermal energy. However, the system requires an extremely high temperature for a reasonable yield that is typically over 5000 K [42]. This degree of temperature cannot be readily provided by the different thermal sources and seems much unrealistic. Decomposition of the water by the direct thermal process not only needs the building of the reactors to work at very high temperature and below the atmospheric pressure but also requires sophisticated technologies for the separation of product gases. This is while the hydrogen should be compressed at the higher pressures that along with the extreme temperature requirement have made the aforementioned process more difficult. However, it is expected that in the future, developing the technologies in constructing suitable types of rectors imposes more energy costs on the system.
- *Photoelectrolysis*: In this process, by immersing the semiconductor electrodes in an aqueous solution and exposing them to sunlight, the electric current along with the voltage is produced and used in the electrolysis process [43]. This process is successfully tested at the laboratory scale, but some challenges should be

solved for its implementation in practice. The decomposition technology needs to be adequately developed for physical separation of the water. This is because mixing the hydrogen and oxygen after the separation process in the electrolyte seriously increases an explosion hazard in the system.

After successfully decomposing the water into hydrogen and oxygen, we need to store it in the hydrogen energy storage for future use. The following section is aimed at discussing the different ways of storing hydrogen energy in hydrogen storage.

3.3.8.1 Hydrogen Storing Techniques

Research on hydrogen energy and its storage systems is increasingly developing to provide cost-effective, compact, reliable, and safe materials aiming to be effectively used for fuel cell technology. For a viable hydrogen economy, one of the technological problems facing the system is the lack of reliable and cost-effective solutions for storing hydrogen, which remains as an undeniable challenge [39]. In this regard, different methods are considered for storing the hydrogen energy that are mainly based on the compression, liquefaction, and solidification of the hydrogen and are described in the following subsections.

3.3.8.1.1 Compressed Hydrogen

Because the hydrogen has a lower amount of volumetric energy density, it needs a large storage tank for storing that takes up a lot of space and imposes more cost on the system. One of the solutions for this challenge is to store hydrogen in the gas format in the storage tank. For this aim, the proper level of the gas pressure should be provided for the hydrogen that is typically up to 10,000 pounds per square in [39]. Providing this range of pressure not only is a costly process but also the storage tank with sophisticated technologies is required to be reliably able to withstand this amount of pressure that is usually expensive. On the other hand, working in the high-pressure ranges seriously increases the explosion hazard in the system. Because hydrogen energy is considered for storing in a smaller space, this method is known as one of the beneficial ways for fuel purposes due to retaining the effectiveness of the energy [44]. Despite that the technology of this method is simple, the related process is gravimetrically and volumetrically inefficient [45].

3.3.8.1.2 Liquid Hydrogen

Hydrogen can be found in the liquid form at very cold temperatures. In this method, more cost should be spent on providing the required energy to convert the hydrogen to the liquid form by reducing its temperature in the desirable range. Afterward, liquid hydrogen, which is known as slush hydrogen, needs to be stored at 20 K in

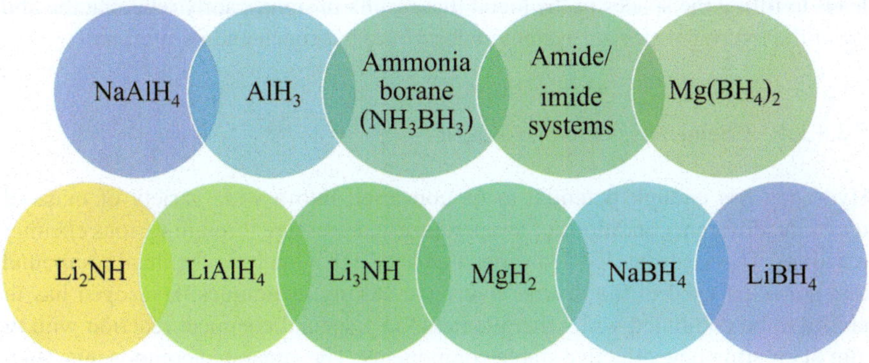

Fig. 3.10 Some of the hydrides that can realize the 9 wt% gravimetric goal in the storage system [39]

the well-insulated storage tanks aiming to maintain the allowable sub-zero temperature [39]. In recent years, the technology of composite tank materials is effectively developed, which results in stronger and lighter tanks for storing liquid hydrogen [44]. Although research is centered around providing promising ways for storing liquid hydrogen, the technology needs to be significantly developed in a way to solve the challenges of dealing with the release and uptake of hydrogen, large energy losses in high hydrogen liquefaction rates, high cost of the hydrogen tank, and hydrogen boil-off [46].

3.3.8.1.3 Metal Hydride Tanks

Metal hydrides are the combination of the metallic alloys that have a great ability to absorb hydrogen and releasing it later. These metals can do this work through heating of the tank or at room temperature and act like the sponges soaking up water. Hydrogen storing capacity for the metal hydrides is 5–7 weight percent (wt%), which can be realized only at 2500 °C or higher temperatures [47]. Currently, there are only some hydrides that can realize the 9 wt% gravimetric goal, which are illustrated in Fig. 3.10 [39].

Although the percentage of the absorbed gas to volume in the metal hydrides is relatively low, this method is famous for its unique ability in the hydrogen absorption process and offers a significant solution for storing hydrogen. One of the valuable advantages of this method is that metal hydrides can safely deliver the hydrogen at the constant pressure and the risk of explosion due to high-pressure hydrogen gas does not threaten the system. Another special benefit of this method is that the alloys act as a sponge in absorbing hydrogen and due to absorbing any impurities of hydrogen in this process, it can deliver the extremely pure hydrogen to the system. This feature eliminates the need for any state-of-the-art method to use special purifiers for releasing pure hydrogen. However, this feature is also considered as one of the disadvantages of this method. This is because absorbing the impurities in the tank leads to filling the spaces in the metal that results in significantly reducing the ability of the hydrogen storage system in absorbing hydrogen and its lifetime.

3.3.8.1.4 Chemically Stored Hydrogen

Hydrogen has enough potential to be converted into a wide variety of materials under the different conditions of temperature and pressure through various chemical reactions. This issue has led to an appearance of it in numerous chemical compounds that are often used in the diverse hydrogen storing techniques. Hydrogen has the ability to be combined with other elements in a special chemical reaction with the aim of creating new effective stable components. The generated components can be used for various objectives and when the hydrogen energy is needed in the system, the second chemical reaction occurs for releasing the hydrogen to be used as a fuel in the fuel cell or stored in the hydrogen storage for later use. This feature of the hydrogen has made it a valuable element that can be contributed to the production of various components, which substantially increases the flexibility of the energy networks in adopting various technologies and storage systems. Some of the materials that can store hydrogen through chemical reactions are demonstrated in Fig. 3.11 [39]. The complete descriptions of these materials and their chemical reactions can be found in Ref. [39].

3.3.8.1.5 Carbon Nanotubes

Carbon nanotubes consist of microscopic tubes of carbon for storing hydrogen, and two nanometers are across in the structure of the storage system for this aim [48]. In this method, carbon is stored in microscopic pores on the tubes, and its application is similar to the metal hydrides method. However, this method can store a larger amount of hydrogen in comparison with the metal hydrides technique. Carbon nanotubes can store hydrogen relatively from 4.2% to 65% of their weight and can be operated anywhere for storing hydrogen [49]. The research in the field of carbon nanotubes is centered on developing manufacturing techniques and improving its economic aspects with the aim of commercialization.

3.3.8.1.6 Glass Microsphere

Glass microsphere is another way for storing hydrogen that does this work at high pressure of around 350–700 bar at 300 °C. Due to control of the release of hydrogen in the storage system, the spheres are reheated again at 200–300 °C and transferred to the lower pressure vehicle tanks. In this method, the materials need high pressure for filling and possess low volumetric capacity [50].

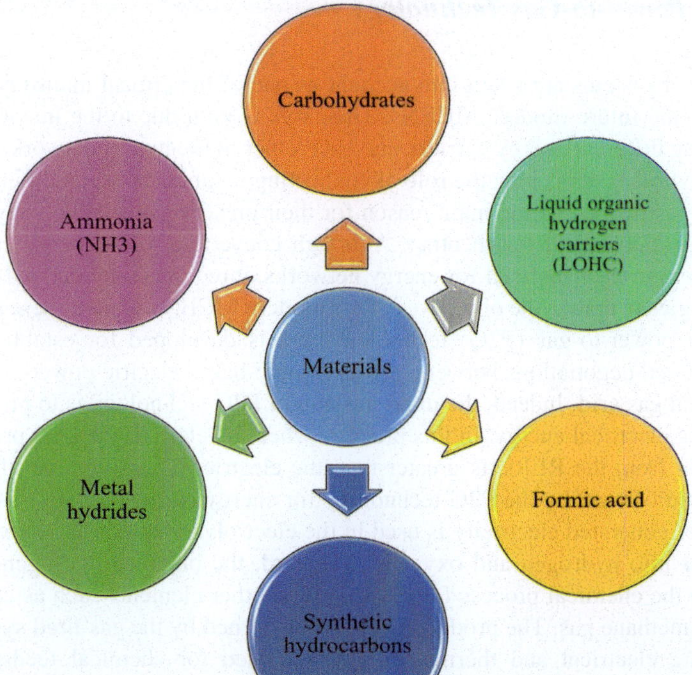

Fig. 3.11 Some of the materials that can store hydrogen through chemical reactions

3.3.8.1.7 Liquid Carrier Storage

In this method, hydrogen is targeted to be used for creating different forms of fossil fuels, which is a common way in today's society. Whenever hydrogen is required in the system, fossil fuels can be reformed as the source of hydrogen for producing it. In this process, the reformation of fossil fuels has led to the separation of hydrogen from the fossil fuels' component. Because carbon is a serious threat for operating certain types of fuel cells and can poison them, the reformed hydrogen needs to be cleaned of surplus carbon monoxide to be used as the fuel for the fuel cell systems.

In the aforementioned subsections, the different techniques for storing hydrogen were briefly described by evaluating their features. After storing hydrogen in the storage system, it can be used for various goals in the system, which is depicted in Fig. 3.11. One of the common usages of hydrogen is to use it as the fuel for the fuel cell systems, aiming to generate clean electrical energy in the system. We are going to get acquainted with fuel cell systems and their different types in the following section.

3.3.9 Power-to-Gas Technology

The hybrid systems are taken into account as one of the critical mainstays in constructing the future modern MCENs. Their role is vital due to the inevitability of converting different energy carriers into each other in the hybrid networks. In other words, hybrid systems play the role of connecting columns between different types of energy networks, so the main reason for their presence is to effectively connect the energy networks to each other. Although converting multi-energy carriers to each other is very beneficial for energy networks, it requires efficient and powerful technologies to realize the objectives of the modern MCENs. One of these technologies is a power-to-gas (P2G) technology that is developed for establishing the energy-based negotiations between the renewable-based electric power system and the natural gas grid. Indeed, the main mission of P2G technology is to produce gas fuel using electrical energy. In this process, when the electrical energy production, especially from the RERs, is greater than the electricity consumption, the excess energy can be used by the P2G technology for energy conversion. To this end, the surplus of generated electricity is used in the electrolyzer system for decomposing the water into hydrogen and oxygen. Afterward, the obtained hydrogen is transferred to the chemical process for reacting with other elements such as carbon for creating methane gas. The produced gas can be burned by the gas-fired systems for generating electrical and thermal energy, intended for chemical feedstocks, or injected to the natural gas networks. In this way, the P2G technology allows hybrid networks to be more flexible than ever before to reliably meet the required energy. Although this technology is significant in providing diverse benefits for energy networks, converting the energy in several steps into the different forms has led to the loss of a considerable amount of energy and reducing the efficiency of the system. Therefore, effective scheduling is essential for the usage of this technology in the future modern MCENs' infrastructure in a way to reduce the conversion steps of the energy to each other aiming to possibly improve the system efficiency and avoiding more energy losses in the system.

3.3.10 Power-to-Hydrogen Technology

The process of the power-to-hydrogen (P2H) technology is a part of the P2G technology, but the objectives may be different. Similar to the P2G technology, the P2H is targeted to use the surplus of electrical energy generated from the pollutant-free energy production units as stimulus energy for the electrolyzer system to break the water for achieving the oxygen and hydrogen as the main products. In the P2G technology, the main objective is to generate natural gas for different applications. However, the produced hydrogen by the P2H technology is targeted for broader applications, including its usage in the carbon economy and for decarbonization purposes.

3.3.11 *Power-to-Heat Technology*

A flexible coupling of electrical and heating energy sectors can contribute to both decarbonization and integration of the RERs [51]. In this regard, a synthesis of achievements demonstrates that power-to-heat technologies can economically and efficiently contribute to decarbonization, renewable integration, and fossil fuel substitution [51]. These technologies play a major role in facilitating the continuous heating energy supply in response to the threat of any demand blocks, especially in the end-user sectors. There are different ways for converting electrical energy to the heating one, and most important of them are categorized in Fig. 3.12 [51].

According to Fig. 3.12, power-to-heat technology makes hybrid networks more flexible and reliable in supplying energy for consumers in the end-user sector. This issue indicates the interdependency between the electric power system and the DHNs in realizing the goals of the future hybrid networks. In this regard, power-to-heat technology can seriously contribute to the prohibition of the synergies between the hybrid elements in line with the future modern MCEN guidelines. Given the coupled structure of the future modern MCENs, the power-to-heat technology will allow for adopting emerging hybrid systems to increase the reliability of heating energy supply while improving the overall reliability, resiliency, stability, and flexibility of the hybrid networks in the presence of the 100% RERs.

3.4 Fuel Cells

Fuel cells are the electrochemical-based systems that operate like battery storage systems and are used for clean energy production by using oxygen and hydrogen energy as a fuel [52]. These devices use the electrochemical process for generating electrical energy and only emit water and heating energy. The fuel cell consists of four main parts including anode and cathode as the catalysts and proton exchange membrane. In the fuel cell structure, catalysts are the special materials that play a key role in facilitating the reactions between hydrogen and oxygen. The very thinly platinum powder coated onto carbon cloth or paper is usually used for making the catalysts. The proton exchange membrane plays a role of electrolyte and is built in a way to allow the passing of only positively charged ions. Indeed, it is a specially treated material, which blocks the passing of the electrons from the membrane. In the fuel cell system, initially, hydrogen gas (H^2) enters the fuel cell from the anode catalyst section. In the anode catalyst, hydrogen is broken down into a positive ion (H^+) and a negative ion (e^-). Since the electrolyte does not allow negative ions or electrons to pass through, only the positive ions cross through the electrolyte and the electrons exit through the connection wire to the anode catalyst, which is the same as electrical energy. On the other hand, due to the fact that it is an anode catalyst, it absorbs positive ions, getting away negative ions, and facilitates the

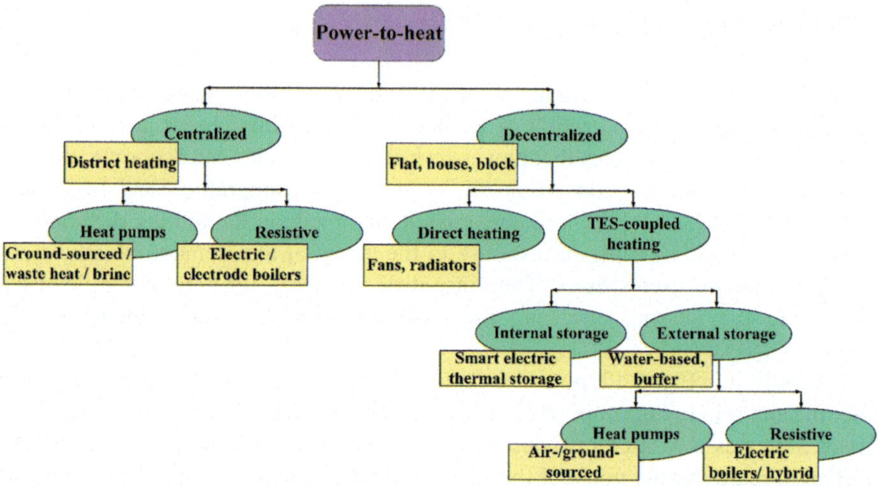

Fig. 3.12 Categorization of different options for power-to-heat technology [51]

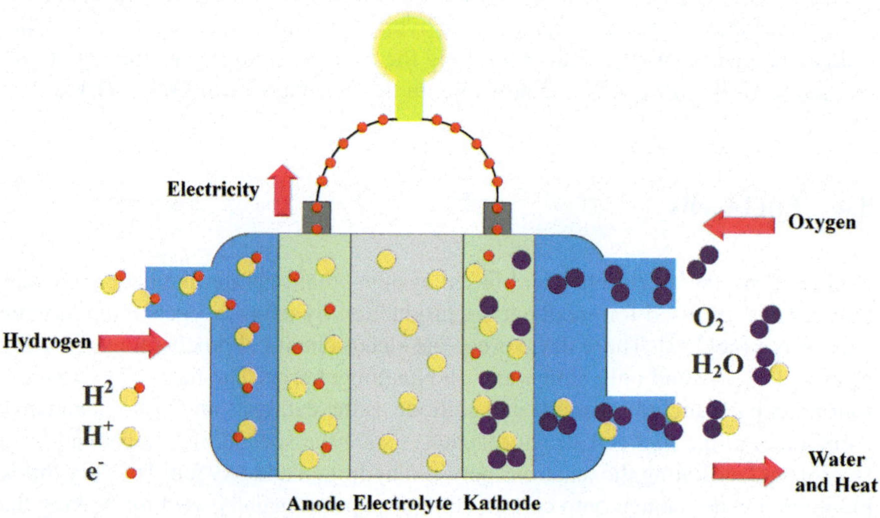

Fig. 3.13 The schematic of the fuel cell system

production of electrical energy. After passing through the electrolyte, the positive ion delivers to the cathode catalyst and reacts chemically with electrons and oxygen (O_2) that have entered the field from other pathways. The result of this reaction is the production of water and heating energy. The schematic of the fuel cell system is illustrated in Fig. 3.13.

The fuel cell system is developed in different types, considering its application and structure, which are discussed next.

3.4.1 Alkaline Fuel Cell

This fuel cell is structured to provide the required electricity and drinking water in the spacecraft. The type of electrolyte in this fuel cell is the aqueous solution of alkaline potassium hydroxide. This fuel cell can produce electrical energy in the range of 300 W to 5 kW. Although the efficiency of the power generation is almost 70%, the alkaline fuel cell is too expensive for commercial applications.

3.4.2 Phosphoric Acid Fuel Cell

This fuel cell is developed to be used in nursing homes, hospitals, and all commercial objectives. The types of electrolyte and catalyst in this fuel cell are the liquid phosphoric acid and platinum, respectively. This fuel cell can produce electrical energy in the range of 200 kW to 1 MW. One of the effective advantages of this fuel cell is that it uses impure hydrogen as a fuel and 85% of the produced steam can be reused for cogeneration. Low power and current, large size and weight, and using the expensive platinum as a catalyst are some important disadvantages of the phosphoric acid fuel cell.

3.4.3 Proton Exchange Membrane Fuel Cell

This fuel cell is also called solid polymers and is targeted to be used for a quick startup in light-duty vehicles and automobiles, and potentially to replace rechargeable batteries. The type of electrolyte and catalyst in this fuel cell is the solid organic polymer, poly-perfluoro sulfonic acid, and metals, usually platinum, respectively. This fuel cell can produce electrical energy in the range of 50–250 kW. Solid polymers use the solid electrolyte as a membrane that reduces management problems and corrosion. One of the special drawbacks of this fuel cell is that it is very sensitive to fuel impurities and only pure hydrogen should be used for this fuel cell as the fuel.

3.4.4 Molten Carbonate Fuel Cell

This fuel cell is majorly considered for electric utility applications. The type of electrolyte is the potassium and/or sodium carbonates and liquid solution of lithium. Instead of platinum, this fuel cell can use inexpensive metals as the catalyst. The molten carbonate fuel cell can produce electrical energy in the range of 10 kW to 1 MW and its efficiency varies between 45% and 60%. Working at high temperatures enhances the breakage and corrosion of cell components. Meanwhile, this fuel cell has significant advantages and are as follows.

- Working at the high temperatures enables the fuel cell to use the inexpensive catalysts.
- A large amount of efficiency in electrical energy production.
- Higher flexibility in using various forms of fuels such as marine, monoxide, carbon, and propane gas due to operating at the high temperatures.

3.4.5 Solid Oxide Fuel Cell (SOFC)

This fuel cell is known as one of the highly promising fuel cells and can be used in large power applications including large-scale central electrical energy producing stations and industrial sectors. The efficiency of this fuel cell in power generation can be reached at 60% and 85% ranges. The SOFC can generate electricity around 220 kW. Utilizing an array of meter-long tubes as well as a compressed disc are two variations in the SOFC fuel cell that have made it closer to commercialization.

3.4.6 Direct Methanol Fuel Cell

The structure of this fuel cell is similar to the proton exchange membrane fuel cell in terms of using a polymer membrane as the electrolyte. The anode catalyst is built in a way to directly draw the hydrogen from the liquid methanol without requiring a fuel reformer. This fuel cell can typically operate at a temperature in the range of 120–190 F and an efficiency of 40%. This is while the higher efficiency can be achieved at higher temperatures. The direct methanol fuel cell is mainly used for low power range applications such as laptops, cellular phones, and mid-sized applications. One of the major problems of this fuel cell is that the process of crossing fuel from the anode catalyst to the cathode is carried out without generating electrical energy.

3.5 Integration of the DEPUs in the Future Modern MCENs

3.5.1 Integration of the DEPUs with a High Share of RERs

Nowadays, there is widespread consensus that the proliferation of RERs in the energy generation sector can act as strong leverage to overcome the concerns regarding climate change. Although the operation of the RERs is a promising way of yielding the economic and environmental objectives, their dependency on the climate changes in energy production has led to intermittences that are a great threat to the reliable operation of the energy networks. One of the reliable ways to overcome the aforementioned challenge is to use the different technologies of the energy storage systems and almost all of them are described in the previous subsections. However, efforts in this research area are being made continuously to find other solutions with respect to the limitations of using the storage systems in some cases. One of the effective solutions for addressing the stochastic nature of the RERs in energy generation is their integration throughout the energy networks. In this respect, the first theory is focused on integrating the RERs with the other controllable power generation systems, which are known as the integration of the DERs with a high share of RERs. According to this theory, the integration includes the operation of some controllable systems such as DGs along with the RERs for ensuring the reliable energy supply [53]. This strategy is accepted as one of the confident ways that enables the hybrid networks to use both the RERs for more clean energy production and non-renewable energy resources' capabilities to support the stochastic producers for ensuring the continuous energy supply for the consumers. However, by the emerging concepts of the modern MCENs, the effects of the first theory are dwindling, and the result of these evolutions has led to the creation of a second theory for the integration of the DEPUs. This is because fossil-fuel-based systems are not only the main components of the energy production sector in this strategy, but also it does not focus on using the maximum potential of the environment to produce clean energy. Therefore, the second theory is developed in accordance with the concept of the future modern MCENs that offers the integration of 100% RERs with the aim of using the maximum potential of the natural energies in producing clean energy.

3.5.2 Integration of the Full Share (100%) of RERs

As mentioned in the previous section, some drawbacks of the first strategy regarding the optimal integration of the RERs have led to the introduction of the second strategy that meets the objectives of the future modern MCENs in terms of clean energy production. The second strategy allows the integration of 100% RERs in the grid to yield the cost-effective and environmentally friendly goals of the future modern MCENs in the energy generation sector. Unlike the first strategy, the second strategy uses the capability of the controllable decentralized energy production units only at times when the system is faced with the challenge of establishing a dynamic energy balance between supply and demand. This means that the gas-fired types of decentralized energy generation units are intended as the backup system to support the stochastic producers only in the emergency conditions [54]. Given the developments in the energy conversion technologies and the related chemical reactions, the CO_2 emissions released by burning natural gas fuels can be considered in the special chemical processes for breaking the CO_2 and producing the carbon and oxygen. The oxygen can be released into the atmosphere while the carbon can be used to combine with the hydrogen energy using the P2H technology for generating the methane gas. Indeed, this strategy not only allows the future modern MCENs to be clean by the integration of 100% RERs but also offers flexible ways for increasing the reliability of the energy supply along with the appropriate usage of the generated greenhouse gas emissions for producing new valuable products. This is while simultaneously realizing these objectives requires the amazing technologies, which will be discussed in Chap. 5 of this book.

3.6 Demand-Side Energy Management in the Future Modern MCENs

The concept of modern energy networks is designed in such a way that it is not relied upon to a few ways to guarantee energy supply with confidence. In other words, modern energy networks provide a set of safe and practical solutions to achieve the goals of grid modernization and are not limited to a few techniques and technologies. Some of these solutions are related to the end-user sector, one of the most famous of which is demand-side energy management (DSEM). Indeed, DSEM refers to a set of energy management schemes that seeks to control and manage the energy consumption of the consumers by improving their energy consumption patterns by suggesting some incentive options. The DSEM is mainly categorized into two broad programs, including demand response programs and energy efficiency programs, which are illustrated in Fig. 3.14 [55].

From the modern MCENs' point of view, the DSEM should focus on proposing more options for the consumers to modify their undesirable behaviors in energy consumption rather than limiting or forcing them to participate in some schemes. Therefore, the DSEM in the modern structure of the MCENs offers additional choices for the consumers not only with the aim of managing energy consumption in the prosumer-sector but also improving the level of people's well-being and conveniences.

3.6.1 Demand Response Programs

Due to day by day growing energy consumption, proliferating of mushrooming the stochastic producers, changes in the consumers' energy consumption patterns, considerably expanding the energy networks, emerging new players and sophisticated technologies, uncertain events, etc., the energy networks face various challenges such as high energy losses, high possibility of market power exercise, low efficiency, reliability and stability issues, and high emissions. In this regard, in addition to other solutions, demand response (DR) programs propose practical solutions for effectively addressing some challengeable problems in the energy networks, such as

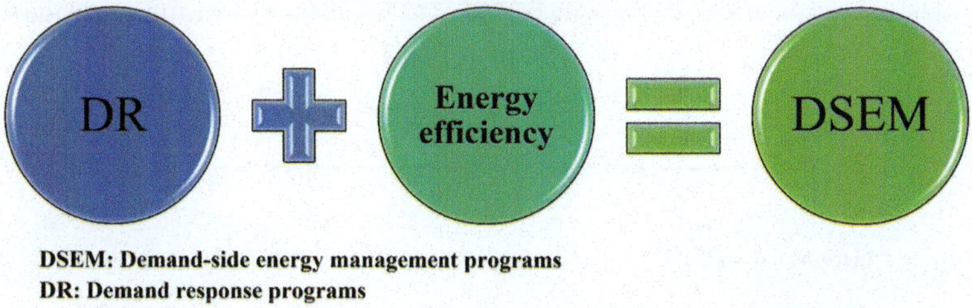

Fig. 3.14 Two main classifications of the DSEM programs [55]

congestion in energy production, transmission and distribution parts of the system, reliability issues, high emissions, high demand's peak to average ratio, and high production cost [55]. The DR is defined as "changes in electrical energy consumption by end-use customers from their normal consumption patterns in response to changes in the price of energy over time, or to incentive payments designed to induce lower energy use when system reliability is jeopardized or at times of high wholesale market prices" [56, pp. ix, 6]. The DR programs are divided into two general classifications: price-based programs and incentive-based programs. In the first category (price-based programs), energy supply cost is considered for creating the retail energy tariffs to charge the consumers with different rates at various consumption times. However, the incentive-based programs offer predesigned incentives for energy consumers to change their energy consumption patterns by proposing different plans. Each of the aforementioned programs is divided into the sub-categories that are demonstrated in Fig. 3.15.

3.6.1.1 Price-Based Demand Response Programs

The price-based DR programs are designed with the aim of changing the energy consumption patterns by applying regulated coefficients on the energy price at different hours during a day. For example, increasing tariffs during contingency hours and peak times by the utilities can result in a remarkable reduction in energy consumption. The different types of price-based DR programs are:

- *Time of use (TOU) pricing*: In this program, the energy price is determined depending on the various time intervals during a day. For this aim, three time intervals are considered during a day including peak, mid-peak, and off-peak intervals. This program's aim is to charge consumers to reduce or shift their energy consumption at peak times.
- *Critical peak pricing (CPP)*: This program is designed to maintain the system balance when the reliability of the energy network is jeopardized. For this aim, the energy price at peak times is increased to a very high price to persuade consumers to reduce energy consumption. This program is usually required to employ only at limited hours per year for maintaining the reliability of the system at an acceptable level.

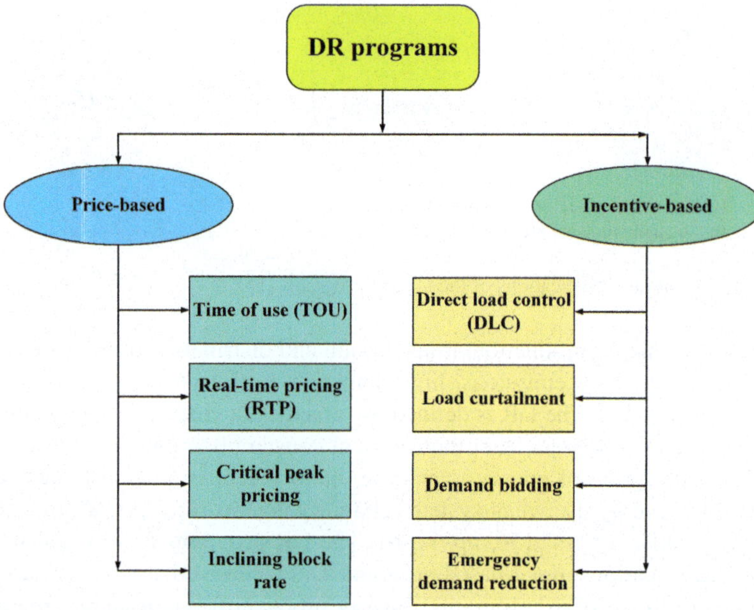

Fig. 3.15 Different classifications for DR programs [55]

- *Real-time pricing (RTP)*: In this program, we witness the dynamical changes in energy prices during a day. Based on this program, the consumers are notified to modify their energy consumption in each hour of a day under the day-ahead and hour-ahead basis. This program is one of the popular and useful DR schemes that is implemented in the system with successful results.
- *Inclining block rate (IBR)*: This program is structured based on the two-level price, which is determined considering the total energy consumption of the consumers. In this program, reaching the threshold energy consumption leads to increasing the energy price. Under this program, unnecessary investments are significantly reduced in the system.

3.6.1.2 Incentive-Based Demand Response Programs

The incentive-based DR program offers different types of incentives for motivating consumers to participate in the DR programs with the aim of energy management consumption. This program includes several schemes that are described below.

- *Direct load control (DLC)*: In this program, consumers receive a certain reward to allow the utility to cut off some predetermined appliances or cycle them when needed (especially at peak times). With this agreement between the consumer and the electricity utility, in cases where the network needs, the utility will curtail the predetermined equipment without informing the consumer.

- *Load curtailment*: This program allows the consumers to register and participate in the scheme that is aimed to pay incentives for them in return for curtailing their energy consumption according to the utility's requests. In this program, failure to comply with the agreement will result in penalties for consumers [56].
- *Demand bidding*: This program makes the participation of large-scale consumers (>1 MW) in the DSEM schemes possible by allowing them to curtail a certain portion of their energy consumption during peak times of contingencies to receive various incentives [56, 57].
- *Emergency demand reduction*: In severe contingencies in the energy networks, this program pays a considerable amount of incentives for interested consumers to curtail their energy consumption for maintaining the continuous energy supply in the system [55].

3.6.1.3 Advantages of Demand Response Programs

DR techniques offer diverse advantages for the energy networks, consumers, utilities, and retailers and as follows [55]:

- They reduce the peak to average ratio of demand.
- They reduce the operation level of the non-renewable energy resources for supporting peak load by reducing the peak demand that is led to a reduction in releasing the emissions in the system.
- They decrease the stress on the energy network by eliminating the contingencies using various programs such as emergency load reduction and direct load control for reducing the energy consumption level in the system.
- They decrease the probability of price spike occurrence in the wholesale electricity market.
- They provide a useful opportunity for effectively using different energy resources in the system by making the retail tariffs dependent on the wholesale market price.
- They increase market efficiency by subtracting the possibility of market energy exercise by energy production companies.
- They offer a suitable opportunity for consumers to save energy costs by adopting an effective strategy in their energy consumption.
- They increase the potential of the energy networks in adopting more RERs by providing various schemes for reducing energy consumption.

3.6.1.4 Challenges and Drawbacks of Demand Response Programs

In addition to the aforementioned advantages, which of course are more advantageous for the current structure of networks, there are some substantial challenges for DR programs. Perhaps the most important challenge in using DR is the inconsistency of most of its programs with the goals of modern networks. Given the features of the different types of DR programs, some of them are not suitable to be used in the future modern MCENs. One of the main missions of the future modern grids is to provide continuous energy to the consumers without creating any limitations in their energy consumption in various ways. Indeed, curtailing a portion of the consumers' energy consumption or controlling some appliances for directly shutting them down in the uncertain times is not in line with the modern grids' policies. The future MECNs are targeted to be modern and is also called modern because it is expected to employ innovative solutions for solving the challenges of the hybrid networks like the DSEM rather than referring to the consumers for accepting some limitations in return for receiving certain incentives. In other words, one of the key attributes of modern grids is to maintain the hybrid network in normal conditions without creating any disruptions in the well-being and convenience of the people. Therefore, among the DR programs, only the RTP program is appropriate for the future modern MCENs to notify the consumers to effectively participate in the energy market interactions. Shifting load in the prosumer-side of the modern grids is also possible in a way that the building control system automatically notifies the occupants by sending them an appropriate message and they will have the option to accept or reject their shifting load offer anywhere. The smart building/home energy management system also has the ability that residents can only receive similar messages if they wish. In addition to the mentioned challenges, most of the benefits of the DR programs rely on the current structure of the energy network that non-renewable energy resources play a key role in it. This is while the future modern MCENs are going to be designed in a way to equip with 100% RERs.

3.6.2 i-Energy

Generally, smart systems can be described using two concepts: the first one is the "*i-Energy*" that focuses on the smart demand-side energy management while the second concept refers to the smart grid [58]. Unlike the smart grid that seeks to manage energy from the suppliers' viewpoint, *i-Energy* is intended for energy management objectives from the consumers' point of view [59]. The *i-Energy* provides the opportunity of using communication and information technologies with the aim of improving energy management techniques in the end-user sector [58]. Indeed, the *i-Energy* systems are developed to automatically manage the controllable appliances and distributed energy production and storage devices at the demand-side [59]. Four steps can be considered for realizing the concept of the *i-Energy* and are as follows [59]:

- *Smart tap network*: In this step, the individual appliances' energy consumption patterns are considered to be monitored in detail along with the dynamic activities of people in factories, offices, and homes.
- *Energy on demand protocol*: This step is aimed to realize the automatic ceiling mechanism for energy consumption in both watt-hours (Wh) and watts (W).

Moreover, it focuses on realizing an energy supply mechanism aiming to best act considering the priorities in the system.

- *Power flow coloring*: This step is structured to allow holistic controls on versatile energy flow considering the costs and types of energy sources.
- *Smart community*: This step offers bidirectional energy exchanging between factories, offices, and households in a local community.

The complete descriptions regarding each of the aforementioned steps and the widespread applications of the *i-Energy* can be found in Refs. [58, 59]. Although the current concept of the *i-Energy* cannot completely meet the requirements of the future modern MCENs, scrutinizing its different aspects can achieve useful results for energy management at the demand-side. Therefore, the *i-Energy* concept can be an introduction for developing the current technologies with the aim of satisfying the future modern MCENs' objectives in terms of energy management in the prosumer-side.

3.7 Flexibility of the Future Modern MCENs

There are many factors that can affect the flexibility of the energy networks. The flexibility of a system reflects its ability to deal safely with a variety of events in the system and to respond to them appropriately. The flexibility of the future modern MCENs more relies on the interactions between the different energy networks. Indeed, the interconnected structure of the hybrid networks is considered for proliferating the synergies among the hybrid systems aiming to enable the system for sufficient reactions to the certain and unexpected events in the system. In this regard, hybrid systems are famous for their ability to convert the multi-carrier energies to each other, so they play a critical role in establishing flexible modern MCENs. In addition, deploying diverse technologies of the energy storage systems, innovative technologies in controlling the supply-side, and full automated energy management schemes in the prosumer-side; integrating the hybrid systems with the full renewable-based structure of the modern grid; and utilizing the potential of the controllable devices as the backup systems are only some prominent actions that can be considered to make the future modern MCENs more flexible than ever before.

3.8 Food-Energy-Water Nexus in the Future Modern MCENs

The increasing cost of multi-carrier energy production processes, dwindling sources of water and non-renewable energies, and shortages in the food cycle inspire the fact that we have to accelerate the transition of the current world to the sustainable modern society [60]. Due to the increasing growth of the world's population along with the level of their expectations and well-being, we can witness a significant increment of pressure on the food, energy, and water (FEW) demands. This condition reflects a great threat of FEW shortages that is increasingly being felt, which requires serious actions in all areas, especially energy. The need for the interconnected system as well as the developed technologies for effectively pursuing the practical solutions to improve the hybrid system performance has led to the appearance of the FEW nexus concepts [60]. Given that the future modern MCENs are a combination of different energy networks, which have direct effects on each other, the FEW nexus systems are expected to play a key role in making the modern networks more sustainable than ever before. In other words, the MCENs engage various cycles such as water and energy in the network's interactions that have made them more dependent on each other. The availability of water and energy directly affects the food cycle and can have positive or negative impacts on the food production cycle. This issue indicates that realizing the goals of modern grids by implementing them in the near future can seriously improve the performance of the FEW nexus systems with the aim of building a sustainable society.

3.9 Summary

This chapter has tried to facilitate the transition from the current structure of the energy networks to the future modern MCENs by investigating their structures from the different viewpoints and proposing effective solutions for addressing ongoing and future challenges. Indeed, this chapter is focused on accelerating the switch procedure to the modern grids by developing the concepts and theories in accordance with the goals of the future modern MCENs. For this aim, the energy production sector is scrutinized by fully analyzing the different types of renewable and non-renewable energy resources and their potentials in determining the priority of their presence in the modern MCENs' structure. The operation of the multi-generation systems is a necessary step for expanding the variety of the MCENs. Energy storage systems are recognized as essential devices for increasing the reliability of renewable-based systems. Thus, all the energy storage systems are assessed with their features to give a suitable overview of them for adopting effective strategies regarding their operation in the modern grids. The integration of the DEPUs as another reliable way for developing the system with a full share of the RERs is considered for evaluating the possible techniques to effectively do this huge duty. This resulted that the future modern MCENs need innovative technologies to integrate 100% RERs in a reliable and sustainable manner. We will discuss one of these techniques in Chap. 5. The demand-side energy management approaches are investigated, by highlighting the advantages and disadvantages of the demand response programs, as well as the realization steps of the *i-Energy* and its potential in the energy networks. This concluded that we need to have significant developments on the prosumer-side for making this part a full automated section aiming to provide satisfactory conditions for the consumers. In order to make the modern grids more sustainable than before, the potential of the food-energy-water nexus systems is analyzed with the aim of developing them to be suitable for participation in the future modern MCENs' interactions.

3.10 Questions

After careful reading and evaluating this chapter, we are expected to be able to answer the following questions.

3.10.1 Energy Production Sector

1. What are the critical aspects that should be analyzed for future modern MCENs and why they are important?
2. How many ways are there in general for energy production in hybrid networks?
3. What are the features of centralized and decentralized energy production units?
4. How will centralized and decentralized energy production units be present in future modern MCENs?
5. Among the centralized and decentralized energy production units, which of them is suitable to be considered as the backup system for the modern grids and why?
6. Why do we need to accelerate the switching from the uni- or bi-dimensional energy generation system to the multi-generation technologies?
7. What are the advantages of centralized and decentralized systems over each other?
8. What are the different types of centralized energy sources and what are the characteristics of each of them?
9. What are the different types of decentralized energy sources and what are the characteristics of each of them?
10. What is the classification of renewable and non-renewable energy resources and what will be the role of each in the future modern MCENs?

3.10.2 Energy Storage Sector

11. What are the different types of energy storage systems and why they are essential for the future modern MCENs?
12. What is the classification of electrical storage systems and what are the characteristics of each of them to participate in the future modern MCENs' interactions?
13. What are the distinguishing features of the battery energy storage systems and what types of them are available in the energy networks?

14. Why the sodium-sulfur battery can be a special option for different goals in the future modern grids?
15. What are the different types of mechanical storage systems and how they can assist in satisfying the goals of the future modern grids?
16. What are the various categories and technologies of the pumped-storage systems and how each of them can be operated in the hybrid networks?
17. What are the various classifications of the compressed air energy storage system and how can each of them play an effective role in the hybrid networks?
18. What are the distinguishing features of the thermal energy storage systems?
19. Which types of thermal energy storage systems are available in the energy grids and what are their attributes?
20. How many operating strategies are there for thermal energy storage and what are their goals?
21. Why the presence of natural gas and water storage systems is necessary for the infrastructure of the future modern MCENs?
22. What is the function of a hydrogen storage system and what are its different types?
23. How many methods are there to separate water for hydrogen energy production and how do they work?
24. What are the techniques for storing hydrogen energy and how they can store hydrogen for which applications?
25. What is the mission of the fuel cell systems, and different types of them are developed for which objectives?
26. How power-to-gas technology can be used in the future modern MCENs and for which goals?
27. How power-to-hydrogen technology can be used in the future modern MCENs and for which goals?
28. How power-to-heat technology can be used in the future modern MCENs and for which goals?
29. How many ways are there for converting electricity to the heating energy and what are their features?

3.10.3 Integration of Energy Generation Resources and DSEM Sectors

30. Why do we need to integrate the decentralized energy generation systems in the hybrid networks?
31. What are the differences between integrating a high level of RERs and the 100% RERs with the modern grids?
32. Why should there be demand-side energy management in energy networks and what are the common methods for doing so?
33. What is the demand response concept and which types of them are considered in the energy networks?

34. What are the features of the price-based demand response programs?
35. What are the features of the incentive-based demand response programs?
36. What are the advantages and challenges of the demand response programs in the energy networks?
37. Why most of the demand response programs are not suitable for the future modern MCENs?
38. What is the concept of the *i-Energy* and which steps are required for realizing it in the energy networks?
39. From which ways the flexibility of the future modern MCENs can be effectively provided?
40. What is the concept of the food-energy-water nexus and how can it play an effective role in the future modern MCENs?

3.11 Suggestions

In this section, some suggestions are provided related to this chapter that can give a good overview of some substantial issues in the future modern MCENs.

1. Employing the full level of renewable-based decentralized energy production units can economically make our society green and beautiful, which is directly in line with the objectives of the future modern MCENs.
2. Switching from uni- or bi-generation systems to the multi-generation units will certainly increase the flexibility of the future modern MCENs in reliably meeting the multi-carriers' energy demands.
3. Deploying the different types of energy storage systems will make hybrid modern networks more reliable than ever before in the presence of 100% RERs.
4. Developing energy conversion technologies can certainly increase the ability of the modern grids in continuous energy supply for different types of consumers.
5. Integration of the 100% RERs in the future modern MCENs is one of the reliable ways for increasing the sustainability of the energy networks.
6. Developing the full automated demand-side energy management systems in the prosumer-side of the modern MCENs can effectively provide the conveniences of the occupants.
7. Developing the food-energy-water nexus systems can enable modern grids to achieve an acceptable level of sustainability.

3.12 Future Trends and Discussion Topics

Here, future trends and some discussion topics are proposed for the interested readers regarding this chapter that can be useful for developing the MCENs' modernization process.

1. Developing clean energy production technologies for maximum usage of the environment's potential in providing free source of natural energies.
2. Analyzing the possible progress in the energy generation technologies for upgrading them to the multi-generation systems.
3. Economically developing the energy storage technologies for increasing the reliability of the energy supply in the presence of 100% RERs.
4. Improving the process of multi-carrier energy conversions for reducing costs and energy losses.
5. Providing innovative technologies for effective integration of the RERs in the future modern MCENs.
6. Developing intelligent systems for improving energy management protocols with the aim of increasing people's well-being and comfortability.

References

1. M. Daneshvar, B.M. Ivatloo, M. Abapour, S. Asadi, R. Khanjani, Distributionally robust chance constrained transactive energy framework for coupled electrical and gas microgrids. IEEE Trans. Ind. Electron, **68**(1), 347–357 (2021)
2. M. Daneshvar, B. Mohammadi-Ivatloo, K. Zare, S. Asadi, Two-stage robust stochastic scheduling model for transactive energy based renewable microgrids. IEEE Trans. Ind. Inf, **16**(11), 6857–6867 (2020)
3. M. Daneshvar, B. Mohammadi-ivatloo, K. Zare, S. Asadi, A. Anvari-Moghaddam, A novel operational model for interconnected microgrids participation in transactive energy market: A hybrid IGDT/stochastic approach. IEEE Trans. Ind. Inf, (2020). doi: 10.1109/TII.2020.3012446 28. July 2020
4. M. Daneshvar, B. Mohammadi-Ivatloo, M. Abapour, S. Asadi, Energy exchange control in multiple microgrids with transactive energy management. J. Mod. Power Syst. Clean Energy, **8**(4), 1–8 (2020)
5. R. Ambrosio, Transactive energy systems. IEEE Electrific. Mag **4**(4), 4–7 (2016)
6. D.M. Tagare, *Electricity Power Generation: The Changing Dimensions* (Wiley, Hoboken, 2011)
7. M. Daneshvar, S. Asadi, CPS-based transactive energy technology for smart grids, in *Cyber-Physical Systems in the Built Environment*, (Cham, Springer, 2020), pp. 323–338
8. M. Daneshvar et al., The chance-constrained models for transactive energy management of interconnected microgrids clusters. J. Clean. Prod. **271**, 122177 (2020)
9. Combined heat and power for buildings. (May 12, 2020). Available: https://www.cibse.org/knowledge/knowledge-items/detail?id=a0q20000008I7nsAAC
10. Y. Chen, M. Hu, Balancing collective and individual interests in transactive energy management of interconnected micro-grid clusters. Energy **109**, 1075–1085 (2016)
11. M. Daneshvar, B. Mohammadi-Ivatloo, M. Abapour, The possibility of using blockchain based cryptocurrency in transactive energy markets: Ongoing activities and opportunities ahead, in *The Fourth International Energy Management and Technology Conference*, (Tehran, 2018)
12. M. Daneshvar, B. Mohammadi-Ivatloo, K. Zare, Two-stage optimal robust scheduling of hybrid energy system considering the demand response programs. J. Clean. Prod. **248**, 119267 (2020)
13. M. Daneshvar, B. Mohammadi-ivatloo, S. Asadi, K. Zare, A. Anvari-Moghaddam, Optimal day-ahead scheduling of the renewable based energy hubs considering demand side energy

management, in *2019 International Conference on Smart Energy Systems and Technologies (SEST)*, (2019), pp. 1–6
14. IEA, Renewable electricity capacity additions, 2007–2021, updated IEA forecast, IEA, Paris. (September 3, 2020). Available: https://www.iea.org/data-and-statistics/charts/renewable-electricity-capacity-additions-2007-2021-updated-iea-forecast
15. IEA, Renewable electricity generation by region and scenario, 2018–2040, IEA, Paris. (September 3, 2020). Available: https://www.iea.org/data-and-statistics/charts/renewable-electricity-generation-by-region-and-scenario-2018-2040
16. N.S. Chukwujindu, A comprehensive review of empirical models for estimating global solar radiation in Africa. Renew. Sust. Energ. Rev. **78**, 955–995 (2017)
17. A. Uihlein, D. Magagna, Wave and tidal current energy–a review of the current state of research beyond technology. Renew. Sust. Energ. Rev. **58**, 1070–1081 (2016)
18. A. Brito, J. Villate, Implementing agreement on ocean energy systems, Annual Report, 2014
19. D. Elliott, Marine renewables: A new innovation frontier. Tech. Anal. Strat. Manag. **21**(2), 267–275 (2009)
20. A. Latifi, Tidal Energy and how it Pertains to the Construction Industry, 2017
21. The Environment: Wave and tidal energy. (May 12, 2020). Available: https://www.ducksters.com/science/environment/wave_and_tidal_energy.php
22. A. Wachtel, D. Voege, *Geothermal Energy* (Chelsea Clubhouse, USA, 2010)
23. M.A. Rosen, S. Koohi-Fayegh, *Geothermal Energy: Sustainable Heating and Cooling Using the Ground* (Wiley, Chichester, 2017)
24. T. Abbasi, S. Abbasi, Biomass energy and the environmental impacts associated with its production and utilization. Renew. Sust. Energ. Rev. **14**(3), 919–937 (2010)
25. J. Twidell, T. Weir, *Renewable Energy Resources* (Routledge, London, 2015)
26. R.R. Bomma, J. Jayakumar, T. Bogaraj, Different types of energy storage systems: A literature survey, in *Innovations in Electrical and Electronics Engineering*, (Springer, Singapore, 2020), pp. 515–540
27. M. Hannan, M. Hoque, A. Mohamed, A. Ayob, Review of energy storage systems for electric vehicle applications: Issues and challenges. Renew. Sust. Energ. Rev. **69**, 771–789 (2017)
28. M.S. Guney, Y. Tepe, Classification and assessment of energy storage systems. Renew. Sust. Energ. Rev. **75**, 1187–1197 (2017)
29. S. Vazquez, S.M. Lukic, E. Galvan, L.G. Franquelo, J.M. Carrasco, Energy storage systems for transport and grid applications. IEEE Trans. Ind. Electron. **57**(12), 3881–3895 (2010)
30. S.F. Tie, C.W. Tan, A review of energy sources and energy management system in electric vehicles. Renew. Sust. Energ. Rev. **20**, 82–102 (2013)
31. M. Semadeni, Energy storage as an essential part of sustainable energy systems: a review on applied energy storage technologies, CEPE Working Paper, vol. 24, 2003
32. H.-K. Kim et al., Scalable fabrication of micron-scale graphene nanomeshes for high-performance supercapacitor applications. Energy Environ. Sci. **9**(4), 1270–1281 (2016)
33. T. Cotterman, *WISE-Energy Storage Technologies: Transforming America's Intelligent Electrical Infrastructure* (IEEE USA Books & eBooks, USA, 2013)
34. Z. Wen, J. Cao, Z. Gu, X. Xu, F. Zhang, Z. Lin, Research on sodium sulfur battery for energy storage. Solid State Ionics **179**(27–32), 1697–1701 (2008)
35. H. Chen, T.N. Cong, W. Yang, C. Tan, Y. Li, Y. Ding, Progress in electrical energy storage system: A critical review. Prog. Nat. Sci. **19**(3), 291–312 (2009)
36. N. Pfleger, T. Bauer, C. Martin, M. Eck, A. Wörner, Thermal energy storage–overview and specific insight into nitrate salts for sensible and latent heat storage. Beilstein J. Nanotechnol **6**(1), 1487–1497 (2015)
37. I. Dincer, On thermal energy storage systems and applications in buildings. Energ. Buildings **34**(4), 377–388 (2002)
38. W. Qikun, Z. Changchun, L. Weihua, W. Ting, Hydrogen storage by carbon nanotube and their films under ambient pressure. Int. J. Hydrog. Energy **27**(5), 497–500 (2002)
39. S. Niaz, T. Manzoor, A.H. Pandith, Hydrogen storage: Materials, methods and perspectives. Renew. Sust. Energ. Rev. **50**, 457–469 (2015)

40. A. Züttel, A. Borgschulte, L. Schlapbach, *Hydrogen as a Future Energy Carrier* (Wiley-VCH Verlag, Weinheim, 2008)
41. I. Dincer, M. Rosen, Thermal Energy Storage (TES), in *Thermal Energy Storage: Systems and Applications*, (Wiley, Chichester, 2002), p. 93
42. I. Abe, Hydrogen production from water, in *Energy Carriers Conversion Systems*, vol. 1, (EOLSS Publishers Company Limited, Oxford, 2001), pp. 1–3
43. I. Dincer, M. Rosen, Energy storage systems, in *Thermal Energy Storage: Systems and Applications*, (Wiley, Chichester, 2010)
44. S. Satyapal, J. Petrovic, C. Read, G. Thomas, G. Ordaz, The US Department of Energy's National Hydrogen Storage Project: Progress towards meeting hydrogen-powered vehicle requirements. Catal. Today **120**(3–4), 246–256 (2007)
45. Y. Li, R.T. Yang, Significantly enhanced hydrogen storage in metal– organic frameworks via spillover. J. Am. Chem. Soc. **128**(3), 726–727 (2006)
46. D. Hirscher, M. Dornheim, Tailoring reaction enthalpies of hydrides, in *Handbook of Hydrogen Storage*, (Wiley-VCH Verlag GmbH & Co. KGaA, Weinheim, 2010)
47. R. Schulz, J. Huot, G.X. Liang, S. Boily, A. Van Neste, Structure and hydrogen sorption properties of ball milled Mg dihydride. J. Metastable Nanocryst. Mater **2**, 615–622: Trans Tech Publ (1999)
48. G. Cacciola, V. Antonucci, S. Freni, Technology up date and new strategies on fuel cells. J. Power Sources **100**(1–2), 67–79 (2001)
49. R. Salar, Proton-conducting solid oxide fuel cell operating on a methylcyclohexane fuel as an organic hydrogen-carrier. The University of Manchester (United Kingdom), 2005
50. G. Wicks, L. Heung, R. Schumacher, Microspheres and microworlds. Am. Ceram. Soc. Bull. **87**(6), 23 (2008)
51. A. Bloess, W.-P. Schill, A. Zerrahn, Power-to-heat for renewable energy integration: A review of technologies, modeling approaches, and flexibility potentials. Appl. Energy **212**, 1611–1626 (2018)
52. R. Kaur, A. Marwah, V.A. Chhabra, K.-H. Kim, S. Tripathi, Recent developments on functional nanomaterial-based electrodes for microbial fuel cells. Renew. Sust. Energ. Rev. **119**, 109551 (2019)
53. M. Daneshvar, M. Abapour, B. Mohammadi-ivatloo, S. Asadi, Impact of optimal DG placement and sizing on power reliability and voltage profile of radial distribution networks. Majlesi J. Electr.Eng **13**(2), 91–102 (2019)
54. M. Daneshvar, B. Mohammadi-Ivatloo, K. Zare, S. Asadi, Two-stage stochastic programming model for optimal scheduling of the wind-thermal-hydropower-pumped storage system considering the flexibility assessment. Energy **193**, 116657 (2020)
55. A.R. Jordehi, Optimisation of demand response in electric power systems, a review. Renew. Sust. Energ. Rev. **103**, 308–319 (2019)
56. Q. Qdr, Benefits of demand response in electricity markets and recommendations for achieving them, US Dept. Energy, Washington, DC, USA, Tech. Rep, 2006
57. H.A. Aalami, A. Khatibzadeh, Regulation of market clearing price based on nonlinear models of demand bidding and emergency demand response programs. Int. Trans. Electr. Energy Syst **26**(11), 2463–2478 (2016)
58. M. Eissa, First time real time incentive demand response program in smart grid with "i-Energy" management system with different resources. Appl. Energy **212**, 607–621 (2018)
59. T. Matsuyama, i-Energy: Smart demand-side energy management, in *Smart Grid Applications and Developments*, (Springer, London, 2014), pp. 141–163
60. S. Asadi, B. Mohammadi-Ivatloo, *Food-Energy-Water Nexus Resilience and Sustainable Development* (Springer Nature, Cham, Switzerland.)

Chapter 4
Data Management in Modernizing the Future Multi-Carrier Energy Networks

Abstract Recently, diverse activities and initiatives for upgrading the existing power system to the modern energy network have led to discovering a significant requirement for a great revolution in the structure, design, and development of a communications infrastructure with the aim of creating effective interoperability among the controllable systems. In the meantime, the increasing penetration of an array of information technologies and controllable devices together suggests that the communications networks enact critical roles in the electric power infrastructure. Indeed, reconfiguration of the current structure of the multi-carrier energy networks by incorporating the advanced communication protocols is one of the essential steps in the grid modernization process. Therefore, this chapter is developed to introduce the required communication systems for modernizing future energy networks. In this regard, the key role of the communication platforms in coordinating the grid agents is examined considering the effective participation of the various control centers and intelligent devices throughout the modern grid. Also, this chapter investigates how to implement some effective technologies such as the Internet of Things (IoT) in the modern energy grid by adopting capable technologies such as the Internet of Energy while satisfying the modern energy networks' standards.

Keywords Big data · Smart meters · Internet of Things · Cyber-physical systems · Digital twin · Energy internet · Virtual power plant · Data mining · Machine learning · Deep learning

4.1 Introduction

Over recent decades, the deteriorating problems of energy resources, environment, and climate have prompted the researchers to strive for moving toward the acceleration of the transition from the traditional systems to modern society [1]. In this regard, not only advanced intelligent systems are needed for modernizing the hybrid

© Springer Nature Switzerland AG 2021
M. Daneshvar et al., *Grid Modernization – Future Energy Network Infrastructure*, Power Systems, https://doi.org/10.1007/978-3-030-64099-6_4

energy networks but also the provision of capable platforms for communication between intelligent systems is necessary for developing the interconnected grids. In fact, the development of communication protocols among smart energy systems is one of the essential and undeniable steps to achieve modern multi-carrier energy networks (MCENs). Emerging information technologies such as mobile intelligence, big data analytics, the Internet of Things, and cloud computing along with the combination of them with the communication technologies are reshaping the landscape of energy networks and directly affect the daily human's lifestyle [2]. With the aim of facilitating the achievement of the goals of these technologies in making the energy networks more robust, efficient, intelligent, and digital, modern grids require a vast amount of data for effectively analyzing the interactions for optimally scheduling the system. These data are mostly related to the energy generation, operation, and consumption sectors that the collection of them creates the energy big data [3]. The modern energy networks need capable systems to effectively record these data, collect and categorize them for sending to the control centers [4]. In this respect, smart meters have attracted widespread popularity for their ability to enable the system to collect a huge amount of energy consumption data. The integration of big data and smart meters has created the concept of the smart meter big data that covers the important component of the energy grid big data such as energy consumption, current, and voltage [5]. In this regard, the need for controlling numerous intelligent systems such as smart meters in the confident platform has formed the concept of the Internet of Things (IoT) in the energy management ecosystem of today's energy networks. Indeed, IoT has acquired immense potential due to its contribution to the smart grids' interactions, which virtually included the control of every controllable device in the energy networks [6]. However, the IoT and other intelligent systems require the hierarchical energy management paradigm that can facilitate the implementation of the smart energy transmission and distribution protocols. Due to this, the Energy Internet is introduced as an intelligent energy system for enabling the smart grids to effectively support the negotiations between smart energy devices [7]. In other words, Energy Internet is recognized as a large-scale cyber-physical system in terms of cybersecurity that is targeted to create a deep integration among the energy components and information and communication technology [8]. Developing a cyber-physical system layer is necessary for implementing a flexible energy control paradigm when the energy networks face different uncertain resources such as energy demand [9]. Therefore, communication protocols play a key role in energy networks that needs to consider influential parts such as the smart meter, big data analytics, the IoT, Energy Internet, cyber-physical system to achieve the goal of establishing a comprehensive communication system in future modern MCENs.

Since the future modern MCENs are aimed to equip with a 100% renewable energy resources (RERs) to comprehensively promote the clean energy production in the system, it will undoubtedly involve a high amount of uncertainties that make the controlling of the system challengeable [10]. Due to this, an expert energy system is required for using the potential of smart devices to allow energy networks to incorporate system components aiming to create a holistic communication controlling system. For this aim, smart meters are essential devices in the end-user parts to collect the demand-side data for different purposes. Big data analytics is one of these goals that should be intended for acquiring useful results in developing smart energy systems. However, considering effective plans for obtaining such results for the grid needs capable technologies, which is led to the appearance of the IoT paradigm in the structure of the smart grids. This is when the IoT system depends on the strong communication gateway to establish a negotiation opportunity between the intelligent elements. The Energy Internet concept is introduced to fill this gas by employing cyber-physical systems in the cyber-physical layer. All of these steps inspire us to establish a hierarchical energy management system for a cloud-based community to ease the communication of the smart energy systems. This is the primary necessity for the future modern MCENs that engage powerful technologies for making the communication controlling system stronger than ever before. Despite the usefulness of this expert system, it is not sufficient for the grids with multi-dimensions' energy and hybrid devices. Therefore, communication protocols need to be developed to effectively cover the numerous smart energy system interactions, enable the system to enhance interoperability among the hybrid networks, and allow it to adopt emerging sophisticated technologies for developing communication protocols. This issue has motivated us to examine how the current technologies can create a holistic control system for establishing efficient interactions by developing the communication protocols in this chapter.

4.2 Big Data Analytics

By promoting the smart energy systems in the energy networks' infrastructure, the need for hybrid information and communication technologies has appeared as a fundamental challenge [11]. These technologies are famous for their ability to make the energy grids more intelligent and digital, efficient, and sustainable in the process of their general trend toward modern networks. In this pass, smart devices play a key role in realizing the modern grids' goal and are the main reason for generating the different types of data that are pretty vast and complex [12]. The huge amount of produced data in the smart energy networks comprises the data of various parts of energy grids such as energy generation, transmission, and distribution that are combined to form the energy networks' big data [3]. Energy big data is a critical element in supporting the different management decisions as well as optimization of the energy networks. However, each type of data is not beneficial and the redundant data not only imposes great pressure on the data transfer gateways but also it takes up space in storage systems and has negative impacts on the efficiency of big data analytics. Meanwhile, big data analytics can give a valuable overview of the marketing strategies development as well as the application of smart devices. Therefore, it is necessary for developing the technology of smart energy systems in line with the objectives of the future modern MCENs. Figure 4.1 exhibits the components required for the big data analytics [13].

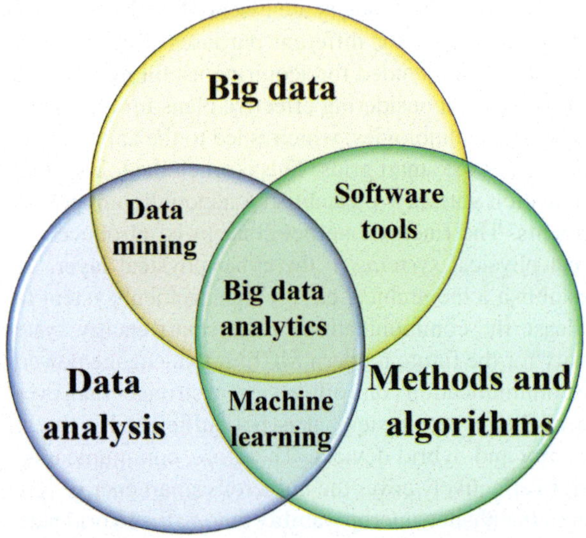

Fig. 4.1 Key components required for big data analytics [13]

4.2.1 Characteristics of Energy Big Data

The modern energy grids are targeted to establish a widely distributed energy trading networks along with the highly automated system that are characterized by the bi-directional mobility of information and energy. In this widespread smart structure, the huge amounts of data produced by intelligent devices such as digital protection devices, energy distribution automated devices, and smart meters form the energy network big data. Generally, energy big data is comprised of business systems data, energy network internal data, user behavior, and user description data [14]. The examples for each of the aforementioned data are depicted in Fig. 4.2.

There are mainly four characteristics of the energy big data, which are as follows [15]:

- *Volume*: The volume of data is one of the effective factors that directly affect the efficiency of the system. In this respect, large-scale energy consumption, transmission, and production data are collected using the potential of thousands of advanced measurement systems in near real-time. Furthermore, the vast amount of other types of data such as geographic information, weather, and temperature data are also targeted to be considered in the big data analytics.
- *Velocity*: The energy networks have to equip with the sophisticated data processing systems to quickly react to the changes in both the energy production and consumption sectors with the aim of establishing a dynamic energy balance. Indeed, possessing rapid data processing devices is essential for quickly collecting, identifying, and recovering energy network failures, load forecasting, and

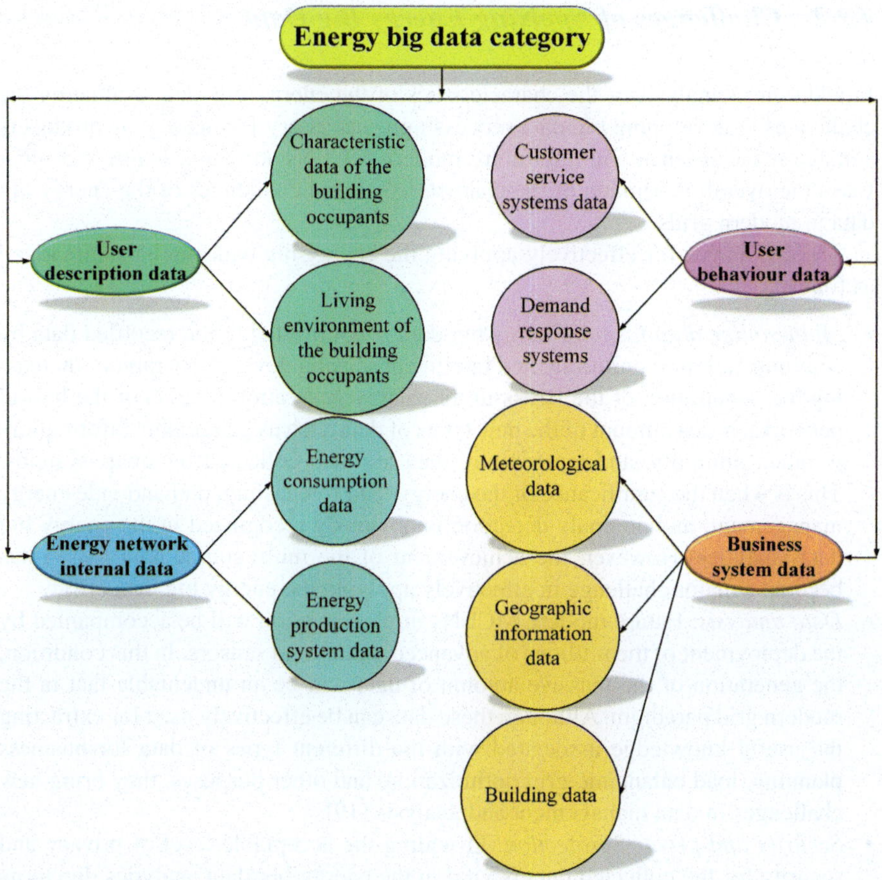

Fig. 4.2 The category for the energy big data

other actions [16]. Therefore, rapid data processing should be intended in developing intelligent systems in the modern structure of the hybrid networks.

- *Variety*: In data processing, we can witness a variety of data with multiple dimensions that are typicaly complex in a high level. In the hybrid networks, the data not only include the gas pressure, heat temperature, water volume, energy of each measurement node, frequency, current, and voltage but also include user data, business data, and numerous energy production data. This is while the varieties of unstructured data such as geographic location, weather images, and network logs information are also becoming more diverse every day [15].
- *Value*: The required knowledge, as well as state of the art technologies for discovering the energy networks data, is a vital issue in supporting the energy generation, transmission, distribution, energy investment, and failure recovery. Additionally, the analysis of the energy network big data contributes to enabling the system to create new business models along with providing personalized energy efficiency services for the consumers [17].

4.2.2 Challenges of Applying Energy Big Data

In addition to analysis of the characteristics of the energy big data, evaluating the challenges in developing big data processing is necessary for making its producers suitable in accordance with the future modern MCENs structure. Figure 4.3 illustrates the overall schematic of the characteristics and challenges of the energy big data in modern grids.

The challenges for effectively applying the energy big data can be summarized as follows:

- *Multi-source data fusion*: Energy big data is not limited to the recorded data by communication, monitoring, and intelligent sensing devices. By rapidly increasing the penetration of the RERs in the energy generation process of the hybrid networks, a vast amount of the new types of data such as geographic information, weather, humidity, and temperature data are also collected by smart systems. This is when the significance of the energy load forecasting, demand-side energy management, and anomaly detection is obviously recognized in the energy big data analytics. However, the achievement of the multi-source data fusion has become a major challenge in effectively applying the energy big data [18].
- *Data analysis*: Future modern MCENs implementation will be accompanied by the deployment of the millions of advanced meters and sensors. In this condition, the generation of the massive amount of data will be an undeniable fact in the modern grid paradigm. Although these data can be effectively used for extracting the useful knowledge associated with the different types of data for business planning, load balancing, grid optimization, and other purposes, they bring new challenges in data management and analysis [19].
- *Security and privacy protection*: Providing the acceptable level of privacy and security for the collected data is vital in the energy big data analytics due to its

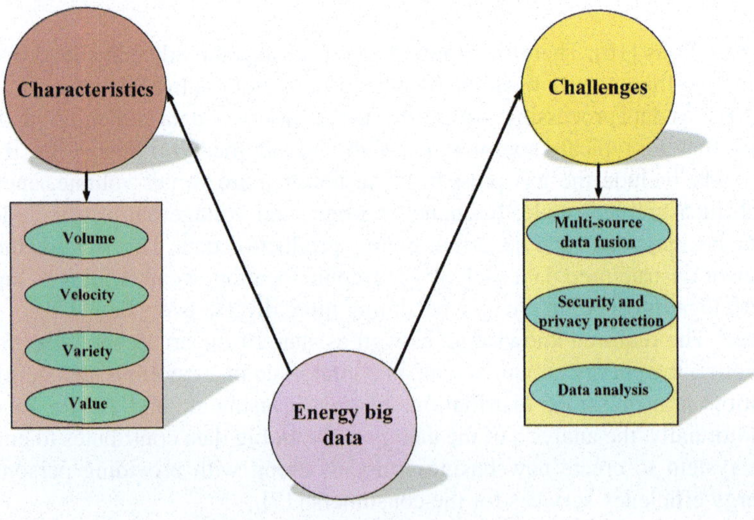

Fig. 4.3 Challenges and characteristics of the energy big data

great importance in employing energy network solutions. In this pass, the expansion of the usage of information technologies is also expected to make energy networks more efficient and intelligent [20]. Growing the appearance level of the information and communication technologies in the modern structure of the hybrid networks promises the emergence of new security and privacy problems. Therefore, there arises a greater need for ensuring a sufficient level of privacy and security for the recorded and collected data in the future modern MCENs.

4.3 Smart Meters

The smart meters have attracted widespread attention so that their popularity has led to the deployment of a large number of them in the energy network that have enabled this grid to collect an immense amount of energy data over the past decade [21]. The deregulation of the energy industry, especially in the energy supply sector, is continuously moving forward worldwide along with increasing the pressure of the shortages in the multi-carrier energies [22]. In this regard, how to record a vast amount of data for enhancing and promoting the sustainability and efficiency of the energy networks has become a pressing issue [23]. In other words, the lack of sophisticated data recording systems can hinder the near real-time optimization of the system for achieving significant benefits like the continuous energy supply. Due to this, the smart meter technologies are rapidly developed to be widely deployed throughout the smart energy networks that can easily allow the system for [15]:

- Optimizing the system operation and energy production sector
- Precisely predicting energy demand
- Dispatching the multi-carrier energy in a timely fashion
- Accurately discovering the energy consumption patterns
- Developing effective energy pricing mechanisms

All of the aforementioned benefits can be reached using the smart meters for recording the different types of energy big data and frequently transmitting them to the data control center in real-time. In the energy networks, smart meters can support different energy big data in the recording process of the networks' important key parameters such as energy consumption, current, voltage [24], gas pressure in the pipelines, heat temperature in the heat nodes, and water volume in the water distribution system.

4.3.1 Characteristics of Smart Meter

At present, numerous smart meters located throughout the smart energy networks are responsible for recording a massive amount of data that needs to be properly processed. In order to effectively use the smart meters' potential, fully focusing on their characteristics is essential. By growing the developments in the context of the information and communication technologies, the technology of the smart meter big data has also matured, which is characterized as follows [15]:

- *High dimensionality of data*: Smart meters are distributed in the load centers for frequently recording the energy consumption information like the amount of electricity consumed, current, voltage. The advanced version of the smart meters can accumulate the recorded data every 5–60 min [15]. In recent years that the technology of smart meters has dramatically grown, we were witnessed the providing of technical solutions for smart meters by the U.S. largest electricity and gas supplier, PECO, which enables the system for collecting a huge amount of data once a minute [25]. This is when this technology is rapidly developing day by day for constructing smart meter faster than before. Nowadays, the tendency to modern MCENs is the introduction for starting the era of smart meter big data that is expected to be accompanied by various opportunities and challenges.
- *Large volume of data*: The development of the smart grids has led to the privilege of the smart meters in accelerating the birth of several start-ups [21]. One of the prominent missions of these start-ups is to expand the deployment of the smart meters throughout the energy networks for energy big data records, collection, and assessment. The huge volume of data recorded by a large number of smart meters has created additional requirements for the smart energy networks. Indeed, smart grids need to be developed as the new special data centers with adequate capacity aiming for storing the high-dimensional recorded data by the smart meters. Therefore, a large volume of recorded data not only needs the additional data centers for data organization but also the technology of the classifying the stored data and analyzing them should be undergone a significant overhaul.
- *High speed of data acquisition and transmission*: The quick reinforcement of the modern grid's leverage in recent years has resulted in significant development in the context of information and communication technologies. These technologies are considered to be incorporated with smart energy systems and intelligent devices to increase the speed of big data acquisition by using intelligent processing programs. The collection and transmission of the energy big data in the near real-time are expected to be conducted at the minimum time by the aforementioned integrated system.
- *High speed of data processing and analysis*: The rapidly analyzing and processing of the smart meter big data provide a great opportunity to the smart energy networks for predicting energy load, determining the location of any failures, detecting the causes of any unnatural events, and so on in a short time. Furthermore, the speed of data processing and analytics has direct influences on the application of the controlling systems in maintaining the reliability and stability of the energy networks.

To sum up, smart meters not only play a key role in recording, collecting, and transmitting the energy big data to the data centers but also they allow the system to effectively use the recorded data for mining the useful information that can be used in developing the different parts of the energy system as well as the data mining techniques.

4.3.2 Benefits of Smart Meters for the Future Modern MCENs

Generally, smart meters are famous for their application in recording the energy big data, which are mostly referred to as the electrical energy parameters such as electricity consumption, current, voltage, and frequency. This is when the future modern MCENs are expected to be an integration of the electric power system, natural gas grid, the district heating network (DHN), and water distribution system. This issue clarifies that the smart meters in the modern MCENs should not be limited to one dimension of the energy but also they are expected to be multi-dimension-based energy metering systems. Therefore, the recording of energy big data is not limited only to the electrical energy parameters like electricity consumption, current, voltage, and frequency. Indeed, in the future modern MCENs, the smart meter in each energy consumption node should be comprised of several parts in which each part is specialized to support a certain dimension of energy. For example, the part associated with the natural gas is responsible for recording the energy big data such as gas consumption, gas pressure, and volume in the pipelines. Figure 4.4 demonstrates different parts of the smart meters and their benefits in recording various types of data in the future modern MCENs structure. It should be noted that the proposed home modern multi-carrier energy management system (HM^2EMS) described in Chap. 2 is targeted to do this work but its duties are not limited only to perform similar works.

The energy big data recorded by the smart meters can be used for achieving valuable benefits such as improving the quality of the energy services, maintaining the reliability and stability of the smart energy networks, and optimizing energy efficiency [26]. For this aim, smart meter big data can be completed by integrating other required data such as weather and energy market data for reaching various purposes such as demand-side energy management, anomaly detection, load

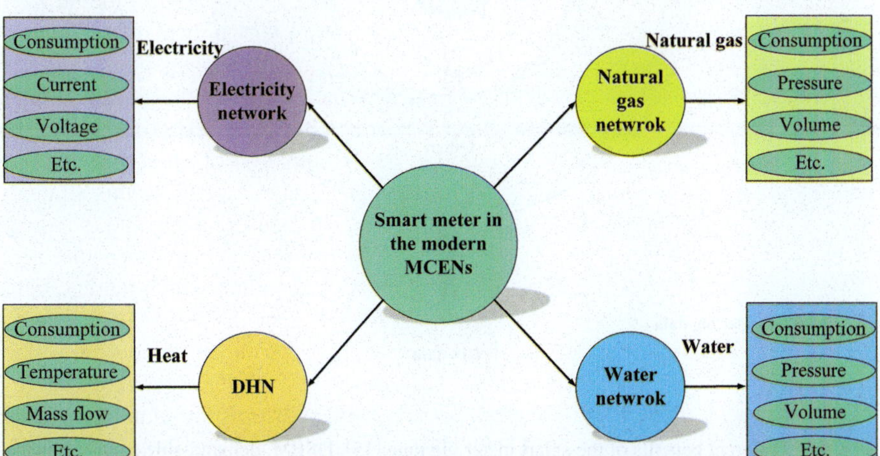

Fig. 4.4 Different parts of the smart meters and their benefits in recording multi-carrier energy

forecasting, and so on. The overall schematic of the smart meter big data in the smart energy networks is illustrated in Fig. 4.5 [15].

Some of the important benefits of the smart meter big data are described below but surely they are not limited to them.

- *Energy load forecasting*: Energy load forecasting is essential for the development planning of energy networks, particularly in the energy supply sectors. Indeed, energy load forecasting supports the system to adopt specific strategies for an operating system more safely, economically, and reliably. Forecasting energy demand can be carried out for different time intervals such as ultrashort-term load forecasting, short-term, medium-term, and long-term load forecasting which can be intended for an hourly, single day, monthly, and annual load forecasting [27]. Until recently, several techniques have been applied for

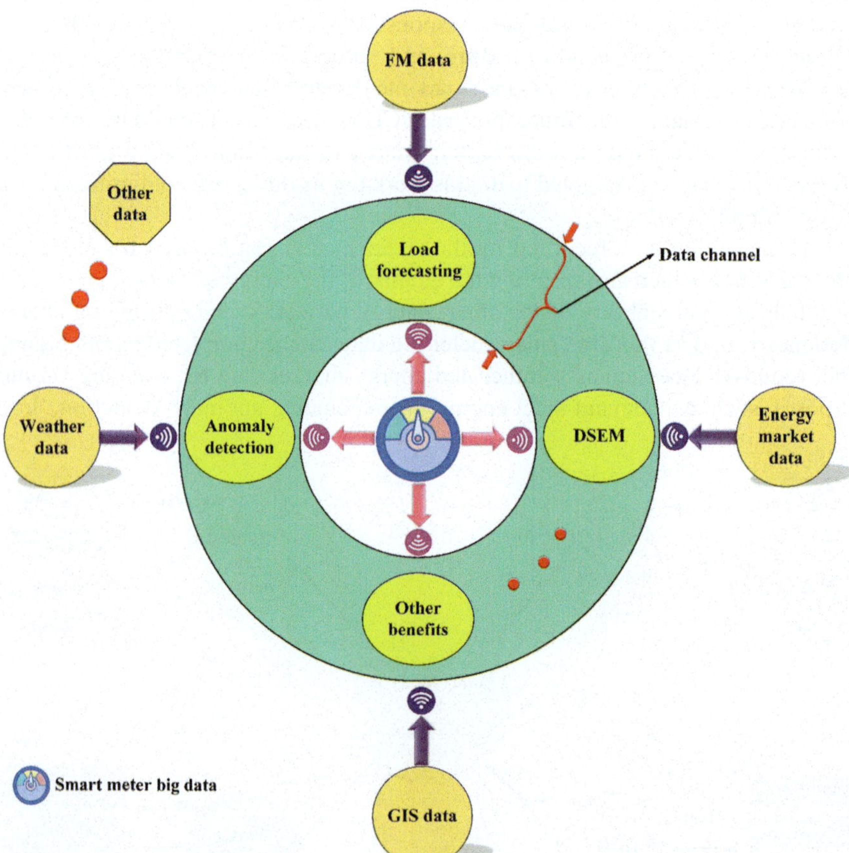

Fig. 4.5 Different benefits of the smart meter big data [15]. DSEM, demand-side energy management; FM, field measurement; GIS, geographic information systems

energy load forecasting such as weighted iteration, exponential smoothing, and regression analysis [15]. Moreover, genetic analysis and neural networks are also widely used for this work along with the improved algorithms such as stochastic time series and adaptive prediction. A vast amount of energy data collected by the smart meters creates the foundation of energy load forecasting. In this respect, to avoid difficulties in using the high-dimensional data in the load forecasting, employing the smart meter big data compressing methods is necessary [15].
- *Anomaly detection of the energy networks*: Anomalies are typically referred to as the variety of faults along with some non-technical energy losses such as energy theft and illegal energy usage in the energy networks [28]. Quickly detecting of the anomalies is critical for timely reacting to them with the aim of maintaining system stability, safety, and reliability. To this end, smart meters are empowered by the intelligent sensors that can effectively detect fault information and send them to the relevant control center. In this process, some other useful information can provide more details for supporting the system to further identify the location and cause of faults such as point of failure information, climate information, and geographic information. At present, several techniques like deep learning and neural networks are used accompanying with the smart meter big data for effectively detecting any faults in the energy networks [15].
- *Demand-side energy management (DSEM)*: The DSEM is a set of techniques that are designed to encourage the consumers to modify their energy consumption patterns to improve the efficiency of the energy networks as well as retaining the reliability and stability of the system while suppressing short-term enhancements in energy prices. In this regard, forecasting of the consumers' energy consumption is a key issue in implementing the DSEM techniques in the prosumer-side and the smart meters are the main devices for providing the required primary information for the energy demand forecasting.

To sum up, some of the most important features of the smart meter big data in the hybrid networks are as follows [29]:

- Providing energy consumption data for utility and consumer
- Remote command (turn on/off) operations
- Outage and failure notification
- Time-based pricing
- Net metering
- Load limiting for DSEM purposes
- Energy theft detection
- Communication with other smart devices
- Energy quality monitoring for the electric power system, natural gas grid, the DHN, and water distribution system as illustrated in Fig. 4.3
- Improving environmental conditions by mitigating environmental pollutants through efficient energy management

4.4 Advanced Metering Infrastructure

In smart energy systems, a hierarchy of sub-systems needs to be realized for achieving intelligent system goals [29]. In this regard, the advanced metering infrastructure (AMI) has experienced considerable changes to be matched with the structure of the smart energy systems [30]. These changes have driven the technology of the AMI to be developed in a way for making it as a configured infrastructure with the incorporation of the diverse technologies in accordance with the smart energy networks' goals. The AMI infrastructure consists of meter data management systems, communication grids in various levels of the infrastructure hierarchy, smart meter big data, and facilities for integrating the accumulated data into software application interfaces and platforms [31]. The schematic of the AMI infrastructure is shown in Fig. 4.6.

According to Fig. 4.6, each of the end-users is equipped with the smart meter big data for real-time collecting data. After collecting data, smart meters can transmit the data to the data processing center through available communication networks such as fixed radio frequency, power line communications, broadband over power line, and public networks such as cellular, paging, and landline [29]. Afterward, data classifying, clustering, analyzing, and data mining are conducted in the data processing and management center. By developing the two-way communication protocol, the AMI enables the control center to directly control the smart devices wherever in the smart energy network.

4.4.1 Communications in AMI

In the AMI infrastructure, the collected data by the smart meters need to be sent to the control center aiming to allow the system to adopt optimal decisions and send them back to the energy management systems. This indicates that the AMI has to

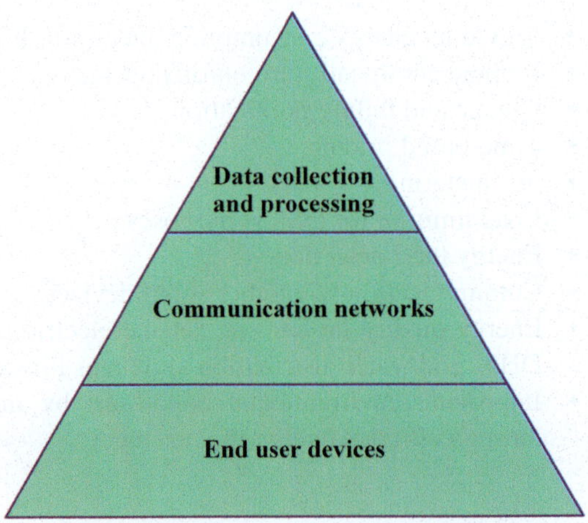

Fig. 4.6 The AMI infrastructure [31]

equip with advanced communication protocols for exchanging various types of data. Given a large number of intelligent devices and smart meters throughout the system, the communication gateway is expected to have sufficient capacity for effectively handling the high volume of data. Therefore, meticulous design and construction are necessary for the communication protocols in the smart energy networks. For this aim, some key factors should be intended, which are as follows [32]:

- Showing status of the grid
- Restriction in accessing data
- Cost-effectiveness
- Supporting future expansion
- Providing complete information regarding consumers' energy consumption
- A huge amount of data transfer
- Confidentiality of sensitive data
- Ability to host modern attributes beyond AMI requirements
- Precision in communication with related device and authenticity of data

For establishing a safe communication in the system, different architectures and topologies can be commonly used in the smart energy networks. The most particular architectures are focused on developing the local data center for collecting the data of group smart meters and transmitting them to the control center through a backhaul channel where the billing and management applications along with the data processing and storing facilities reside [31]. There are different communication and medium technologies for realizing the AMI, which some of them are depicted in Fig. 4.7 [29].

The future modern MCENs are not one-dimension energy-based networks; instead, they are the community-based grids that are targeted to be developed in the form of the combination of multi-carrier energy and several hybrid energy networks. Indeed, because the exchanged data in the AMI infrastructure are not limited only to the electricity network parameters, the communication systems should also be developed in a way to effectively handle the related parameters of the natural gas network, the DHN, and water distribution networks in the future modern MCENs.

4.5 Internet of Things (IoT)

As the smart term has appeared in the philosophy of the energy networks, the hybrid grids have witnessed the increasing number of intelligent devices. These systems include home appliances, vehicles, physical items, and other systems that are embedded with actuators, sensors, software, electronics, and characterized by their connectivity to the Internet. In this digital environment, the IoT has emerged as a network of such smart devices in which they can negotiate with each other pursuing special goals. Due to enhancing the penetration of intelligent devices throughout the smart energy networks, expanding of the IoT in the smart grids' interactions is at the pinnacle stage [6].

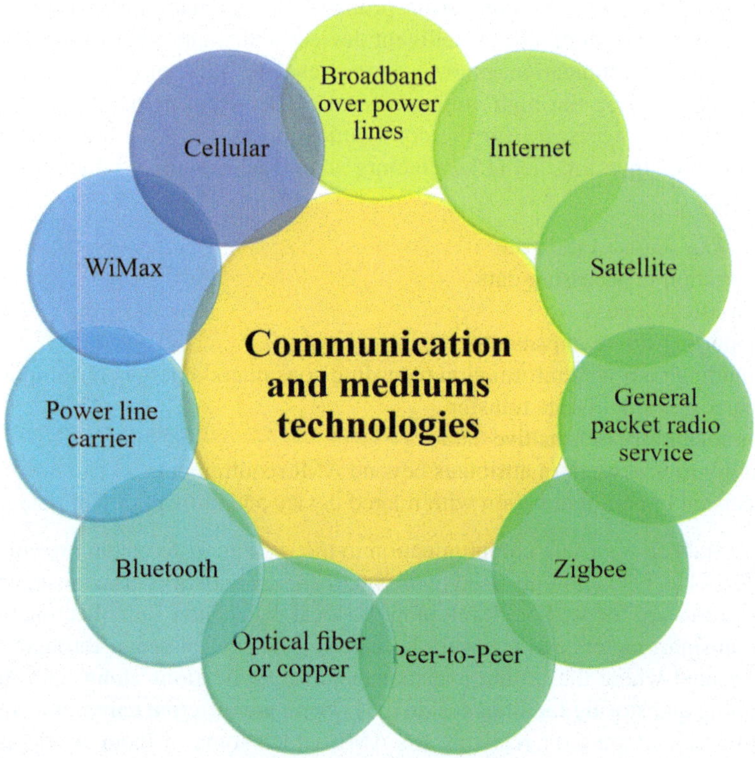

Fig. 4.7 Some of the communication and medium technologies for realizing the AMI [29]

4.5.1 Classification of IoT

In today's world, the IoT has experienced significant changes with manifold definitions due to its widespread application in various fields, which has led to the appearance of its different classification for it. It can be typically classified into the following items [6]:

- *Using Internet technologies*: These technologies are broadly considered for creating interconnections among all the smart objects.
- *Internet services*: Several supporting technologies need to be employed for realizing the perception such as machine-to-machine communication devices, actuators/sensors, radiofrequency identification.
- *The ensemble of services and applications*: Applying the state-of-the-art technologies for creating the new market and business opportunities.

Due to shifting in IoT in the digital network with only one common Internet protocol (IP), it is not necessary to use the concept of three different networks such as neighborhood area network (NAN), wide area network (WAN), and home area network (HAN) in the smart energy networks technology. Indeed, the one IP-based networks allow all sub-departments in any industry to be complemented in accordance with their applications using different services without any hurdle [6].

4.5.2 Characteristics of IoT

The IoT platform can also be developed to make the system flexible enough to allow third parties to effectively use the provision of an application program interface with the aim of developing complex applications [33]. This development is beyond some specialized services and presentation of raw data for the IoT-based systems. In such systems, technology synthesis involves suitable conditions for perceptive technology, artificial intelligence technology, advanced network technology, intelligent systems for uniting with each other aiming to realize the interoperability in the system [6]. According to the complex ecosystem theory regarding the context of the social-economic-natural, the essence of sustainability has been intended as a great mutation for the coordinated development of the life and its habitat, material production environment, culture, society, and work. This mutation is very important in the smart energy networks area because it coincides in line with the goals of the IoT [34]. The general characteristics of IoT are demonstrated in Fig. 4.8 [6].

Given the characteristics exhibited in Fig. 4.8, creating effective interoperability is a key issue throughout the IoT-based system in the presence of a large number of technologies and devices [33]. In such a system, generating a massive amount of data by intelligent devices has made the data extensibility, scalability, and deluge more significant.

4.5.3 Challenges in IoT

Notwithstanding, the IoT systems have adequate capacity for making the energy networks a smarter system. On the other side, they create new concerns that are critical issues in successfully implementing smart energy systems. Some of the most significant of them are as follows [6]:

- *Identification of different objects*: Identification of various intelligent objects is a key challenge in the IoT system that indicates how can each object be identified in the energy networks. In this respect, each object can be uniquely identified considering various specific scenarios for distinguishing them from the class. On the other hand, the identification of any object is possible by means of QR codes [6]. These possibilities are feasible either by developing the characterization of objects or providing their descriptions. These two ways can complement each other and are not mutually exclusive.

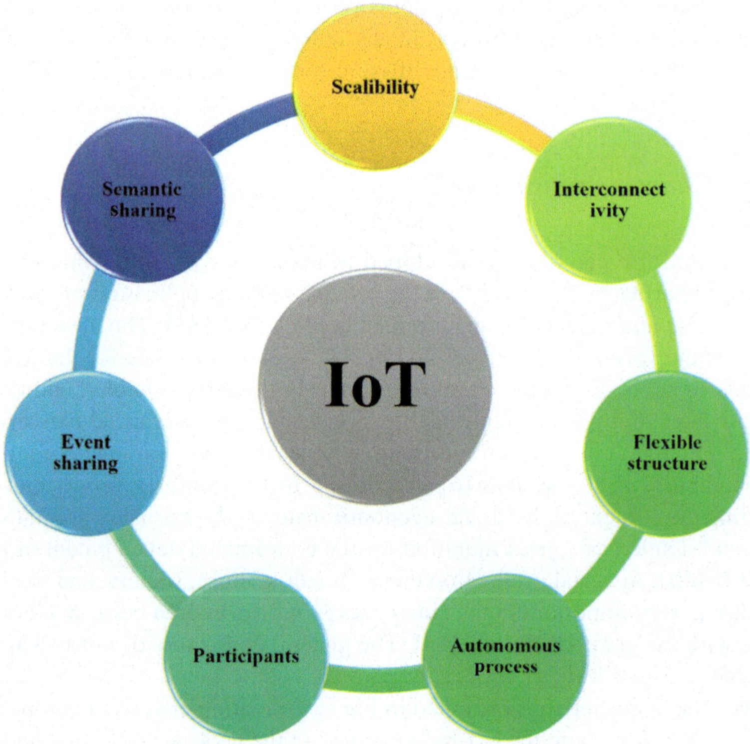

Fig. 4.8 Characteristics of IoT [6]

- *Sensing and actuation in IoT systems*: Sensing and actuation dimensions provoke the fundamental operations that seek to exhibit the pairing and interface among the digital and real domains [35]. Sensor and actuator networks provide research opportunities in the prominent area with ad hoc systems that can be used for developing interactions to pursue a particular task. Indeed, the IoT application is expanded beyond the current smart technologies.
- *Anything communicates*: Anything that is intelligent and enables the IoT-based system to establish wireless communication between the smart devices by developing ad hoc grids with the community-based objects.
- *Anything identifies*: Smart systems can be identified using the information of their relationships with each other in the digital domain without the possibility of establishing physical intervention.
- *Anything interacting*: Smart systems have an ability to create interactions using the actuation and sensing competence, whenever present.
- *Privacy and security*: IoT systems are targeted to provide plenty of benefits as the state-of-the-art technology for future energy networks. The major challenge overlooked in the IoT is the effective provision of sufficient security and privacy. However, advancing the technology along with the appropriate remedial

measures can support the system by bringing new authentication schemes, public key infrastructure, encryption methods as well as by developing standardized application scheme interfaces.

There are several considerable features for the IoT systems that some of them are taken into account as the advantages of the IoT while some others are intended as the disadvantages of the IoT. All of these features are depicted in Fig. 4.9. More descriptions regarding each of these features can be found in [6].

4.5.4 Combination of IoT and Smart Energy Grids

In smart energy networks, the term "smart" inspires a group of intelligent devices that are united with each other to enable the energy networks for monitoring all interactions and operations through automated processes. In such an environment, the generated data are typically much greater than the conventional energy grids due to the necessities of every moment interacting among the smart devices for realizing the goals of the smart energy networks. Indeed, the enormous data processing is inevitable and undeniable due to establishing the unceasing dual communication among the smart meters at the end-users' sector with the utility center. In order to successfully conducting the processing of the huge data in the system, the smart infrastructure along with the capable two-way communication systems is essential. Herein, a significant role of the IoT technology is highlighted by assisting the smart system to narrow down the exchanging of the massive data between the utility and users over an IP. In smart energy networks, IoT technology is responsible for identifying energy shortages or excessive in particular areas and at certain times with the aim of properly managing the systems' interactions. Similar dependencies exhibit the necessity of the combination between the IoT technology and smart energy networks. This combination is developed through several layers that are illustrated in Fig. 4.10 [6].

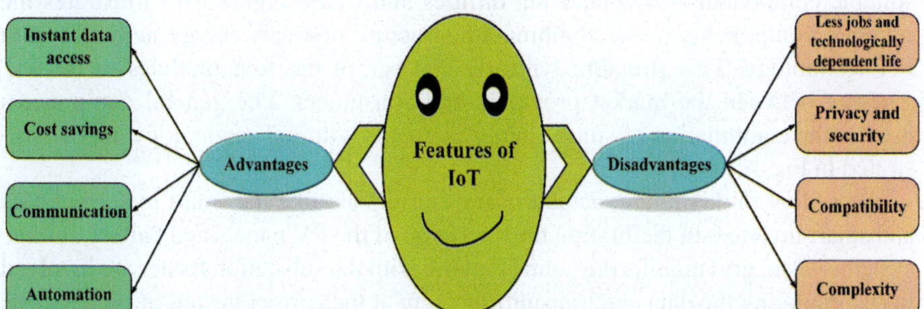

Fig. 4.9 Features of IoT [6]

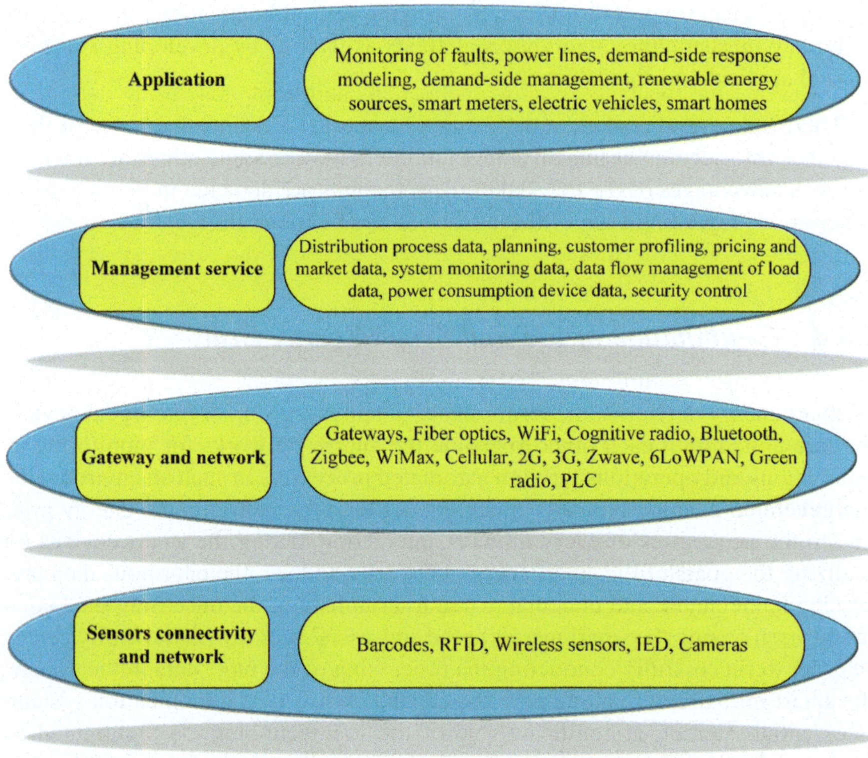

Fig. 4.10 Basic layers in the combination of the IoT technology with the smart grid [6]

4.5.5 Components of IoT with the Smart Energy Grid

The necessity for combining the smart energy networks with IoT technology will surely require the stable connections between the different components. In this regard, successfully realizing smart energy networks directly depends on involving suitable components for connecting utilities and users. Figure 4.11 illustrates the required components in the combination structure of smart energy networks with IoT technology. This structure typically consists of the four modules in the IoT gateway between the market providers and consumers. The general components used in the combination of smart energy networks with IoT technology are demonstrated in Fig. 4.11 [6].

According to this figure, actuators and sensors act like the smart meter big data and smart inverters in facilitating the operation of the PV panels and battery storage systems. Also, grid monitoring controls along with the substation feeder are involved in accumulating the data and transmitting them at the correct instant of time. In this process, the gateway embarks a significant role as the common connection platform with the providers and commercial users in the smart energy networks. The monitoring of the system is conducted by recording and transmitting data by the sensors

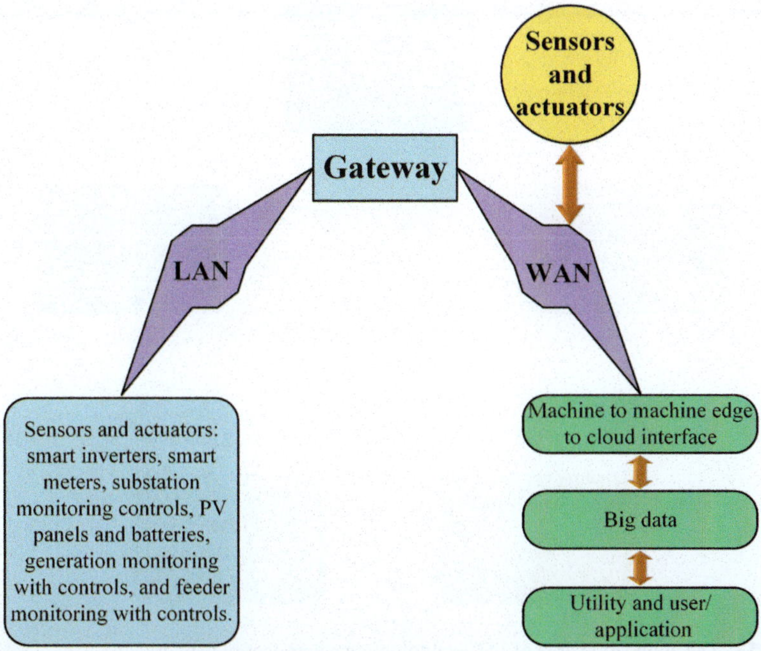

Fig. 4.11 Components used in the combination of smart energy networks with IoT technology [6]. LAN, local area network; WAN, wide area network

to the gateway using Internet protocols [36]. Afterward, temporary decisions have been adopted and the data are transferred to the final data processing. In order to effectively manage the smart devices as well as the collected data, the cloud of the machine-to-machine interface is used as another key area for monitoring the system. In this respect, event processing and appropriate data streaming along with the authentication are carried out in a continuous fashion. Big data analytics is taken into account as another critical component that comprises the continuous series of data [37]. Even the properly streamlining of the unstructured data is also continued for analyzing by means of suitable data analytics solutions [38].

4.5.6 Applications of IoT in Smart Energy Grids

The rapid growth in IoT technology has driven the trend of its progress toward making it as the state-of-the-art technology for global applications in all sectors such as commercial, industrial, and residential. The IoT progress in the industrial sector is more focused on making it most beneficial aiming to promote its widespread usage in various manufacturing sectors. This technology is widely used as a core constituent in the smart meters with two-way communication capability that enables users to establish stable communication with the utility provides.

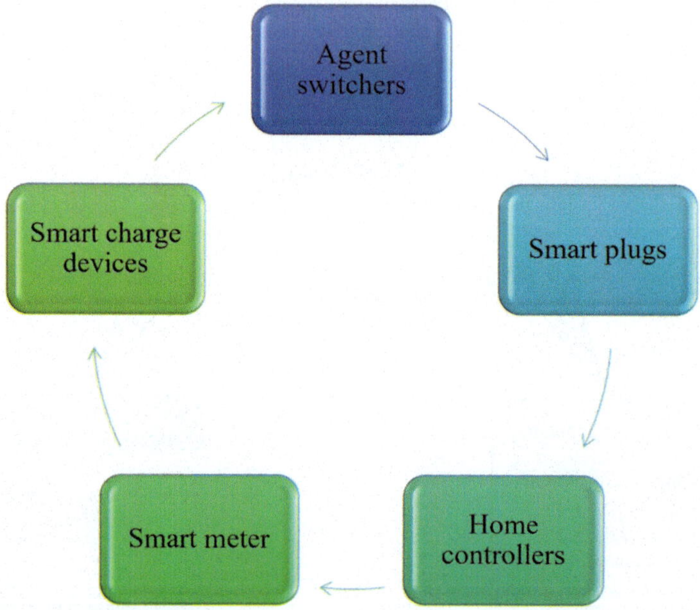

Fig. 4.12 The application area of IoT technology in smart energy networks [6]

The amount of consumers' energy consumption is recorded at certain periods and sent to the utility through the web browser that provides appropriate conditions for the users to inform them about the price of energy at every hour of a day in a more apparent manner. The application area of IoT technology in smart energy networks is shown in Fig. 4.12.

The other arena in the IoT application is the smart charge devices that allow the system to automatically charge the intelligent devices considering the real-time battery array when the system is scheduled at peak times. This system enables the users to know about the real-time energy price using the web browser during a day, especially at peak times. In addition, the IoT technology can make the smart plugs possible in the system for connecting the residential loads with the energy lines. The most benefit of the smart plugs is that they can be properly adjusted by receiving appropriate signals through Internet connectivity to help the stability of the system, especially during heavy demand times. The agent switcher evaluates the received data from the smart meters with the aim of submitting suggestions for the consumers to persuade them to do appropriate actions by providing the information regarding the deferrable and non-deferrable energy demand. Additionally, IoT technology can be applied in home controllers to establish suitable interactions between the residents and utility for adopting appropriate strategies in the prosumer-side.

Although the IoT technology is famous for its capability to develop smart energy systems, it needs significant changes to be suitable for the future modern MCENs structure. The current version of the IoT only supports one-dimension energy in the smart grids' interactions. This is while modern MCENs will be built based on the multi-carrier energy systems. In such environments, privacy and security are the vital issues that their acceptable level needs to be provided throughout the system. However, the IoT technology lacks sufficient privacy and security that can ensure the safety of the system from the different attacks and other remarkable risks. Moreover, the IoT technology increases the complexity of monitoring systems at the wide-area while the innovative technologies are expected to join the modern structure of the future MCENs aiming to facilitate the interoperability in the system with minimum complexity. Therefore, the developed version of the IoT technology can be intended to be used in the future modern MCENs for supporting the system to be more reliable, resilient, stable, flexible, and safer than ever before.

4.6 Energy-Cyber-Physical Systems (ECPSs)

Over recent years, evolutions in human life exhibit that the coming century collectively brings new challenges for societies in the context of unreliable and aging infrastructure, massive urbanization, cyberspace security, and natural resource depletion [39]. In this regard, the burden of enormous pressure in meeting societal demands is more felt on the urbanization parts like the smart energy networks and buildings to solve this challenge by switching from mere physical systems to intelligent ones with the capability of advanced computations. These systems are targeted to serve the human-centric services as well as increasing the system efficiency and resilience. In this regard, Cyber-Physical System (CPS) has become a common effort for achieving the aforementioned expectations by integrating the physical and cyber components using sophisticated computational devices, network-based technologies, and modern sensors [40]. Given the capabilities of the CPSs in effectively addressing the challenges of the energy fields, their incorporation has become an inevitable issue in the smart energy systems. Engaging the CPSs with different aspects of energy sectors has led to an introduction of the new concept of CPSs, which is called Energy-Cyber-Physical Systems (ECPSs) [39]. The ECPSs are physical systems with cyber layer that covers the energy area of the energy generation part to the energy distribution and operation sector. Remarkable advancements in the different dimensions of the ECPSs such as in system optimization, control, learning, and sensing have enabled the smart grids to further unlock their undeniable potential to significantly alleviate the pressure of considerable shortages in the world's energy sector. Successfully achieving such goals undoubtedly needs significant progress in developing new market mechanisms, new privacy- and security-protection mechanisms, new operational and design methodologies, new sensing and data processing solutions, and new technologies at all stages of the energy sector [39]. By pursuing the aforementioned changes regarding the ECPSs, some key benefits are expected to be reached such as resilience and safety of the interdependent critical infrastructures, flexibility in dynamic environments, operational agility, enhancing the system efficiency, cost-effectiveness of resource management, and broadly defined human-centric services like the well-being, health, privacy, and comfort [39].

4.6.1 Features of CPSs

The rapid growth of the CPSs technologies has led to this fact that the concept of CPSs is not limited to the cyber and physical systems but also the recent developments have given rise to new concepts regarding the CPSs. In this respect, a novel computing paradigm brings a new version of the CPSs that is recognized as cyber-physical-social [41]. Indeed, the term "cyber-physical-social" is originated from the cyber-social system and CPSs that includes the signs of people's interactions and participation in the social space [42]. Such developments in the history of the CPS have made it as an appropriate paradigm for constructing smart energy networks with specific features. Some of these distinctive features of CPS systems are given as [43, 44]:

- Ability to adapt and train
- Interaction of physical- and cyber-components
- Ensuring the reliable operation of smart energy networks
- Data flows and cross-domain sensor resources
- Mobile and embedded sensing
- Providing interoperability using Internet protocols such as IoT technology
- Human in/outside the loop
- Ensuring system robustness by automated intellectual control
- Operating the system in a way to be certifiable in most cases
- Presenting common cyberspace that can establish energy exchanging both within the environment and with the system considering the information security in the different types like antiviruses, firewalls, cryptosystems.

4.6.2 Principles of CPSs

The recent substantial advancements of the CPSs have made it the basis for developing some critical areas such as smart cities, smart buildings and infrastructures, smart medicine, smart manufacturing, mobile systems, wearable devices, smart vehicles, meteorology. The vastness of the CPSs applications in the smart energy networks has formed the domain of the CPSs that is illustrated in Fig. 4.13 [45].

In the smartization, the cyber and physical layers are two main layers in the CPS architecture [45]. In the current state of the CPSs, the variables present the data received from the sensors while the control variables are the representative of the control signals [46]. In this respect, a set point is named for the normal value of a parameter in a certain process [45]. The controllers are used for calculating the distance of the process variables with the control ones. After measuring this offset, the controllers recompute the variables using the newly available data and conditions of the system for achieving new setpoints that can retain the system process more closely to the optimal points. In this process, the graphical user interface is used for providing the status of the controlled objects at each time for the system

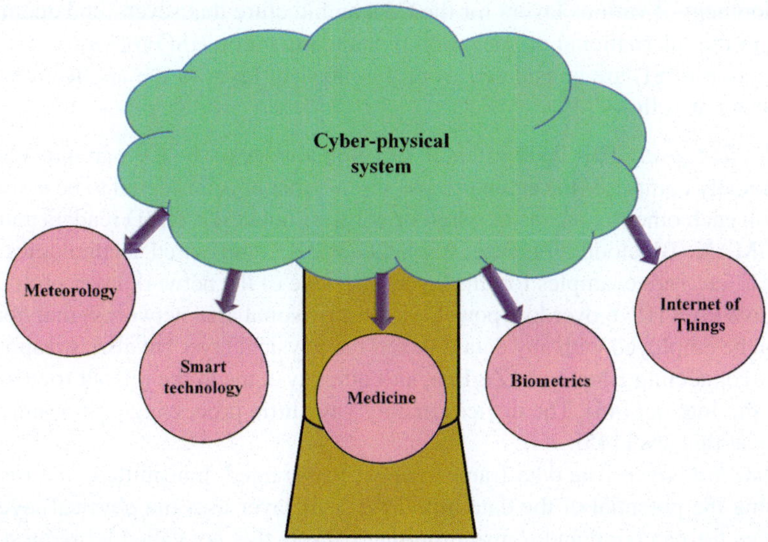

Fig. 4.13 The domain of the CPSs [45]

operators to let them adopt appropriate strategies for better controlling the system. The following stages present the common division of the CPS process [47]:

- Monitoring
- Networking
- Computational processing
- Actuation

In this regard, some industrial protocols such as Modbus, 61,850, and DNP3 are often used for creating communication between the cyber and physical layers [45].

4.6.3 CPSs Architecture

The CPSs are commonly characterized by intelligent decision making, heterogeneous information flow, and cross-domain sensor cooperation, which may comprise multiple mobile/static actuators and sensor networks that are incorporated based on the intelligent decision systems [48]. Under the effective connectivity, various types of CPS components can be integrated to form different combinations of key functions that directly depend on their applications [45]. Mostly the computational components are intended in the CPS by focusing on the information and knowledge of physical processes to determine the characteristics that must be considered and to what extent in accordance with their applications [45]. The realization of all of them directly depends on how CPS architecture should be designed and which factors should be considered in this process. This issue has led to the development of various layers for the CPS architecture that seven fundamentals of them in the international organization of standardization (ISO)/open system interconnection (OSI) are considered from the physical layer to the application layer, which are as follows [49]:

- *Physical layer*: This layer refers to the groundwork of the CPS architecture and typically comprises the actuators and sensors that are considered to be connected with each other by means of wired or wireless networks. RFID readers and tags, WiMAX, Bluetooth, ZigBee, WiFi, 2G/3G/4G, and wired technologies (NC, PLC, etc.) are examples for this layer [45]. One of the network layer protocols is 6LoWPAN (IPv6 over low power wireless personal area networks) that not only can be employed with any data link and the physical layer but also is responsible for connecting Bluetooth, ZigBee, and other systems to the related routers such as the Internet [45]. The devices mainly have little processing power and memory at this level [45].
- *Data link layer*: The data frames are usually created, transmitted, and received using the potential of the data link layer. This layer uses the physical layer services for receiving and transmitting the packets that are called by request from the network layer. This layer is classified into two sublayers: logical channel management and media access control. In this respect, the network layer service is provided by the logical channel management sublayer while the media access control sublayer is responsible for regulating any accesses to a shared physical environment.
- *Network layer*: In this layer, converting media access control addresses to network ones is intended as a common way of routing the packets using the RPL ("Ripple" routing) and IPv4/IPv6 protocols.
- *Transport layer*: This layer is where the packets are targeted to be broken into small fragments. The Internet control message protocol (ICMP), transmission control protocol (TCP), and user datagram protocol (UDP) are the most typical protocols at the transport layer.
- *Session layer*: This layer is aimed to manage the conversation (communication session) by monitoring the order of messages transmitted over the network. When the fault occurs, the labels are inserted into long messages without requiring to start from the beginning again. Session-level protocols are taken into account as a key layer of the functions in the top three layers of the model.
- *Presentation layer*: Transferring data from the external format to internal format as well as the data decryption and encryption are two application processes that enable the presentation level to effectively coordinate the data presentation (syntax). For example, the secure socket layer (SSL) protocol allows the system to exchange a secret message for the application layer protocols of the IP/TCP stack.
- *Application layer*: This layer receives information from previous layers to store, analyze, and update them for various applications in different domains (Fig. 4.13). In this layer, the virtual prototype interface is used to make the visualization of the adopted control decisions possible aiming to protect data privacy as one of the most important issues.

4.6.4 Challenges of CPSs

As the CPS capability has brought significant benefits for the energy sector worldwide, the fast-growing trend of its applications has also faced the CPS-based systems with remarkable challenges, especially in the confidentiality and security areas. In this context, the security of information is the preservation of availability, integrity, and confidentiality of information while the confidentiality refers to those properties of the information that does not allow the information to be available to unauthorized processes, entities, and individuals [45]. Moreover, integrity introduces the completeness and accuracy properties of the information, and availability refers to accessibility and usability upon demand by an authorized entity. Reliability, non-repudiation, accountability, and authenticity are other properties that can be involved in the CPS systems. An effectively providing the aforementioned properties for the CPS systems has been recognized as a fundamental challenge. In other words, the satisfactory level of each of the mentioned attributes should be provided for the smart energy networks to meet the acceptable levels of smartization. For this aim, a large group of intelligent devices needs to be employed in the CPS structure, which can cause some substantial challenges as follows [50]:

- Growing the penetration of the IoT by rapidly rising its devices enhances the vulnerability of such systems to different threats.
- Developing a formal technique for assessing CPS vulnerabilities is essential.
- Designing fault-tolerant and reliable architectures for analyzing of rapidly developing physical and cyber threats.
- The necessity of security threat modeling.

Similar to the IoT technology, the ECPSs also need a significant overhaul to be suitable for the future modern MCENs. The current version of the ECPSs only supports electrical energy interactions, especially in the physical layer. This is while we are going to develop the future modern MCENs with several energy dimensions. This means that both the cyber and physical layers need to be reconfigured by including the multi-carrier energy supports in the new infrastructure. In other words, the term ECPSs should cover all energy dimensions and not just be limited to electrical energy so that physical and cyber layers need to change from single systems to the hybrid ones. In such conditions, the capability and potential of the ECPSs should be enhanced in a way to effectively handle the energy interactions simultaneously in the interconnected environment with the participation of several energy networks. This support not only includes the hybrid system privacy and security but also should cover the multidisciplinary clues of hybrid energy networks such as system reliability, resilience, flexibility, safety. Therefore, the current paradigm of the EPCSs should be changed to the modern ECPSs architecture aiming to pursue the future modern MCENs guidelines in terms of design, validation, implementation, and operation.

4.6.5 *Digital Twin Technology*

As the cyber-physical systems (CPSs) have become a hot topic and popular tool in the energy sector due to their significant advancements, their technology has been faced with great vastness in the smart energy networks. The digital twin is one of the technologies that is developed as one of the components of the CPSs and is the product of the recent advancements in the CPSs. The digital twin is a special type of the CPSs, which refers to a digital replica of both actual and potential physical twin, devices, systems, places, people, and processes that can be developed for particular purposes [51]. The digital twins can effectively enable the data transmission between the virtual and physical worlds, which allow the system to easily monitor, learn, and optimize the functions of physical entities [52]. This technology can also provide continuous feedback for humans by connecting the virtual model or counterpart to the physical model with the aim of improving the quality of life and well-being. In the digital twin technology, the capability of the different types of sensors is used for producing the required real-time data for establishing the connection between the virtual and physical models [53]. In order to structure living digital simulation models for changing and updating in accordance with their physical counterparts' changes, the digital twin technology integrates software analytics, machine learning techniques, artificial intelligence, and IoT with the spatial network graphs [54]. Indeed, receiving various types of data from the related sensors as the primary source enables the digital twin technology to use the special type of the algorithms for effectively learning the patterns hidden in the data and update its application near real-time status corresponding with the targeted work. The digital twin as a formative technology can be used for developing the industrial IoT that seeks to create virtual interactions among the physical objects with the other machines and people [55] by optimizing the operation of the systems, physical assets, and manufacturing processes [56].

4.6.5.1 Characteristics of the Digital Twin Technology

Different characteristics can be considered for this technology in the digital world. These characteristics can be used for distinguishing the digital twin from other technologies in the CPSs. The characteristics of the digital twin technology can be categorized as follows.

4.6.5.1.1 Connectivity

Connectivity is one of the main characteristics of the digital twin technology, which enables it to establish a connection with the diverse intelligent devices at the minimum time. This character of the digital twin is more derived from the feature of communication in the IoT technology and originated from it. Indeed, the digital twin has directly relied on its connectivity and its existence makes sense when it has the ability to create the connection between the physical components and their digital counterpart. The digital twin technology uses the sensors for generating data, special algorithms for integrating them, and capable techniques for increasing the connectivity among the customers, products, and organizations [57].

4.6.5.1.2 Homogenization

Homogenization of data is another key characteristic of the digital twin so that its existence is mainly consequenced from the homogenization of data. The possibility of storing and transmitting any type of information in the digital form has made the decoupling of information from its physical form possible by virtually representing the products in the form of the digital twin [58]. Therefore, the homogenization of data along with the possibility of the information coupling with its physical artifact enables the digital twin to come into existence and emerge as formative technology.

4.6.5.1.3 Reprogrammable and Smart

The digital twin technology can be reprogrammable in an automatic manner by suitably using predictive analytics, artificial intelligence technologies, and sensors on the physical product [59]. The special outcome of this reprogrammable is the functionalities that mean the digital twin can accumulate the special type of data, analyze them for learning, and adjust the related system if needed for creating newer products.

4.6.5.1.4 Digital Traces

One of the observable characteristics of the digital twin technology is that it leaves digital traces. These tracks are valuable for identifying where the error originated by checking the traces of the digital twin when a related system malfunctions [60]. These diagnoses can inform the manufacturer of the system to use the digital twin tracks for improving the designs and eliminating the problematic errors of the system in its new version.

4.6.5.1.5 Modularity

From the manufacturing industry's stand of view, modularity can be expressed as the customization and design of production modules and products [61]. Modularity characteristic allows the manufacturer to have the ability to tweak machines and models. The digital twin technology makes the tracking of the machines possible for the manufacturers to provide appropriate conditions for improving the quality of the systems. Indeed, when the modularity feature is considered in constructing the smart machines, the manufacturers can easily identify the components that cause for poorly working of the system to replace it with a better one aiming to increase the quality of the related process.

4.6.5.2 Benefits and Drawbacks of the Digital Twin for the Future Modern MCENs

As one of the applicable components of the CPSs, digital twin technology can be effectively used for improving the application of the smart systems in the energy networks as well as developing the cyber, physical, and social layers of the CPSs. It can enable the manufacturers to build useful smart products by providing valuable information about the previous applications of the systems. For example, it can improve the construction process of the wind turbines with the aim of effectively following the directions of the wind speed or can be used for making the most of the buildings' appliances smart for increasing the well-being of the occupants. These applications of the digital twin are feasible when the deterministic initial data are available for its process to achieve reliable information from this technology for sensitive decisions in the system. For instance, for a special goal that we know all possible states of realizing it, the digital twin can be effectively used for smartizing the related process for obtaining the mentioned goals. However, it is not a suitable choice for those objectives that their relaxing process is stochastic or we do not have access to their deterministic information in the future modern MCENs. Indeed, in the future modern MCENs, the application of the digital twin technology is limited to those goals that the related data are deterministic or all the possible states for its process are certain and it cannot confidently cover the processes with stochastic sources. Moreover, the algorithms and techniques used in the digital twin need to be seriously developed to minimize the tolerances in the different processes. Also, the time of process for those procedures with a vast amount of data is very high that has made this technology become unsuitable for most of the goals in the future modern MCENs that intensely require fast response technologies. Therefore, the digital twin technology should be developed in a way to significantly reduce the processing time for effectively presenting in the future modern MCENs' interactions.

4.7 Energy Internet

At each time period from the recent century until now, the changes in the energy infrastructure have been made it more and more complex than ever before [62]. This is more due to the emerging of different technologies and innovative systems in all sectors of the energy as well as the rapidly growing demand for various carriers of energy. By developing the energy networks and appearing the intelligent devices with significant advancements, the energy infrastructure has experienced substantial alterations so that business and information flows are also developed in addition to the energy flow in the new evolved energy structure. Energy Internet is a new term named for the new development form of the energy infrastructure [63]. Energy Internet can realize the integration of business and information flows with the energy flow by developing energy services and business models [63]. From the viewpoint of the Rifkin in his book [64], Energy Internet is the new energy paradigm that aims to integrate distributed power plants, RERs, electric vehicles, energy storage systems, and hydrogen with the Internet technology. The core idea for Energy Internet is taken from its emphasis on personalization, equality, interaction, openness, and energy sharing in the whole process of energy consumption and generation [63]. In the digital environment of energy, Energy Internet caused the energy to no longer remain just a kind of commodity but it has also provided a significant opportunity for making it as a technological tool, a generalized resource, and a support platform that enables the system for modifying the lifestyle of individuals as well as operation process of business for the enterprises [63]. Therefore, the innovation of services models and management concepts has a direct impact on innovations in the energy systems.

4.7.1 Key Concepts in Energy Internet

This section is targeted to clarify some key concepts in Energy Internet that are illustrated in Fig. 4.14 [63]. Understanding these concepts is very useful for effective usage of the Energy Internet capability in evaluating business innovations as well as recognizing the business values in the Energy Internet ecosystem. Among the exhibited concepts in Fig. 4.14, the concept of the smart grid, smart energy, microgrid, and prosumer is completely discussed in Chap. 1 of this book. The concept of the virtual power plant (VPP) is also expressed in the following subsection.

4.7.1.1 Virtual Power Plant (VPP)

The virtual power plant (VPP) is one of the innovative operational concepts in smart energy systems. The VPP integrates different types of DERs, electric vehicles, controllable loads, energy storage systems using the advanced metering, control, and communication systems in the coordination of the other sophisticated technologies [65]. The structure of the VPP is demonstrated in Fig. 4.15 [66].

In the VPP, the optimal coordination of the DERs is carried out aiming to achieve the optimal utilization and configuration of the resources by using the higher level of software architecture [63]. In other words, VPP engineering enables the system to deal with the instability defects of various energy production units. It is estimated that by mostly relying on the offshore wind turbines and PV station on the roof of residents, the amount of the global VPP capacity will reach 28 GW [63]. VPP paradigm provides the opportunity of participating in the energy market interactions for the distributed energy production units by enabling them to present as aggregated

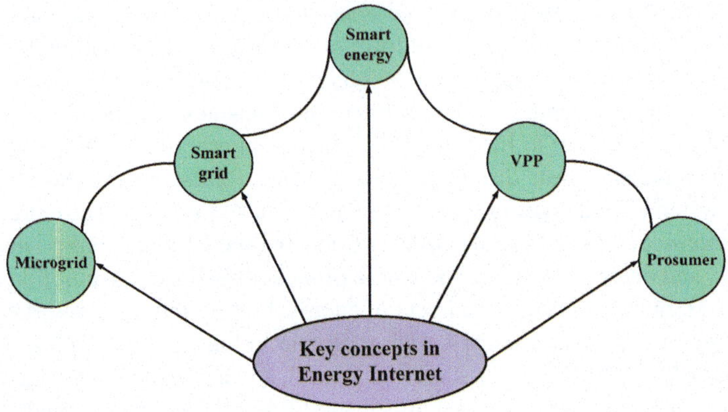

Fig. 4.14 Key concepts in Energy Internet [63]

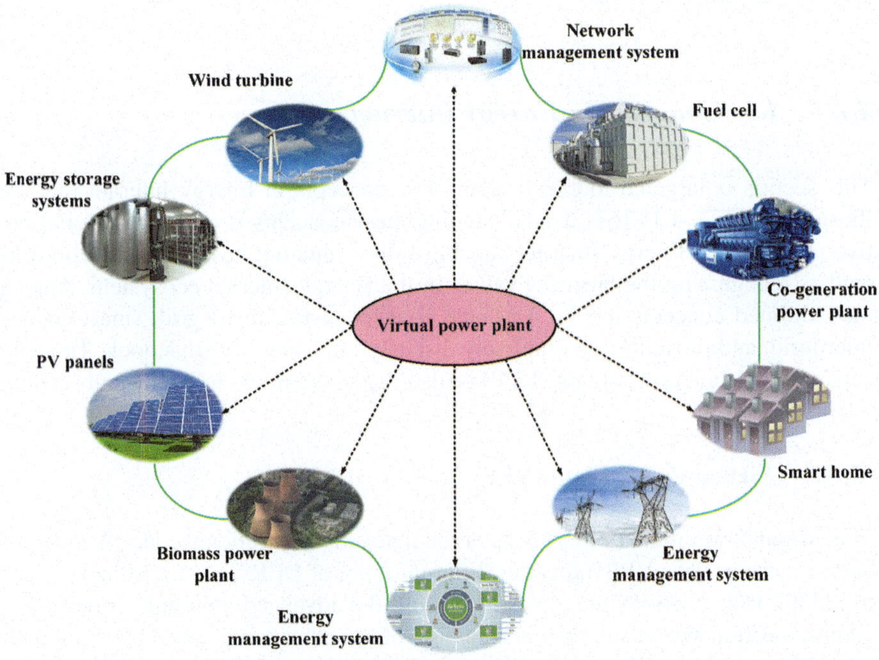

Fig. 4.15 The structure of the VPP [66]

controllable groups for making energy contracts and offer different services to the system operator in the wholesale market. This new business model allows the Energy Internet to inherit most of the DERs merits while makes the achieving of significant benefits for the VPP possible in the competitive energy market. For example, having a short start-up time for the controllable DERs provides efficient flexibility options for the VPP to quickly react for any energy signal in the system.

4.7.2 Architecture of Energy Internet

Energy Internet has emerged as the evolution of the smart grid with the aim of accelerating the transition from the traditional energy environment to the modern energy networks by integrating different forms of energy into a highly efficient and flexible grid to act similar to the Internet in providing energy routing and packing functions [67]. In this new energy environment, one of the main principles is to redesign and develop the energy network structure for engaging diverse DERs technologies aiming for creating a green and reliable society. Although incorporating the DERs especially the clean version of them into the hybrid energy networks can facilitate the making of green, clean, reliable, and highly efficient grids, their rapid growth can also cause substantial challenges in the energy network operation and market regulation. From the system operators' point of view, DERs need to be controllable and visible, especially in the systems with a high share of the RERs while this issue is not the case in some countries [8]. On the other hand, from the market regulators' point of view, the argument of how DERs can incorporate into the existing energy markets and regulatory framework indicates a source of inefficiency or potential conflict [8]. To effectively address such challenges, the architecture of the future modern MCENs needs a significant overhaul, evolving from the traditional structure to the so-called "Energy Internet." As depicted in Fig. 4.16 [8], Energy Internet is representative of the developed architecture that incorporates the smart energy transmission and distribution sectors. In the smart grid, the Energy Internet is

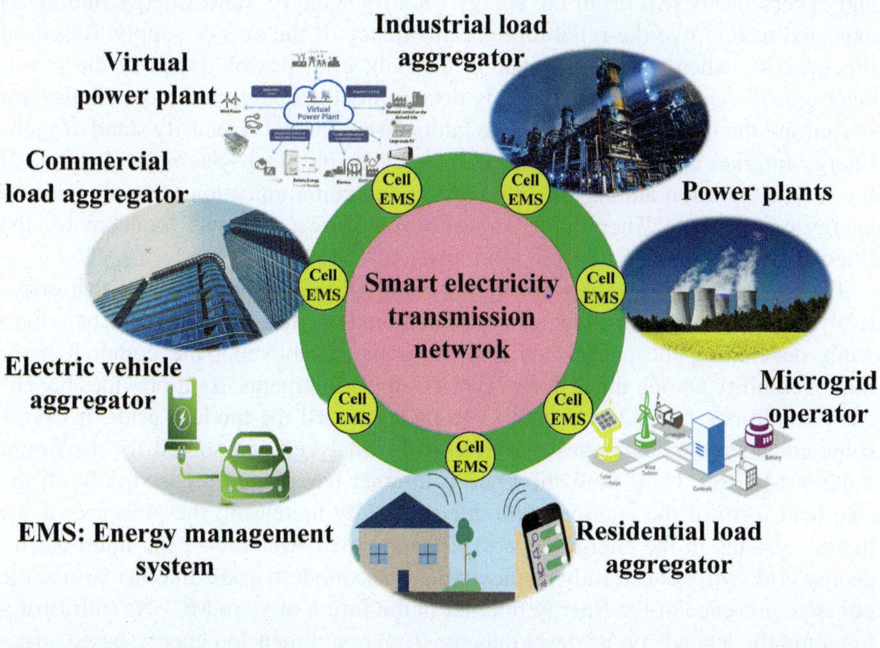

Fig. 4.16 Schematic of the Energy Internet architecture [8]. EMS, energy management system

equipped with a number of "cell" units that support the system by acting as a highly efficient organization form of the controllable units, electric vehicles, and other types of the DERs to the grid [68]. For this support, hierarchical energy management systems for distribution and transmission networks as well as the cell units are expected to be deployed in the Energy Internet architecture. Energy Internet can be characterized as follows [63]:

- Integration of diverse energy consumption and production of prosumers
- Extensive application of information and communication technologies
- Interconnection of multiple energy systems

As future energy networks are expected to appear with a large number of the DERs and EVs, the Energy Internet paradigm will require a more decentralized and flexible energy management pattern that high incorporation of the information and communication technologies would be inevitable in achieving such goals. Due to this, energy security and routing are taken into account as two major concerns for the Energy Internet. Energy routing allows the bi-directional energy flow between different energy systems as well as make the flexible control possible for the energy flow analogous in information routing on the Internet [69]. This process facilitates the optimal integration of the DERs and electric vehicles while enables the system for the interconnecting of various energy systems. For supporting energy routing, energy routers act as the key hardware that are mainly developed based on the information and communication technologies as well as the power electronics technology [70]. Energy Internet security is divided into energy supply security and cybersecurity [8]. From the energy security stand of view, Energy Internet is expected to improve the reliability and resiliency of the energy supply. Based on this process, when the faults occur in the bulk grid, flexible types of the power electronic devices are used for quickly decoupling the Energy Internet with the aim of limiting the harmful impacts of the faults. From the cybersecurity stand of view, Energy Internet is intended as a large-scale cyber-physical system that is targeted for deep integration among the information and communication technologies and energy components. Thereby, cybersecurity is known as a vital concern for the Energy Internet.

Energy Internet is an innovative architecture of the energy networks that effectively enables the system elements to be integrated into the interconnected paradigm while developing the energy markets for easily establishing the comprehensive interoperability among the different types of the participants. As it provides beneficiary conditions for energy networks to move toward the modern grids, it causes some considerable challenges that need to be analyzed and solved for the future modern MCENs. Until now, the Energy Internet has mostly focused only on the electrical form of the energy while due to rapidly increasing the presence of the hybrid systems in the energy networks, the coupled structure of the multi-carrier energy grids will not certainly be inevitable in the modern grids. In other words, the effective presence of the Energy Internet in the future modern MCENs' infrastructure directly depends on its developments from one-dimension energy-based structure to the multi-carrier energy paradigm. This means that, in addition to the electric power system, the Energy Internet should be reconfigured in a way to support all interdependent energy networks' interactions, for example, the natural gas grids and the DHN in a reliable manner. This is a key responsibility for the Energy Internet in the future modern MCENs that is expected to be conducted by considering the adequate level of system security, privacy, reliability, resiliency, flexibility, safety, and stability.

4.8 Machine Learning in the Future Modern MCENs

Over recent decades, the adverse effects of climate change have made the need for moving from the fossil fuel-based systems to pollutant-free and sustainable technologies more critical than before. Indeed, the focus on carbon-free energy production units has become more urgent than other technologies for transiting from conventional systems to modern networks. This issue is more highlighted when we consider the International Energy Agency report that asserts over 65% of the greenhouse gas emissions emanate from the energy sector [71]. In this regard, RERs play a burgeoning leading role in significantly alleviating this percentage under the decarbonization schemes. However, intermittent energy production patterns of RERs caused by meteorological changes have rendered their deep operation significantly difficult, which further efforts for proposing practical solutions are required to make their implementation more reliable. As the optimal operation of different types of energy storage systems is a famous approach to effectively overcome this challenge, its realization in the real world imposes high costs to the system. This issue has created a new research trend in the context of the renewable-based systems that drive the most of endeavors toward the prediction of the energy demand and outputs of the RERs. Indeed, forecasting of the stochastic energy consumption and production with sufficient accuracy is expected to be beneficial in facilitating the RERs technological penetration as well as minimize the operational costs [72]. In this regard, one of the approaches for addressing this issue is the enhancing of the near-term predictability of the RERs aiming to use this knowledge in the smart control systems for optimizing the energy dispatch [73]. To this end, big data analytics has appeared as a strong contender and "enabler" due to its capability in converting real-time data into "actionable knowledge" [73]. The big data generated in the smart grid need comprehensive evaluations aiming to extract valuable information. Without this process, most of the collected data will be unusable and only occupy a lot of space without any benefit. Machine learning has emerged as a useful tool to make the collected data applicable to significant goals in the smart system with numerous intelligent devices. Machine learning is recognized as the final piece of the energy networks that is responsible for assessing the accumulated data for strategic decision-making. Indeed, machine learning techniques are developed in a way to act as logical instruments in pursuing the smart grids' goals by analyzing data in the first step, secondly adopt appropriate decisions, and finally allow the smart grids for realizing diverse functions.

4.8.1 Data Mining and Machine Learning Techniques and Models

Machine learning is a common term that refers to a set of endeavors that receive the collected data for parsing them, learn from data patterns, and use what they have learned to make decisions or predictions without being explicitly programmed to conduct so. The process of discovering the specific patterns in the big data for learning from them to achieve certain objectives is called data mining. Indeed, data mining is not only a key step in the machine learning techniques that enables them to use the useful information of the big data for adopting special decisions and/or predictions but also it is a very applicable and essential step for the statistics and database systems. Machine learning consists of different computer algorithms for scrutinizing the energy collected big data through a set of instructions to give data-driven decisions and/or predictions. It undergoes the rigorous processes for developing applicable algorithms in terms of designing and programming with the expected applications. Some of the most important steps and objectives in machine learning and data mining processes are illustrated in Fig. 4.17 [13].

4.8.1.1 Machine Learning Techniques

By developing technologies in the different fields of science, various smart systems are manufactured for diverse objectives. On the other hand, enhancing the interconnections between all parts of the energy with each other has led to the production of different types of data in the system inevitable. This issue indicates that we face several sets of data that can be used in different problems for various goals. Due to this, machine learning techniques are developed to modify their algorithms consid-

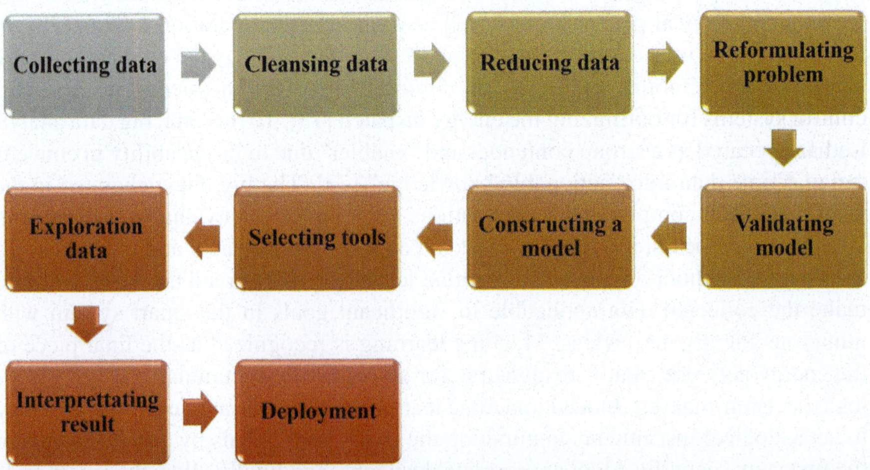

Fig. 4.17 Some of the most important steps and objectives in machine learning and data mining processes [13]

ering the types of data, problems, and objectives with the aim of reaching reasonable achievements in different conditions. Given the type of data as well as the type of problem or task, there are different machine learning approaches as follows:

4.8.1.1.1 Supervised Learning

Supervised learning algorithms use a set of training data related for building the relevant mathematical model [74]. In this algorithm, a set of training examples form training data that each example may have one or more input data and the desired output. In the mathematical model, each of the training examples, known as a supervisory signal, includes an array or vector and a matrix of training data. Based on this process, supervised learning methods learn a function through iterative optimizing the related objective function with the aim of predicting the outputs corresponds to the new inputs. Regression, classification, and active learning are the different types of supervised learning algorithms [75].

4.8.1.1.2 Unsupervised Learning

Unsupervised learning algorithms use a set of input data for predicting the related structure by clustering or grouping of data points. Afterward, the test data that are not labeled, categorized, or classified are used for learning algorithms [76]. Unsupervised learning algorithms focus on identifying commonalities in the data and react considering their absence or presence in the piece of data instead of responding to the related feedbacks. The unsupervised learning technique can be effectively used for density estimation in statistics like detecting the probability density function [77].

4.8.1.1.3 Semi-Supervised Learning

Semi-supervised learning is a learning technique that is the combination of supervised learning with labeled training data and unsupervised learning without any labeled training data [78]. This approach has the ability to simultaneously use a few labeled data and plenty of unlabeled data for pattern recognition. Semi-supervised learning techniques can be roughly divided into the graph-based method, discriminant-based method, difference-based method, and generation-based method [78].

4.8.1.1.4 Reinforcement Learning

Reinforcement learning is one of the applicable areas of machine learning that pursue how software agents should adopt appropriate decisions for maximizing some notion of cumulative reward [79]. Due to its applicability, it is considered in many other disciplines such as genetic and statistics algorithms, swarm intelligence, multi-agent systems, simulation-based optimization, information theory, operations research, control theory, and game theory. Typically, dynamic programming techniques are used in the reinforcement learning algorithms that enable them to be effectively employed in autonomous vehicles [79].

4.8.1.1.5 Self-Learning

Self-learning is a type of machine learning technique that is learning with no external teacher advice and no external rewards. It is introduced along with the crossbar adaptive array (a neural network capable of self-learning) in a crossbar fashion [80] that decisions about emotions and actions are made by the integration between emotion and cognition [81].

4.8.1.1.6 Feature Learning

Machine learning algorithms are often targeted to discover better representations of the input training data [82]. Among them, feature learning algorithms are famous as representation learning algorithms that endeavor to preserve the information of the input data as well as transforming them to make them useful for predictions [82]. The reconstructing of the inputs that belong to the unknown data-producing distribution is possible under feature learning that enables a machine to learn the attributes of the data and use them to conduct a specific task. This technique can be unsupervised or supervised. The features are learned with (e.g., supervised dictionary learning, multilayer perceptrons, and artificial neural networks) and without (e.g., various forms of clustering, matrix factorization, autoencoders, independent component analysis, and dictionary learning) labeled input data in the supervised and unsupervised feature learning, respectively.

4.8.1.1.7 Sparse Dictionary Learning

Sparse dictionary learning is the group of feature learning techniques in which a training example is presented as a linear combination of foundation functions and is intended as a sparse matrix [83]. The main focus of this method is to determine a class that contains a previously unseen training example and can be used in several contexts.

4.8.1.1.8 Anomaly Detection

Anomaly or outlier detection is a set of efforts that seeks to identify rare observations, events, and items that enhance the suspicions by differing the majority of the data [84]. Anomalies typically refer to as exceptions, deviations, noise, novelties, and outliers [85]. Indeed, this method allows the system to detect the deviations from the expected amount for timely adopting appropriate strategies aiming to prevent the system from different anomalies.

4.8.1.1.9 Robot Learning

Robot learning algorithms make significantly developing the robotics possible by allowing them to have their own sequences of learning experiences (curriculum) with the aim of acquiring new skills through social interactions with humans as well as self-guided exploration [86]. Motor synergies and imitation, maturation, and active learning can be commonly used as the guidance mechanisms for this type of robot.

4.8.1.1.10 Association Rule Learning

The need for discovering various key relations among variables in large databases propose a new type of rule-based machine learning that is called association rule learning [87]. In this method, those groups of rules that can satisfy both a minimum confidence threshold and minimum support threshold are the priority to be selected as the interesting rules among the set of all possible rules [87]. Association rule learning can be generally carried out in the two steps as follows:

- Frequent pattern mining: in this step, each of the rule sets will satisfy only the minimum support threshold.
- Strong association rule generation: those groups of rules that are created from sets with minimum support should satisfy a minimum confidence constraint.

4.8.1.2 Machine Learning Models

In order to effectively use the data for predictions, machine learning needs to be involved in creating effective models for enabling the training of data for specific goals. This issue has led to the creation of the different models for machine learning that are described as follows.

4.8.1.2.1 Artificial Neural Networks

Artificial neural networks are the special types of machine learning models that are widely used for forecasting the building loads due to their significant features in the prediction [88]. The effectiveness of the artificial neural networks' models is more highlighted when dealing with nonlinear, complex problems. These models are very applicable for estimating sub-level components optimization and operation, total electricity consumption, and cooling/heating loads [89].

4.8.1.2.2 Support Vector Machines

Support vector machines are very famous for their capability as one of the effective machine learning models that can solve the nonlinear problems with small quantities of training data [90]. These models can be applied for regression and classification problems where the latter case is named support vector regression [89]. Effectively locating global minima instead of local minima in the solution space is one of the most significant advantages of the support vector machines over other machine learning techniques like artificial neural networks [91].

4.8.1.2.3 Regression Trees

Regression trees are the other machine learning models that have specific features for modeling data interacted in nonlinear and complicated ways [89]. Due to such features, these machine learning approaches are strongly preferred over linear regression models for problems with the aforementioned conditions. Generally, a global predictive formula is used in the linear regression models that hold the entire data-space [89]. However, space is sub-divided into the further partition of the subdivisions and smaller regions for assigning to its nodes with the aim of conducting more manageable interactions. A tree form regressing is mainly adopted by the regression tree models for regressing the decisions, which starts from the root node down to a leaf node included with responses [92].

4.8.1.2.4 Decision Trees

Decision tree learning is intended as one of the predictive modeling methods that is commonly used in machine learning, statistics, and data mining. In this method, a decision tree is considered as a predictive model for continuing the process from observations about items to get conclusions about their target value [93]. In this regard, classification trees are those tree models that a discrete set of values can be taken by the target variable. They are called regression trees if the continuous values are taken rather than a discrete set of values.

4.8.1.2.5 Bayesian Networks

A Bayesian network is a probabilistic graphical model that indicates a set of random variables and is also known as a directed acyclic graphical model or belief network [94]. In a Bayesian network model, a directed acyclic graph is also intended for the conditional independence of a set of random variables. This model can present the probabilistic relationships between symptoms and diseases [94]. Decision problems under uncertainty can be solved using the Bayesian networks considering the different uncertainty sets.

4.8.1.2.6 Genetic Algorithms

A genetic algorithm is a heuristic technique and search algorithm that mimics the natural selection process to produce new genotypes using the mutation and crossover techniques. Typically, machine learning methods are used for improving the performance of evolutionary algorithms like the genetic algorithm [95].

4.8.2 Machine Learning Applications and Limitations

Machine learning has widespread applications in several fields but most of the significant ones from the energy sector point of view are as follows:

- Predictions of energy consumption, production, and price
- Future optimum schedule
- Fault detection
- Adaptive control, sizing, and detection of network intruders during a data breach
- Wind, solar, and marine energy forecasting
- Optimization of the energy networks
- Telecommunication
- Energy network security objectives

From the grid security stand of point, machine learning can be used for various applications that are demonstrated in Fig. 4.18 [13].

Although machine learning techniques are widely used in most of the fields, they often fail to reach the expected results [96]. This can be due to various reasons such as evaluation problems, lack of resources, wrong tools, and people, badly chosen algorithms and tasks, privacy problems, data bias, lack of access to the data, and lack of suitable data [97]. Because they use the previously recorded data for

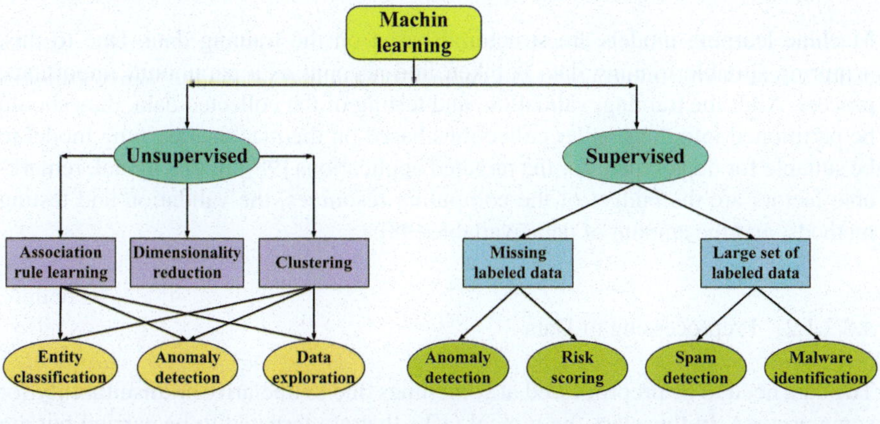

Fig. 4.18 Application of machine learning in the security of smart energy networks [13]

pursuing a specific process for achieving goals, they cannot be effectively used for solving the challenges of the energy networks with a high share of RERs.

4.8.3 Autonomic Machine Learning

This section is aimed to examine the design choices in machine learning for evaluating the requirements to make autonomic machine learning possible. The developers or experts can adopt design choices based on developing machine learning applications.

4.8.3.1 Design Factors of Autonomic Machine Learning

Although several machine learning tools are available for various objectives, effectively using them in the actual problems directly depends on the different factors that need to be taken by an expert [98]. In the autonomic machine learning model, characteristics of the data are considered in their generation and optimization processes without the intervention of expert knowledge. This issue not only incurs high costs for generating different types of numerous data but also it takes a remarkable amount of time in the process. Hence, in order to reduce the cost and time in properly building the autonomic machine learning platform as well as minimizing expert intervention, the appropriate selection of the different design factors is essential [98]. Figure 4.19 shows various design factors in the autonomic machine learning model [98]. The characteristics of generic design factors used in the machine learning techniques are described in the following subsections.

4.8.3.1.1 Acquisition of Training Data and Partitioning

Machine learning models are structured based on the training data. Due to this, acquiring suitable training data is taken into account as a paramount importance process. After the training, validation, and testing of the collected data, they should be partitioned into the smaller collections based on the machine learning model to be suitable for deployment for the targeted applications [98]. In this model, remarkable factors are the budget of the computing resources, the validation and testing methods, and the amount of data available [98].

4.8.3.1.2 Preprocessing of Data

The data need to be preprocessed at sometimes due to unclarity or unsuitability for some process. In this step, the typical tasks that are targeted to be carried out are data cleansing, normalization, and transformation [98]. This is because most of

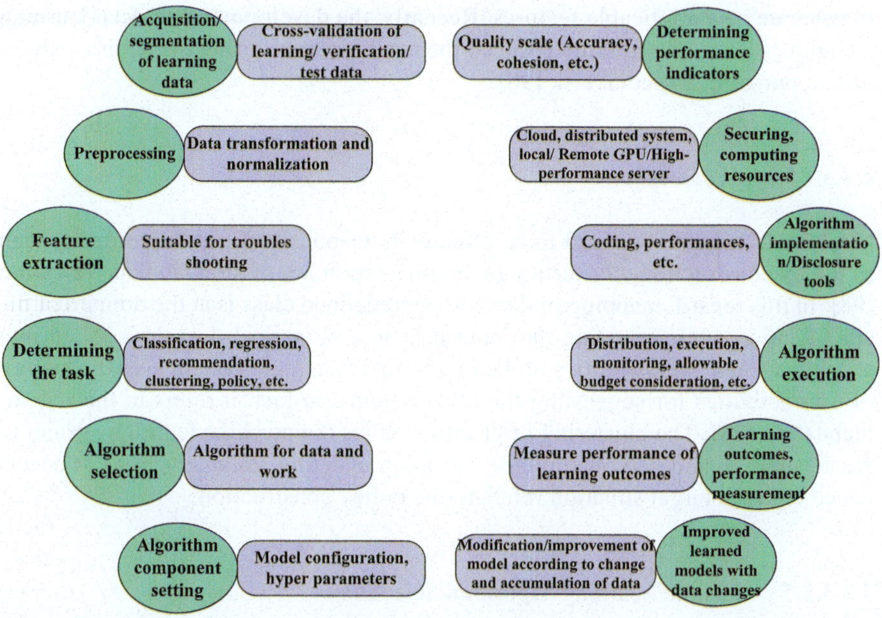

Fig. 4.19 Different design factors in the machine learning process [98]

the machine learning algorithms are not dependent on the input data format and raw data are not acceptable for their process and format. Thus, the possibility of converting the data into an amenable format is considered in the data transformation step [98]. The differences in data characteristics have made normalization one of the most fundamental steps during the data preparation process. Indeed, normalization is conducted for standardizing the data and making their scale uniform [98].

4.8.3.1.3 Feature Extraction

Machine learning algorithms are designed in a way that has enough capability to extract the features of data while learning and using these attributes for a designated task. The quality of the data features is significantly important because it seriously influences the performance of the machine learning model [98]. The selection of quality features is occasionally a time-consuming work and needs the expertise to be effectively done. Feature generation and selection methods are two common processes in each feature extraction technique that the first one is focused on extracting the useful attributes of data while the second one is responsible to choose certain features from the available data [98]. Moreover, the feature generation techniques are also empowered to combine the available attributes to generate new applicable features. Recently, the developments in deep learning techniques have substantially reduced the need for considering the feature extraction in pursuing a special task [98].

4.8.3.1.4 Task Decision

Machine learning is targeted to be effectively responsive for different tasks such as policy construction, clustering, recommendation, regression, and classification [98]. In this regard, mapping of data into a predefined class is at the domain of the classification while mapping the input data in a numerical domain is taken into account as the responsibilities of the regression. When the system uses the history of user activities for suggesting the related items, in fact, it refers to the recommendation [98]. The clustering of groups enables the machine learning system to create subgroups based on similarity. Finally, selecting suitable actions in accordance with the target situation refers to the policy construction.

4.8.3.1.5 Machine Learning Algorithm Selection

As mentioned in the previous sections, machine learning techniques are broadly classified into three categories including reinforcement learning, supervised learning, and unsupervised learning. The diversity of machine learning techniques is enough widespread that allows the system to select appropriate ones considering their weaknesses and strengths. An expert can choose the proper algorithm based on the given type of the problem domain.

4.8.3.1.6 Machine Learning Algorithm Configuration

Most of the machine learning algorithms need configuration endeavors like hyperparameter setting and model configuration for adopting a proper strategy of problem-solving considering the details of the algorithms and problems. In the neural networks and deep learning, the performance of the machine learning applications directly depends on the network configuration [98]. Among various parameters, hyperparameters allow the developers to use their own insights and experiences for setting the system [98].

4.8.3.1.7 Performance Measure Selection

Machine learning algorithms are developed in a way to effectively improve the related models given the specific performance measures [98]. They refine candidates of their performance measures. The performance measure needs to be selected by the developers for usage in the machine learning applications. Developers should also determine the criteria for accepting the modified model.

4.8.3.1.8 Computing Resource Acquirement

Machine learning algorithms are expected to face a large volume of data that will certainly impose heavy computations and will need a considerable amount of computing sources [98]. In this regard, the candidates are examined for training and identifying a satisfactory model which has made the existence of some trial and test processes inevitable. Cloud services are recognized as the most accessible computing systems and easiest types of enabling technology for developing machine learning applications.

4.8.3.1.9 Algorithm Implementation and Public Tool Usage

Machine learning algorithms should be converted as a code to be implementable by a developer in the system. For this aim, different excellent frameworks and tools are developed to make machine learning applicable for public use and several types of them are also commercially available [98]. This is the responsibility of the developer to determine which tools need to be developed independently and used. Although this self-development imposes remarkable costs and time resources, it makes the system more flexible than before [98]. Because commercial and public tools possess independent usage methodologies, a developer needs to learn how to use which of them for building the execution.

4.8.3.1.10 Execution of Machine Learning Algorithms

The time to execute varies for each of the machine learning algorithms due to differences in their attributes and applications. In order to develop a special type of machine learning application, different machine learning models are typically examined with the aim of finding an excellent one [98]. Developers should control the monitoring of the model training tasks to put an end to the execution of those algorithms that seem doomed to failure.

4.8.3.1.11 Evaluations of Trained Models

Several machine learning models are targeted for training with various hyperparameters and configurations. In this respect, effective evaluations should be conducted for choosing the most appropriate model based on the test data and certain evaluation criteria [98]. Developers must prepare a schedule and/or producer for this type of model evaluation.

4.8.3.1.12 Model Updates for Data Distribution Changes in the Problem Domain

At times, data distribution may be faced with some changes in the problem domain. Given the volume of the collected data is rapidly increasing day by day, machine learning models should be improved in a way to have sufficient potential for handling a massive amount of data for learning [98]. Accordingly, developers need to undertake appropriate actions with the aim of improving the models in line with the new volume of the accumulated data.

4.8.3.2 Requirements of an Autonomic Machine Learning Platform

An autonomic machine learning platform is targeted to make an environment that allows the non-experts to develop the desired applications using the appropriate type of the machine learning technique. Providing such a platform is subject to several requirements as listed in the following [98]:

- The platform should be general at the interface by providing public tools and frameworks for the usage of machine learning.
- The platform should be easily understandable for even non-experts to facilitate the development of machine learning applications for them.
- The platform should be developed in a way to not only make the use of external computing resources possible but also enables the system for executing learning.
- The platform needs to include a scripting language for enabling the system to engage autonomic machine learning tasks in a parallel and distributed manner.

4.8.4 Deep Learning

Deep learning is one of the branches of machine learning that includes deeper inner hidden layers that connect it to the network [99]. The main goal of deep learning is to develop the machines to act similarly to the computer in the execution of commands by human beings and think and comprehend as human thinks by mimicking the network of the human brain connection [99]. Indeed, deep learning is structured to use a logical structure for continually analyzing data analogous to how a human would draw conclusions. For this aim, an artificial neural network is employed as the layered structure of algorithms in the deep learning applications. The biological neural network related to the human brain is used for designing the artificial neural network. This type of learning provides a specific learning system that acts as far more capable than other standard machine learning models. Similar to the other machine learning models, a deep learning model also needs lots of training to correctly get the required learning processes. This is while deep learning acts as a scientific marvel when it is targeted to do a certain task, which has made it as the backbone of true artificial intelligence.

4.8.4.1 Deep Learning Methods

As data science has attracted significant attention along with the substantial developments in artificial intelligence, deep learning has experienced remarkable overhauls in its history over recent years. The advancements in deep learning have led to the introduction of different methods for them that each of them is developed for diverse goals in certain problems. The descriptions for the different deep learning methods are provided as follows.

4.8.4.1.1 Autoencoder

Autoencoder is taken into account as the member of feed-forward neural network groups that are aimed to copy input neurons to output ones considering the pass of a hidden layer [100]. Indeed, Autoencoder is a learning algorithm that consists of three layers input, output, and hidden layers, and is typically used for corrupted data reconstruction, feature learning, and dimensionality reduction [99]. Denoising Autoencoder, Sparse Autoencoder, and Undercomplete Autoencode are the Autoencoder models that are respectively used for the aforementioned kinds of problems [100].

4.8.4.1.2 Recurrent Neural Network

The recurrent neural network is one of the common sequential data neural network processors that is equipped with internal memory for updating the state of each neuron in the previously updated network [99]. Backpropagation is typically used for training this type of neural network. Vanishing gradient descent is one of the great weaknesses of the recurrent neural network that causes it to fail at usual [101]. However, the deep recurrent neural network creates more layers into the architecture of this neural network that brings several significant benefits [102].

4.8.4.1.3 Long Short-Term Memory

Long short-term memory is designed with the aim of filling the gap of the recurrent neural network in vanishing gradient decent and makes a longer-term memory possible for the system [99]. It contains the internal self-loops in its model for storing information and makes a profit of existence of five key elements in its computational graph including [99]:

1. Input gate
2. Forget gate
3. Output gate
4. Cell
5. State output

The gates operations like erasing, reading, and writing are conducted in the cell memory state [101].

4.8.4.1.4 Convolutional Neural Network

The convolutional neural network is a feedforward neural network that typically emulates the human neurons in the relevant platform. This feature of deep learning convolutional neural network has made it suitable for widely applying in the audio and visual processing, natural language processing, and video recognition [100]. This type of deep learning method is also intended to be used for processing grid data topology and at least in one of its layers, the convolution mathematical operation is applied for supporting the learning process [100]. In order to build the network architecture, the convolutional neural network consists of three basic layers that the first layer possesses multiple parallel convolution layers, the second one is equipped with the rectified linear activation for detection operations, and eventually the third layer is the pooling function stage [100].

4.8.4.1.5 Restricted Boltzmann Machine

One of the most reputed deep probabilistic techniques is the restricted Boltzmann machine that is recognized as the undirected probabilistic graphical models [100]. The restricted Boltzmann machine possesses two basic layers, which the first one is based on the visible inputs while the second one works with hidden variables. This machine is designed in a way to be deeper by stacking one layer on top of the other ones [99].

4.8.4.1.6 Deep Belief Network

Deep belief network architecture is a special type of the deep learning that possesses multiple layers of hidden units famous as stacked restricted Boltzmann machine that are trained by employing the Backpropagation algorithm [103]. In this architecture, the units between in one layer with other units in the next layers create the connections of the deep belief network architecture without intra-connection for each layer unit [100]. Indeed, the deep belief network possesses more hidden layers in comparison with the restricted Boltzmann machine with only one hidden layer [100].

4.8.4.1.7 Deep Boltzmann Machine

Deep Boltzmann machine also acts similarly to the restricted Boltzmann machine with more hidden layers and variables than restricted Boltzmann machine [99]. On the other hand, unlike the deep belief network, deep Boltzmann machines possess entirely undirected connections in its architecture that are placed among variables within all layers [100].

4.8.5 Differences Between Machine Learning and Deep Learning

From a practical stand of view, deep learning is taken into account just as a subset of machine learning and technology acts in a similar way. However, capabilities of the deep learning are different. Although the machine learning models provide significant benefits and their capabilities are progressively becoming better, they face some basic limitations that need to be effectively addressed. In the machine learning architecture, if an artificial intelligence algorithm returns an inaccurate prediction, the direct human reactions are required for stepping and making the appropriate adjustments [104]. This issue has made the possibility of implementing the full automated system challenge. This issue is partly addressed in the deep learning models by employing the algorithms that work based on the neural network for determining that one prediction is accurate or not. Indeed, machine learning techniques use the algorithms to assess the received data, learn from them, and adopt informed decisions considering what they have learned. This is while the deep learning models are aimed to create the artificial neural network by structuring algorithms in layers and it mostly relies on the human-like artificial intelligence [104].

4.8.6 Presence of the Machine Learning and Deep Learning in the Future Modern MCENs

In the previous sections, machine learning and deep learning issues are widely discussed. The main purpose of this investigation is to determine how machine learning and deep learning models can play an applicable role in future modern MCENs interactions. Both machine learning and deep learning techniques generally receive the collected special data, examine them to discover different patterns in them, learn from extracted behavioral patterns, and ultimately make decisions based on what they have learned, or predict. As mentioned before, one of the applications of machine learning techniques is to forecast the energy demand, price, and production form RERs. In this regard, forecasting the outputs of the RERs is a key issue that is commonly paid specific attention in the machine learning applications in the energy networks. The future modern MCENs is designed in a way to equip with a full share of RERs for fully clean energy production. This means that we will witness the full RERs in the future modern MCENs that certainly impose the energy network with a large amount of uncertainties. The main challenge here is how the system with 100% RERs can establish a dynamic energy balance for providing a reliable energy supply in the system. In this regard, as forecasting the outputs of the RERs can be more applicable in optimal operation and scheduling of the system, it will be beneficial when the forecasting is done accurately. This is obvious that, until now, there is not any method that can exactly predict the energy generation of the RERs and it seems that it is not practical to discover such a method, especially in the near future. This is because that numerous information has affected the results of one forecasting while all information is not available from the RERs such as wind speed and some of the effective factors are detected but some other may not be detected that can affect the behavior of the wind speed. Therefore, we can never claim to have discovered all the factors influencing RERs behavior and can learn from them for future predictions and decisions. As a result, the output of these predictions will always be accompanied by an error, and sometimes the amount of this error is very large. For this reason, the use of forecasting methods cannot be a reliable solution to solve the main challenge of future modern MCENs. Machine learning techniques have also used the collected data for analyzing their behaviors to extract some specific patterns, learn from these detected patterns, and then use the learned things for making different decisions and/or predictions. Indeed, when we want to use machine learning methods for predicting the outputs of RERs such as wind or solar power production, the vast amount of the related data should be delivered to the machine learning system, which these data are already recorded for each RERs. In other words, wind speed, for example, is recorded in the special region for every hour, day, or year that is mainly stochastic. Under these circumstances, these data, whose source is completely random, are to be given as input information to machine learning techniques to be evaluated for extracting their behaviors and then learned for doing predictions and/or other decisions. The output of these systems can be severely erroneous, especially in the systems with 100% RERs, because they are based on preliminary data that could not be repeated in the future due to completely random conditions. This can be very worrying, especially for future modern networks with a full share of RERs, due to their high uncertainty. Therefore, the use of machine techniques for forecasting the outputs of the RERs cannot be more applicable and reliable for the future modern MCENs. On the other hand, the deep learning algorithms also require more information to be able to learn in-depth the different behaviors that lie in the data so that they can take appropriate actions without human intervention. In the renewable-based systems, for example, more information (different types of data) should be provided for the deep learning algorithms to enable them for analyzing a vast amount of data to pursue the deep learning processes. But the main question here is how much information is available about renewable energy sources and how reliable they are. For wind speed, we only access the wind speed recorded in every hour while for accurate prediction, deep learning needs more information regarding the factors that have a direct effect on wind velocity. Some of these factors may have been discovered by now, and many of them may not have been discovered yet, and they may remain undiscovered forever, and remain as mysteries hidden in the universe that affect some things, such as wind speed, but we are unaware of them. This is while deep learning needs them for upgrading its procedure for more learning to accurately predict uncertain parameters like the wind speed. However, the algorithms not only are incomplete but also they do not have complete input information to effectively do such tasks.

From another viewpoint, receiving a vast amount of data, effectively analyzing them, extracting some behavioral patterns from them, learning these patterns, and performing some predictions and/or making required decisions is a very time-consuming process. As mentioned, machine learning and deep learning techniques are not designed only to do some predictions but also they are targeted to adopt appropriate decisions if needed. Because their process is time-consuming, so it may not be suitable for real-time operation of the even current networks that need the fast response systems to analyze the system conditions and take the proper strategies at the minimum time for maintaining the stability of the system. These conditions are expected to be much more acute in future modern MCENs, where rapid action in the presence of 100% RERs is critical to maintaining the system. Also, because the machine learning and deep learning systems should support the multiple energy dimensions and data of the several energy networks in addition to the electric power system, the volume of data not only will increase dramatically but also more time will be needed for pursuing the related process. Although the future modern MCENs are in dire need of rapid response and operation over a shorter period of time, the above reasons have made it riskier to use machine learning and deep learning techniques in the sensitive parts of this type of network, which have made them unreasonable to use in the operation of the future modern MCENs. However, they can play a suitable role in planning-based problems. In other words, machine learning and deep learning algorithms can be effectively used for predicting some uncertain parameters that more information of them is almost available for their process. Energy demand is one of these parameters that we can access almost all the influential factors by focusing on its pattern and collect the relevant data and use deep learning methods for more accurate predictions. These predictions are more indicative of their use in planning schemes so that they can be used to give an overview of the energy consumption situation in the future. Although this is anticipation, it can provide at least a good perspective for system planners to develop the energy systems.

Although these techniques do not seem to be very useful for network-based analysis of the future modern MCENs based on the above descriptions, they can have a major impact on the grid modernization. For example, deep learning techniques can be used to make the most of buildings parts intelligent or some urban interactions can also be effectively automatic based on artificial intelligence to minimize human involvement in certain processes.

4.9 Summary

This chapter is focused on scrutinizing the advanced communication protocols in modernizing the future modern MCENs, identifying their challenges and barriers, and proposing practical solutions to effectively address them as much as possible. We witness this fact that developing intelligent systems are rapidly increasing, their presence in the hybrid networks is accelerating day by day, and the relevant scientific trends are eagerly moving toward significantly change the functions of the today's world to the modern universe. This issue obviously inspires that the future modern MCENs will face the massive amount of data that is transferred between smart elements at each moment for implementing grid modernization goals. Indeed, the generation of the different types of big data will be inevitable in the smart hybrid energy networks. Due to this, big data analytics is investigated for clarifying the volume of data and their features and effectiveness in the future modern grids. To assess how they can be collected, analyzed, implemented, and effective in the future modern MCENs, the role of the smart meters, Internet of Things, energy cyber-physical systems, Energy Internet is deeply discussed along with their characteristics. Afterward, the potential of machine learning and deep learning techniques are investigated to clarify their strangeness and weakness for presenting in the future modern energy networks' interactions. The results indicated the creation of some opportunities and remarkable challenges by the machine learning methods for the future modern MCENs. In other words, due to some special challenges and limitations of them, they can be intended only in some insensitive parts of the future modern MCENs for developing the autonomous feature of the energy systems.

4.10 Questions

After careful reading and evaluating this chapter, we are expected to be able to answer the following questions:

4.10.1 Big Data Analytics Sector

1. Why do the future modern MCENs require advanced communication protocols in their infrastructures?
2. What is the energy big data in the smart energy systems?
3. How the energy big data are created in hybrid energy networks?
4. Why big data analytics is essential for future modern MCENs?
5. What are the key components required for big data analytics?
6. What are the characteristics of energy big data in the smart energy grids?
7. Which type of data can be supported by big data analytics?

8. What are the categories for energy big data?
9. When can the applying of big data be challenging in the system?

4.10.2 *Smart Meters and Advanced Metering Infrastructure Sectors*

10. What is the key role of the smart meters in the energy networks?
11. Smart meters allow the system to do what things?
12. What are the characteristics of the smart meters in the system?
13. Which benefits can be provided by the smart meters for the future modern MCENs?
14. What are the different parts of the smart meters?
15. Why the smart meters can be beneficial for the smart energy networks?
16. What features do the smart meters need to implement in the future modern MCENs?
17. Why the presence of the advanced metering infrastructure is necessary for future modern MCENs?
18. The paradigm of the advanced metering infrastructure should include which components?
19. Which factors should be intended in designing and constructing of the advanced metering infrastructure?
20. What are the communication and medium technologies for realizing the advanced metering infrastructure?

4.10.3 *Internet of Things (IoT) Sector*

21. What is the Internet of Things and why it is important in smart energy networks?
22. Which features of the Internet of Things have made it a key part of the smart energy networks?
23. What is the classification of the Internet of Things?
24. What are the characteristics of the Internet of Things?
25. What are the challenges of the Internet of Things?
26. How the Internet of Things can communicate with smart energy grids?
27. Which layers are needed for the combination of the Internet of Things technology with the smart grid?
28. Which components are used for the combination of the Internet of Things technology with the smart grid?
29. What are the considerable applications of the Internet of Things for the future modern MCENs?

4.10.4 Energy Cyber-Physical Systems (ECPSs) Sector

30. What are the responsibilities of the energy cyber-physical systems in smart energy networks?
31. What are the distinguishing features of the energy cyber-physical systems in smart energy networks?
32. What are the principles of the energy cyber-physical systems in smart energy networks?
33. What is the domain of the cyber-physical system in hybrid energy networks?
34. Which guidelines should be considered in the architecture of the energy cyber-physical system?
35. Which layers are typically used for structuring the architecture of the energy cyber-physical system?
36. What are the challenges of the energy cyber-physical system?
37. How the energy cyber-physical systems can be applied for the future modern MCENs?
38. What is the digital twin technology and why it is important in the CPSs?
39. What are the features of the characteristics of the digital twin technology in smart energy networks?
40. What are the benefits and drawbacks of the digital twin for the future modern MCENs?

4.10.5 Energy Internet Sector

41. What is the role of the Energy Internet in the future modern MCENs?
42. What are the key concepts in the Energy Internet?
43. How the virtual power plant can play an effective role in the Energy Internet area?
44. What is the architecture of Energy Internet in the smart energy networks?
45. Which components are used for structuring the virtual power plant?

4.10.6 Machine Learning and Deep Learning Sectors

46. What is machine learning and why it is vital for smart energy systems?
47. What is data mining and how it works?
48. What are the machine learning techniques and models?
49. What are the most important steps and objectives in machine learning and data mining processes?
50. What are the features of the machine learning methods and how they can be used in smart energy networks?

51. What are the features of the machine learning models and how they can be applied in smart energy networks?
52. What are the machine learning applications and limitations in hybrid energy networks?
53. Why the autonomous machine learning is essential for smart energy systems?
54. What are the design factors of autonomic machine learning?
55. What are the features of the different design factors of autonomic machine learning and how they can be used in designing the autonomic machine learning?
56. What are the key requirements of an autonomic machine learning platform?
57. What is deep learning and what are its relationships with machine learning?
58. What are the features of deep learning methods and how they can be employed in smart energy networks?
59. What is the role of artificial intelligence in the smartization of energy networks?
60. What are the differences between machine learning and deep learning?
61. How can the machine learning and deep learning be presented in the future modern MCENs?
62. What are the challenges and opportunities of machine learning and deep learning methods for the future modern MCENs?

4.11 Suggestions

In this section, some suggestions are provided related to this chapter that can give a good overview of some substantial issues in the future modern MCENs.

1. An effective collection of the different types of data along with the big data analytics can achieve valuable information for developing the smart energy systems.
2. Upgrading the structure of the smart meters and advanced metering infrastructure from one-dimension energy to the multi-carrier energy systems can provide a significant opportunity for deploying different intelligent devices in the smart energy networks.
3. Developing the technology of the Internet of Things can be very beneficial for implementing future modern MCENs.
4. Developing energy cyber-physical systems (ECPSs) in cyber, physical, and social layers can make the future modern MCENs more efficient, reliable, flexible, secure, resilient, and stable.
5. Promoting the Energy Internet paradigm throughout the smart energy networks can facilitate the establishment of confident communications among intelligent energy devices.
6. Developing the algorithms of machine learning and deep learning methods can significantly facilitate the implementation of artificial intelligence in the smart energy systems aiming to promote automatic interactions in the system.

4.12 Future Trends and Discussion Topics

Here, future trends and some discussion topics are proposed for the interested readers with regard to this chapter that can be useful for developing the MCENs modernization process.

1. Developing big data analytics for extracting useful information of the collected data and avoiding the usage of the invalid data in the decision-making process.
2. Developing smart meter systems for enabling the system for simultaneously recording a massive amount of diverse data and transmitting them to the data control center.
3. Improving the paradigm of the advanced metering infrastructure for supporting the multi-carrier energy in the future modern MCENs.
4. Effectively addressing the challenges of the Internet of Things for making them suitable for presence in the future modern MCENs interactions.
5. Developing the layers of energy cyber-physical systems from one-dimension-based support systems to the multi-carrier energy-based layers for covering the security and privacy requirements in the future modern MCENs.
6. Improving the Energy Internet architecture for effectively satisfying the communication protocols of the future modern MCENs.
7. Significantly developing machine learning and deep learning techniques for increasing their fastness and accuracy with the aim of making reliable decisions and perform confident predictions.

References

1. J. Rogelj et al., Paris agreement climate proposals need a boost to keep warming well below 2 C. Nature **534**(7609), 631–639 (2016)
2. V.C. Gungor et al., A survey on smart grid potential applications and communication requirements. IEEE Trans. Industr. Inform. **9**(1), 28–42 (2012)
3. B. Saha, D. Srivastava, Data quality: The other face of big data, in *2014 IEEE 30th International Conference on Data Engineering*, (IEEE, USA, 2014), pp. 1294–1297
4. M. Daneshvar, B. Mohammadi-Ivatloo, M. Abapour, S. Asadi, Energy exchange control in multiple microgrids with transactive energy management. J. Mod. Power Syst. Clean Energy **8**, 719–726 (2020)
5. J. Schleich, C. Faure, M. Klobasa, Persistence of the effects of providing feedback alongside smart metering devices on household electricity demand. Energy Policy **107**, 225–233 (2017)
6. S.S. Reka, T. Dragicevic, Future effectual role of energy delivery: A comprehensive review of Internet of Things and smart grid. Renew. Sust. Energ. Rev. **91**, 90–108 (2018)
7. M. Daneshvar, M. Pesaran, B. Mohammadi-Ivatloo, Transactive energy in future smart homes, in *The Energy Internet*, (Elsevier, Cambridge, England, 2019), pp. 153–179
8. B. Hong, W. Zhang, Y. Zhou, J. Chen, Y. Xiang, Y. Mu, Energy-Internet-oriented microgrid energy management system architecture and its application in China. Appl. Energy **228**, 2153–2164 (2018)
9. H. Zhao et al., Improvement of fuel sources and energy products flexibility in coal power plants via energy-cyber-physical-systems approach. Appl. Energy **254**, 113554 (2019)

10. M. Daneshvar, B. Mohammadi-Ivatloo, K. Zare, S. Asadi, Two-stage robust stochastic scheduling model for transactive energy based renewable microgrids. IEEE Trans. Industr. Inform. **16**, 6857–6867 (2020)
11. M. Daneshvar, B. Mohammadi-Ivatloo, S. Asadi, M. Abapour, A. Anvari-Moghaddam, A Transactive Energy Management Framework for Regional Network of Microgrids, in *2019 International Conference on Smart Energy Systems and Technologies (SEST)*, (IEEE, USA, 2019), pp. 1–6
12. H.-y. Zheng, N. Jin, C. Ji, Z. Xiong, K. Li, Analysis technology and typical scenario application of electricity big data of power consumers. Power Syst. Technol. **39**(11), 3147–3152 (2015)
13. E. Hossain, I. Khan, F. Un-Noor, S.S. Sikander, M.S.H. Sunny, Application of big data and machine learning in smart grid, and associated security concerns: A review. IEEE Access **7**, 13960–13988 (2019)
14. X. Wang, Y. Shi, J. Zhang, L. Bo, C. Cheng, Computation services and applications of electricity big data based on Hadoop. Power Syst. Technol. **11**(39), 3128–3133 (2015)
15. L. Wen, K. Zhou, S. Yang, L. Li, Compression of smart meter big data: A survey. Renew. Sust. Energ. Rev. **91**, 59–69 (2018)
16. Y. Huang, X. Zhou, Knowledge model for electric power big data based on ontology and semantic web. CSEE J. Power Energy Syst. **1**(1), 19–27 (2015)
17. X. Peng, D. Deng, S. Cheng, J. Wen, Z. Li, L. Niu, Key technologies of electric power big data and its application prospects in smart grid. Proc. CSEE **35**(3), 503–511 (2015)
18. A. Ukil, R. Zivanovic, Automated analysis of power systems disturbance records: Smart grid big data perspective, in *2014 IEEE Innovative Smart Grid Technologies-Asia (ISGT ASIA)*, (IEEE, Piscataway, 2014), pp. 126–131
19. F. McLoughlin, A. Duffy, M. Conlon, A clustering approach to domestic electricity load profile characterisation using smart metering data. Appl. Energy **141**, 190–199 (2015)
20. Y.-J. Kim, M. Thottan, V. Kolesnikov, W. Lee, A secure decentralized data-centric information infrastructure for smart grid. IEEE Commun. Mag. **48**(11), 58–65 (2010)
21. Y. Wang, Q. Chen, T. Hong, C. Kang, Review of smart meter data analytics: Applications, methodologies, and challenges. IEEE Trans. Smart Grid **10**(3), 3125–3148 (2018)
22. M. Daneshvar et al., The chance-constrained models for transactive energy management of interconnected microgrids clusters. J. Clean. Prod. **271**, 122177 (2020)
23. M. Daneshvar, S. Asadi, CPS-based transactive energy technology for smart grids, in *Cyber-Physical Systems in the Built Environment*, (Springer, Cham, 2020), pp. 323–338
24. K. Reddy, M. Kumar, T. Mallick, H. Sharon, S. Lokeswaran, A review of Integration, Control, Communication and Metering (ICCM) of renewable energy based smart grid. Renew. Sust. Energ. Rev. **38**, 180–192 (2014)
25. A. Molina-Markham, P. Shenoy, K. Fu, E. Cecchet, D. Irwin, Private memoirs of a smart meter, in *Proceedings of the 2nd ACM Workshop on Embedded Sensing Systems for Energy-Efficiency in Building*, (ACM Digital Library, Zurich, Switzerland, 2010), pp. 61–66
26. Z. Guo, Z.J. Wang, A. Kashani, Home appliance load modeling from aggregated smart meter data. IEEE Trans. Power Syst. **30**(1), 254–262 (2014)
27. K. Zhou, S. Yang, Understanding household energy consumption behavior: The contribution of energy big data analytics. Renew. Sust. Energ. Rev. **56**, 810–819 (2016)
28. T. Ahmad, Non-technical loss analysis and prevention using smart meters. Renew. Sust. Energ. Rev. **72**, 573–589 (2017)
29. R.R. Mohassel, A. Fung, F. Mohammadi, K. Raahemifar, A survey on advanced metering infrastructure. Int. J. Electr. Power Energy Syst. **63**, 473–484 (2014)
30. A. Aziz, S. Khalid, M. Mustafa, H. Shareef, G. Aliyu, Artificial intelligent meter development based on advanced metering infrastructure technology. Renew. Sust. Energ. Rev. **27**, 191–197 (2013)
31. NETL Modern Grid Strategy, Advanced metering infrastructure, in *US Department of Energy Office of Electricity and Energy Reliability*, 2008

32. S.S.S.R. Depuru, L. Wang, V. Devabhaktuni, N. Gudi, Smart meters for power grid—Challenges, issues, advantages and status, in *2011 IEEE/PES Power Systems Conference and Exposition*, (IEEE, USA, 2011), pp. 1–7
33. R. Roman, J. Zhou, J. Lopez, On the features and challenges of security and privacy in distributed internet of things. Comput. Netw. **57**(10), 2266–2279 (2013)
34. W. Rui-gang, Main character and basic theory for Internet of Things. Comput. Sci. **39**(6A), 201–206 (2012)
35. H.J. La, A conceptual framework for trajectory-based medical analytics with IoT contexts. J. Comput. Syst. Sci. **82**(4), 610–626 (2016)
36. S. Tonyali, K. Akkaya, N. Saputro, A.S. Uluagac, M. Nojoumian, Privacy-preserving protocols for secure and reliable data aggregation in IoT-enabled Smart Metering systems. Futur. Gener. Comput. Syst. **78**, 547–557 (2018)
37. C. Tu, X. He, Z. Shuai, F. Jiang, Big data issues in smart grid: A review. Renew. Sust. Energ. Rev. **79**, 1099–1107 (2017)
38. M.L. Tuballa, M.L. Abundo, A review of the development of Smart Grid technologies. Renew. Sust. Energ. Rev. **59**, 710–725 (2016)
39. M. Jin et al., Energy-cyber-physical systems. Appl. Energy **256**, 113939 (2019)
40. S. Zeadally, N. Jabeur, *Cyber-Physical System Design with Sensor Networking Technologies* (Institution of Engineering and Technology, London, 2016)
41. A. Sheth, P. Anantharam, C. Henson, Physical-cyber-social computing: An early 21st century approach. IEEE Intell. Syst. **28**(1), 78–82 (2013)
42. J. Zeng, L.T. Yang, M. Lin, H. Ning, J. Ma, A survey: Cyber-physical-social systems and their system-level design methodology. Futur. Gener. Comput. Syst. **105**, 1028–1042 (2020)
43. L. Wang, X.V. Wang, *Cloud-Based Cyber-Physical Systems in Manufacturing* (Springer, Cham, 2018)
44. P. Sobhrajan, S.Y. Nikam, D. Pimpri, P.D. Pimpri, Comparative study of abstraction in cyber physical system. Int. J. Comput. Sci. Inf. Technol. **5**, 466–469 (2014)
45. R. Alguliyev, Y. Imamverdiyev, L. Sukhostat, Cyber-physical systems and their security issues. Comput. Ind. **100**, 212–223 (2018)
46. M. Krotofil, A.A. Cárdenas, Resilience of process control systems to cyber-physical attacks, in *Nordic Conference on Secure IT Systems*, (Springer, Berlin, Heidelberg, 2013), pp. 166–182
47. A. Hahn, R.K. Thomas, I. Lozano, A. Cardenas, A multi-layered and kill-chain based security analysis framework for cyber-physical systems. Int. J. Crit. Infrastruct. Prot. **11**, 39–50 (2015)
48. F.-J. Wu, Y.-F. Kao, Y.-C. Tseng, From wireless sensor networks towards cyber physical systems. Pervasive Mob. Comput. **7**(4), 397–413 (2011)
49. H. Li, L. Lai, H.V. Poor, Multicast routing for decentralized control of cyber physical systems with an application in smart grid. IEEE J. Sel. Areas Commun. **30**(6), 1097–1107 (2012)
50. C.H. Liu, Y. Zhang, *Cyber Physical Systems: Architectures, Protocols and Applications* (CRC Press, Boca Raton, 2015)
51. J. Vitorino et al., Industry 4.0-digital twin applied to direct digital manufacturing. Appl. Mech. Mater. **890**, 54–60 (2019).: Trans Tech Publications
52. A. El Saddik, Digital twins: The convergence of multimedia technologies. IEEE MultiMedia **25**(2), 87–92 (2018)
53. G. Bacchiega, Creating an embedded digital twin: monitor, understand and predict device health failure, in *Inn4mech-Mechatronics and Industry*, (Inn4mech, Italy, 2018), vol. 4
54. S. Malakuti, J. Schmitt, M. Platenius-Mohr, S. Grüner, R. Gitzel, P. Bihani, A four-layer architecture pattern for constructing and managing digital twins, in *European Conference on Software Architecture*, (Springer, Cham, 2019), pp. 231–246
55. M. Bacidore, Digital twin to enable asset optimization. *Smart Industry*, 2015
56. D. Bolton, *What Are Digital Twins and Why Will They Be Integral to the Internet of Things*, 2018

57. M.E. Porter, J.E. Heppelmann, How smart, connected products are transforming companies. Harv. Bus. Rev. **93**(10), 96–114 (2015)
58. D. Tilson, K. Lyytinen, C. Sørensen, Research commentary—Digital infrastructures: The missing IS research agenda. Inf. Syst. Res. **21**(4), 748–759 (2010)
59. D. Hamilton, Seeing double: Why IoT digital twins will change the face of manufacturing (28 May 2020). Available: https://www.networkworld.com/article/3219847/seeing-double-why-iot-digital-twins-will-change-the-face-of-manufacturing.html
60. Y. Cai, B. Starly, P. Cohen, Y.-S. Lee, Sensor data and information fusion to construct digital-twins virtual machine tools for cyber-physical manufacturing. Procedia Manuf. **10**, 1031–1042 (2017)
61. R. Rosen, G. Von Wichert, G. Lo, K.D. Bettenhausen, About the importance of autonomy and digital twins for the future of manufacturing. IFAC-PapersOnLine **48**(3), 567–572 (2015)
62. M. Daneshvar, B. Mohammadi-Ivatloo, K. Zare, S. Asadi, A. Anvari-Moghaddam, A novel operational model for interconnected microgrids participation in transactive energy market: A hybrid IGDT/stochastic approach. IEEE Trans. Industr. Inform., 1 (2020). https://doi.org/10.1109/TII.2020.3012446
63. K. Zhou, S. Yang, Z. Shao, Energy internet: The business perspective. Appl. Energy **178**, 212–222 (2016)
64. J. Rifkin, *The Third Industrial Revolution: How Lateral Power Is Transforming Energy, the Economy, and the World* (Macmillan, New York, 2011)
65. Z. Wei, S. Yu, G. Sun, Y. Sun, Y. Yuan, D. Wang, Concept and development of virtual power plant. Autom. Electr. Power Syst. **37**(13), 1–9 (2013)
66. X. Zhang, J. Li, H. Fu, Distribution power & energy internet: from virtual power plants to virtual power systems. Proc. CSEE **35**(14), 3532–3540 (2015)
67. K. Wang, X. Hu, H. Li, P. Li, D. Zeng, S. Guo, A survey on energy internet communications for sustainability. IEEE Trans. Sustain. Comput. **2**(3), 231–254 (2017)
68. F.F. Wu, P.P. Varaiya, R.S. Hui, Smart grids with intelligent periphery: An architecture for the energy internet. Engineering **1**(4), 436–446 (2015)
69. A.Q. Huang, M.L. Crow, G.T. Heydt, J.P. Zheng, S.J. Dale, The future renewable electric energy delivery and management (FREEDM) system: The energy internet. Proc. IEEE **99**(1), 133–148 (2010)
70. M. Gao, K. Wang, L. He, Probabilistic model checking and scheduling implementation of an energy router system in energy internet for green cities. IEEE Trans. Industr. Inform. **14**(4), 1501–1510 (2018)
71. International Energy and Climate Change, *World Energy Outlook Special*, ed. Rep, 2015
72. D. Tayal, Achieving high renewable energy penetration in Western Australia using data digitisation and machine learning. Renew. Sust. Energ. Rev. **80**, 1537–1543 (2017)
73. M. Sharifzadeh, A. Sikinioti-Lock, N. Shah, Machine-learning methods for integrated renewable power generation: A comparative study of artificial neural networks, support vector regression, and Gaussian Process Regression. Renew. Sust. Energ. Rev. **108**, 513–538 (2019)
74. S. Russell, P. Norvig, *Artificial Intelligence: A Modern Approach* (Prentice Hall, Upper Saddle River, 2002)
75. E. Alpaydin, *Introduction to Machine Learning* (MIT Press, Cambridge, MA, 2020)
76. R. Sathya, A. Abraham, Comparison of supervised and unsupervised learning algorithms for pattern classification. Int. J. Adv. Res. Artif. Intell. **2**(2), 34–38 (2013)
77. A.B. Tucker, *Computer Science Handbook* (CRC Press, Boca Raton, 2004)
78. H. Zhao, J. Zheng, W. Deng, Y. Song, Semi-supervised broad learning system based on manifold regularization and broad network. IEEE Trans. Circuits Syst. I Regul. Pap. **67**(3), 983–994 (2020)
79. M. Wiering, M. Van Otterlo, *Reinforcement Learning*, Adaptation, Learning, and Optimization, vol 12 (Springer, Berlin, Heidelberg/New York, 2012), p. 3
80. S. Bozinovski, A self-learning system using secondary reinforcement, in *Cybernetics and Systems Research*, (North-Holland, Amsterdam, 1982), pp. 397–402

81. S. Bozinovski, Modeling mechanisms of cognition-emotion interaction in artificial neural networks, since 1981. Procedia Comput. Sci. **41**, 255–263 (2014)
82. Y. Bengio, A. Courville, P. Vincent, Representation learning: A review and new perspectives. IEEE Trans. Pattern Anal. Mach. Intell. **35**(8), 1798–1828 (2013)
83. F. Meng, X. Yang, C. Zhou, Z. Li, A sparse dictionary learning-based adaptive patch inpainting method for thick clouds removal from high-spatial resolution remote sensing imagery. *Sensors* **17**(9), 2130 (2017)
84. L. Liu, M.T. Özsu, *Encyclopedia of Database Systems* (Springer, New York, 2009)
85. V. Hodge, J. Austin, A survey of outlier detection methodologies. Artif. Intell. Rev. **22**(2), 85–126 (2004)
86. O. Kroemer, S. Niekum, G. Konidaris, A review of robot learning for manipulation: Challenges, representations, and algorithms. *arXiv:1907.03146* (2019)
87. S. Ozawa, T. Ban, N. Hashimoto, J. Nakazato, J. Shimamura, A study of IoT malware activities using association rule learning for darknet sensor data. Int. J. Inf. Secur. **19**(1), 83–92 (2020)
88. A. Foucquier, S. Robert, F. Suard, L. Stéphan, A. Jay, State of the art in building modelling and energy performances prediction: A review. Renew. Sust. Energ. Rev. **23**, 272–288 (2013)
89. B. Yildiz, J.I. Bilbao, A.B. Sproul, A review and analysis of regression and machine learning models on commercial building electricity load forecasting. Renew. Sust. Energ. Rev. **73**, 1104–1122 (2017)
90. H.-X. Zhao, F. Magoulès, A review on the prediction of building energy consumption. Renew. Sust. Energ. Rev. **16**(6), 3586–3592 (2012)
91. S.M. Clarke, J.H. Griebsch, T.W. Simpson, Analysis of support vector regression for approximation of complex engineering analyses. J. Mech. Des. **127**(6), 1077–1087 (2005)
92. J. Lafferty, L. Wasserman, Challenges in statistical machine learning. Stat. Sin. **16**(2), 307 (2006)
93. R. Blanquero, E. Carrizosa, C. Molero-Río, D.R. Morales, Sparsity in optimal randomized classification trees. Eur. J. Oper. Res. **284**(1), 255–272 (2020)
94. B. Cai, Y. Liu, Z. Liu, Y. Chang, L. Jiang, Risk analysis of subsea blowout preventer by mapping GO models into Bayesian networks, in *Bayesian Networks for Reliability Engineering*, (Springer, Singapore, 2020), pp. 165–187
95. J. Zhang et al., Evolutionary computation meets machine learning: A survey. IEEE Comput. Intell. Mag. **6**(4), 68–75 (2011)
96. K. Hosanagar, A. Saxena, The first wave of corporate AI is doomed to fail. *Harvard Business Review*, 18 Apr 2017
97. A. Artasanchez, 9 Reasons why your machine learning project will fail (23 May 2020). Available: https://www.kdnuggets.com/2018/07/why-machine-learning-project-fail.html
98. K.M. Lee, J. Yoo, S.-W. Kim, J.-H. Lee, J. Hong, Autonomic machine learning platform. Int. J. Inf. Manag. **49**, 491–501 (2019)
99. A. Almalaq, G. Edwards, A review of deep learning methods applied on load forecasting, in *2017 16th IEEE International Conference on Machine Learning and Applications (ICMLA)*, (IEEE, USA, 2017), pp. 511–516
100. I. Goodfellow, Y. Bengio, A. Courville, *Deep Learning* (MIT Press, Cambridge, MA, 2016)
101. D.L. Marino, K. Amarasinghe, M. Manic, Building energy load forecasting using deep neural networks, in *IECON 2016-42nd Annual Conference of the IEEE Industrial Electronics Society*, (IEEE, Piscataway, 2016), pp. 7046–7051
102. P.-F. Pai, W.-C. Hong, Support vector machines with simulated annealing algorithms in electricity load forecasting. Energy Convers. Manag. **46**(17), 2669–2688 (2005)
103. S. Ryu, J. Noh, H. Kim, Deep neural network based demand side short term load forecasting. *Energies* **10**(1), 3 (2017)
104. B. Grossfeld, Deep learning vs machine learning: A simple way to understand the difference (23 May 2020). Available: https://www.zendesk.com/blog/machine-learning-and-deep-learning/

Chapter 5
Energy Trading Possibilities in the Modern Multi-Carrier Energy Networks

Abstract Nowadays, innovations in communication and information technologies along with the massive deployment of renewable energy resources (RERs) have developed the energy grid structure from a hierarchical system to a network-based deregulated environment with the aim of making it more appropriate for the modern architecture of the energy networks. This transformation raises significant challenges regarding the energy markets and operational issues due to the uncontrollable factors of the stochastic producers. Thereby, there is a pressing need for innovative solutions more than ever to overcome the mentioned challenges, which these solutions can be effectively adapted to the modern paradigm of the system. In this regard, energy trading possibilities between various entities provide an exceptional opportunity for vanquishing the high level of uncertainties caused by the ever-increasing penetration of RERs. Therefore, this chapter is focused on evaluating the different types of energy trading mechanisms for the future structure of hybrid energy networks. In this respect, different types of energy trading platforms are effectively evaluated, which encourage direct energy exchanging between peers considering the key technologies and elements involved in the energy sharing process. Finally, transactive multi-carrier energy technology is proposed as the end-to-end energy trading paradigm for the future modern multi-carrier energy networks to establish a dynamic energy balance in a reliable and sustainable manner. In the end, the Grid Modernization-Version I (GM-V1) model is proposed for the future modern multi-carrier energy networks (MCENs).

Keywords Transactive energy · Multi-carrier energy trading · Peer to peer energy trading · Blockchain technology · Reinforcement learning · Modern energy models

© Springer Nature Switzerland AG 2021

M. Daneshvar et al., *Grid Modernization – Future Energy Network Infrastructure*, Power Systems, https://doi.org/10.1007/978-3-030-64099-6_5

5.1 Introduction

In recent decades, the energy sector has witnessed revolutionary changes due to rapidly growing pressure on the multi-carrier energy demand, the advancements in the multidisciplinary energy technologies and expert systems, and proliferation in mushrooming the decentralized energy systems [1]. The recent developments in energy production techniques and substantial advancements in their technologies promise extraordinary improvements in developing energy systems with an acceptable level of energy efficiency [2]. In this regard, the end-user part can effectively involve in the decentralized bilateral energy exchanging with the aim of fulfilling the part of their energy demand without further stressing the energy network. To this end, creating an energy trading possibility among the energy market participants is gradually becoming a key option for developing a sustainable energy environment [3]. The significant achievements in the context of energy trading have accelerated the mutation of the research movements toward the improvement of technologies and techniques that typically leads to making hybrid networks more flexible than ever before [4]. The developments in energy trading technologies and algorithms have resulted in an appearance of diverse techniques for energy exchanging in accordance with the different conditions of the energy environment. All of these techniques have united in reaching the goal of developing sustainable, reliable, resilient, stable, and flexible energy networks [5]. The value of the energy trading techniques can be more highlighted when they are applied in the networks with a high penetration of the uncontrollable energy production units. The energy trading possibility can effectively reduce the need for a large number of energy production units, energy storage systems, and energy conversion process with a high amount of losses. This is due to its capability in creating the energy sharing possibility that allows producers and consumers to exchange energy with each other instead of just producing, storing, or converting energy. As it can occur in different ranges and at each time of the day, it can effectively increase the reliability, resiliency, sustainability, stability, and flexibility of the system particularly in renewable-based energy networks [6]. On the other hand, the future modern MCENs are planned to be equipped with a 100% RERs for fully clean energy production. Energy trading possibility is one of the vital ways for achieving the acceptable level of system reliability, resiliency, stability, sustainability, and flexibility in modern grids. Therefore, employing the effective energy trading mechanisms will be essential in the future modern MCENs. This is while the energy sharing mechanisms are not clearly examined yet to analyze their suitability for the structure of the future modern MCENs. In other words, the lack of effective techniques for establishing strong energy exchanging environment has created serious concerns regarding the reliable interconnectivity of the future modern hybrid energy networks. Therefore, to effectively fill such gaps, we have deeply focused on scrutinizing the diverse energy trading techniques by examining their suitability for the future modern MCENs. In this regard, the new transactive multi-carrier energy technology is proposed with the aim of effectively supporting multiple dimensions of the energy sharing in the interconnected structure of the future modern MCENs. Moreover, the new GM-V1 model is proposed for the future modern MCENs to facilitate the success of the grid modernization process.

5.2 Why Energy Trading Is Essential?

As energy popularity has been spread worldwide, the energy systems have experienced significant changes and have faced with substantial changes in energy consumption [7]. This has made the energy grids' schedulers think about how to deal with such evolutions in the energy sector so that the system is developed in a way to maintain the system reliability and sustainability at the acceptable level whereas to satisfy the techno-economic aspects [8]. In this respect, the recent advancements in the technology of the energy production systems have made the decentralized energy generation units economically and technically viable for distributed energy supply throughout the hybrid networks. On the other hand, the environmental and economic concerns for the traditional fossil fuel-based systems have been mainly led to a lack of new ideas for their development [9]. The DERs, especially the renewable type of them, have attracted remarkable focuses for being at the center of the energy production system. As the future modern MCENs are aimed to implement the full renewable-based energy production system, different types of energy storage systems have revived the expectations for constructing the energy systems with 100% RERs. In other words, the maturing of the various energy storage systems technologies has altered the world of low carbon-based energy systems to be effectively operated in hybrid networks. However, one of the significant challenges that is usually brought along with RERs is their uncontrollable feature in energy production. Indeed, the intermittences in the RERs outputs have posed the concern of reliably establishing a dynamic energy balance in the optimal scheduling of the energy networks. To effectively address this challenge, some efficient ways are proposed that have already showed satisfactory results. In Chap. 2 of this book, operating the different forms of the energy storage systems and optimal integration of the RERs are discussed as two effective ways for dealing with the uncertainties of the stochastic producers. However, there are some considerable concerns regarding their application in solving the uncertainty problem of RERs. The operation and investment costs of the energy storage system are often expensive and direct utilization of most of them depends on the geographical conditions. This is while the optimal integration of the high level of RERs is mostly relied on applying the capable technologies that need to be economically and technically viable. In other words, the lack of appropriate technologies has limited the optimal integration of the stochastic producers in the deregulated environment. At the end of this chapter, we have introduced powerful technology for optimal integration of the RERs in the system. Given the aforementioned challenges regarding the energy storage systems and optimal integration of RERs, they are not much suitable to be intended in the renewable-based systems alone. Therefore, the research trends have been focused on finding other appropriate ways to support the renewable-based systems to supply energy reliably. Providing the energy trading possibility between the eligible elements of energy networks is recognized as one of the effective solutions for balancing energy in the presence of a high level of environmentally friendly energy production units. Therefore, energy trading frameworks are developed to be effectively employed in the energy market's interactions. The success of these frameworks directly depends on the availability of enabling technologies as well as the active participation of end-users [10]. Herein, the biggest motivation for the participation of the end-users in the energy market interactions commonly comes in the form of profits and cost savings. The incorporation of the DERs, energy storage systems, and energy sharing possibility with each other can typically bring the following benefits for the system [10]:

- Substantially alleviate greenhouse gas emissions.
- Improve the overall efficiency of energy networks.
- Minimize the operation cost of the system.

In the context of energy trading, the enabling technologies are required for developing the energy exchanging frameworks with the aim of achieving desired outcomes. A simple overview of the energy sharing components is depicted in Fig. 5.1 [10].

5.3 Energy Trading Benefits

Energy trading, as one of the efficient routes, is designed to make the realization of the sustainable carbon-free energy networks possible. The benefits stem from the energy exchanging frameworks that can definitely increase the chance of setting up the pollutant-free hybrid energy networks. Some of the most significant benefits are described below [10].

5.3.1 Improving System Efficiency

Recently, the energy networks have moved toward becoming strongly hybrid by advancing different types of DERs, energy storage systems, hybrid devices, and electric vehicles. Such hybrid energy grids enable the system operator to effectively provide various ancillary services with the aim of substantially improving their efficiency. In this respect, energy trading in the local area can drastically reduce the use of high capacity energy production systems and subsequently their costs and greenhouse gas emissions. Adopting such strategies will reduce the congestions on energy transmission lines and energy losses, especially in the heat format. As a valuable result, the need for system upgrades will be gradually deferred to a wider time hori-

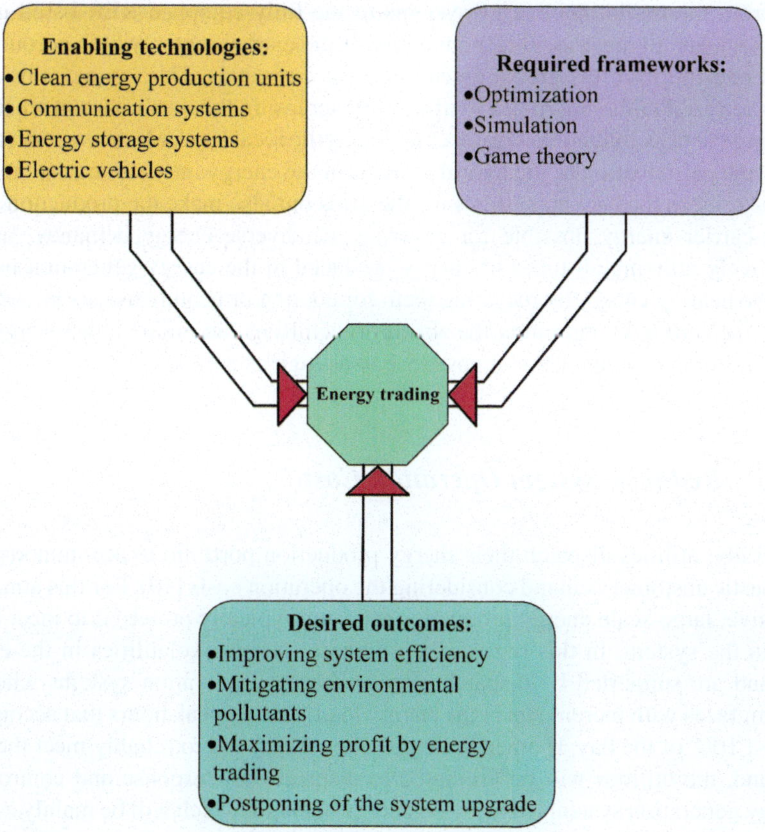

Fig. 5.1 A simple overview of the energy sharing components

zon [10]. Furthermore, these improvements will increase the overall reliability of the system in energy supply and minimize the interruption cost.

5.3.2 Mitigating Environmental Pollutants

Over recent decades, the governments' policies in the context of environmental efforts around the globe have been directed to a significant reduction of greenhouse gas emissions, especially by focusing on energy production from cost-effective and environmentally friendly energy generation units rather than traditional systems. For instance, European Union countries have decided to adopt suitable strategies for reducing greenhouse gas emissions by 20% by 2020 [10]. One of the most important of these strategies is the maximum exploitation of the clean energy resources in the midterm period to the full exploitation of them in the long-term period run in future modern energy networks. In this regard, driving the energy production system from the traditional energy systems to the fully equipped RER-based grid is necessary due to the critical responsibility of the energy sector, which accounts for about one-fourth of the global greenhouse gas emissions. This issue can be effectively realized in the microgrid platform by increasing the mushroom of RERs and creating the possibility of energy exchanging at the local level. Moreover, microgrids allow the participation of the hybrid systems in the energy interactions that not only can be used as the backup systems for the grids but also make the production of the multi-carrier energy possible for covering the diverse energy demands. Indeed, effectively utilizing all types of energy produced in the energy generation process like the heating energy by recycling them for cooling or heating usages in buildings and refrigeration by operating the absorption chilling can improve the energy efficiencies for consumers from almost 35% to as high as 85% [10].

5.3.3 Reduced System Operation Cost

Generally, utilities dispatch their energy production portfolio to accommodate the stochastic customer demand considering the operation costs [10]. For this aim, cost-effective, large-scale energy production units are typically preferred to meet baseload in the system. In this regard, the minute-by-minute volatilities in the energy demand are supported by dispatching more energy generation systems when the system faces with increments in the energy load. During peak hours that account for almost 10% of the day, if other strategies are not pursued to reliably meet the peak demand, the utilities will be prompted to dispatch fast response and controllable energy generation systems to meet the energy demand, which will be mainly accompanied with the release of environmental pollutants. Energy trading is famous for its capability in reducing the peak-to-average demand ratio by enabling the eligible participants for energy sharing with the main grid at the local level. This allows the system to compensate for a considerable portion of the energy shortage through the energy sharing possibilities instead of dispatching more units of the fossil fuel-based energy generation systems. Thereby, the operation and investment costs of the system are substantially reduced.

5.3.4 Energy Profiling

The microgrid generators can achieve the greatest amount of energy efficiency when they have energy interactions in the form of the networked microgrids for profiling the user's energy load and subsequently control the energy production and flow. Similar to the bulk energy network, smart microgrids can produce and distribute energy to the consumers, but at the local level. Local networks allow smart microgrids to offer much lower line losses rather than imposing high losses in energy transmission over long distance. The operation of smart microgrids in the coupled format enables the system to optimally integrate the RERs for reliably establishing energy balance between energy demand and supply near the point of use as well as allowing for customer participation in the energy market. In this regard, energy trading has emerged as the necessary tool for constructing the building blocks of the future perfect energy networks [10]. The energy networks can effectively rely on the networked microgrids in rapidly enlarging the overall energy systems rather than on utility companies.

5.3.5 *Energy Trading in the Future Modern MCENs*

Undoubtedly, the dependency of the globe to the energy is increasing more and more simultaneously by emerging of intelligent systems to respond to increasing living standards. The research trends in the field of energy are mostly centered on finding practical solutions for extracting different forms of energy from the cost-effective and environmentally friendly energy resources. In this respect, RERs are recognized as popular energy resources to play a central role in energy production due to the economic and environmental concerns of the traditional fossil fuel-based systems. The cost-effective and pollutant-free features of the RERs have made them be considered as the main energy production units in the future modern MCENs. In other words, the 100% RERs are planned for the future modern MCENs to fully meet the energy demand from the carbon-free energy production units. As mentioned earlier, optimal integration of the RERs and operation of the various energy storage systems are two effective ways of alleviating the stochastic behaviors of the RERs in energy generation. However, using them alone cannot be an appropriate strategy technically, and economically won't be reliable. Thus, using other useful solutions is necessary for increasing the reliability of the continuous energy supply in the presence of the full stochastic producers. Herein, the energy trading possibility as one of the effective solutions is introduced for the future modern MCENs to assist the assurance of the reliability of the energy supply. Since the future modern MCENs are targeted to equip with 100% RERs, the energy trading at a large scale is essential for creating a dynamic energy balance in the system. Redesigning the prosumer-side rather than an absolute consumer-side in the future modern MCENs enables the end-users for active participation in the energy trading market that allowing the system for structuring a large-scale energy sharing environment. The existence of the large-scale energy exchanging environment is essential for the future modern MCENs due to its capability in real-time energy management, especially in the systems with a full share of RERs. On the other hand, energy trading in the future modern MCENs not only is limited only to the special type of energy like the electricity but also makes the multi-carrier energy exchanging possible throughout the hybrid networks. In other words, diverse dimensions of energy, for example, the natural gas energy, heating and cooling energy can be easily intended for energy sharing in the multi-carrier energy market under the future modern MCENs paradigm. Given the multi-carrier energy trading possibility at the local level, the modern grid can be more flexible than ever before by operating the hybrid systems that enable the hybrid network to share energy among the different grids and convert them to each type of energy that is required for maintaining the stability of the system. Thereby, the multi-carrier energy trading is critical for the sustainability of the future modern MCENs. As energy trading possibility needs capable technologies to securely implement the related frameworks in the modern grids, it can bring several significant benefits for the hybrid energy networks. This is while the weight of the suitability of these benefits is far more than its challenges, which have mobilized many efforts for solving its challenges. Some of the most prominent of these benefits are as follows:

- Eliminate the need for a large number of fossil fuel-based energy generation systems that are scheduled to work only during peak times with more investment and operation costs.
- Reduce the release of the environmental pollutants considering the limiting of operating the carbon-diffusion based energy production systems.
- Reduce the probability of load shedding, especially at peak times for maintaining the stability of the system.
- Provide the opportunity of participation of the end-users in the energy market interactions to improve their economic benefits.
- Facilitate the employing of different energy management schemes on the prosumer-side.
- Improve the hosting capacity of the system with a full share of RERs.
- Enable the system for effective use of the potential of the environment in clean energy production.
- Increase the capability of the modern MCENs in quickly responding to any changes in the system for maintaining the reliability and stability of the grid. This issue is essential in the full renewable-based energy generation systems for balancing energy.

5.4 Real-Time Energy Trading Mechanism

In the smart grid area, one of the key goals is to develop a two-way communication infrastructure for making the energy networks smarter and secure than before. The distribution network operator is empowered by the combination of the information exchange and sophisticated energy management and monitoring systems to perform automated operations in near real-time. Monitoring of the system in near real-time along with the expansion of the operation of intelligent devices can bring the following significant benefits [11]:

- Enables electric appliances to be more efficiently used in households with the aim of increasing the economic benefits of the residents
- Reduces the energy bill payment for the load aggregators
- Provides higher flexibility for the distribution network operator

- Reduces the operation costs of the generators
- Enhances the quality of the technical operation of the system

In the modern hybrid energy networks with the full share of the RERs, realizing the aforementioned benefits mostly relies on deploying intelligent systems as well as capable technologies. This is while the intermittent nature of the RERs has made the employment of those technologies more difficult in realizing some sensitive objectives, such as establishing a dynamic energy balance in the system. In such conditions, real-time energy trading is taken into account as one of the effective ways for developing the energy trading frameworks aiming to assist the real-time balancing energy. In other words, real-time energy trading allows the eligible network's elements to trade energy with each other in real-time to not only increase their economic benefits but also support the system to effectively maintain its reliability, stability, and sustainability at the permissible level. Therefore, the developing of real-time-based energy trading frameworks is necessary for the future modern MCENs for ensuring the reliable energy supply in the presence of the 100% stochastic producers.

5.5 Peer-to-Peer Energy Trading Mechanism

Recently, the developments in the DER technologies along with the demand-side energy management techniques have paved the way for the transformation of the traditional consumers into the prosumers. Prosumers are the new players that allow for realizing the two roles, that is, consumer and producer in the energy networks. Indeed, prosumers can produce energy for meeting their energy demand and sell its excess to the consumers, store it for later use, or export it back to the energy network [12]. The direct energy trading between the prosumers and consumers is called Peer-to-Peer (P2P) energy trading [12]. The P2P energy trading is developed based on the concept of sharing economy or P2P economy [13], which is typically used for energy sharing within a local electrical energy distribution system. In the P2P energy trading process, a peer refers to one or a group of prosumers, consumers, and generators, which are enabled by the P2P energy trading mechanism for direct selling or buying energy with each other without intermediation by traditional energy suppliers [14]. The information and communication technologies are usually used for developing P2P energy trading based on online services [13].

Generally, the conventional energy system gives a unidirectional energy trading possibility that energy is transmitted from the large-scale centralized energy production units to the load centers over long distances. In contrast, P2P energy exchanging develops a direct energy sharing among peers or multidirectional energy trading in which the energy produced from the small-scale DERs in prosumer-side such as factories, offices, and dwellings is traded between local energy consumers and producers within a local geographical area [12]. In this regard, Piclo, as an online platform, is typically used for performing P2P energy trading for business consumers and generators [12]. The Piclo employs matching algorithms to match local energy consumption and production, then the data analytics and visualization are prepared for consumers. The consumer preferences information, generator pricing, and smart meter data are accumulated to be used in matching the energy supply and demand every half an hour.

5.5.1 Peer-to-Peer Energy Trading Platform

Recently, the P2P economy is proposed for the energy exchanging arrangements in local distribution grids. The P2P energy trading enables the peers, for example, suppliers, prosumers, generators, or even consumers, to decide on which other peers to trade energy with while they meet their goals such as the most reliable energy supply, minimum pollution, maximum profits, and minimum costs [12]. A software platform is an integral part of the P2P energy trading mechanism that makes the information exchange possible among peers with the aim of facilitating the monitoring and controlling of the distribution network for the system operator. In the P2P energy trading platform, various energy sharing rules regulated in the platform have direct impacts on the decision-making by peers in their local energy trading interactions with other peers. Different software platforms are developed aiming to facilitate P2P energy trading. For example, the "Elecbay" P2P energy trading platform is proposed in [12] for creating the energy sharing possibility in the grid-connected microgrid. Some key players typically participate in the P2P energy trading platform that their interactions in the energy trading environment are demonstrated in Fig. 5.2 [12].

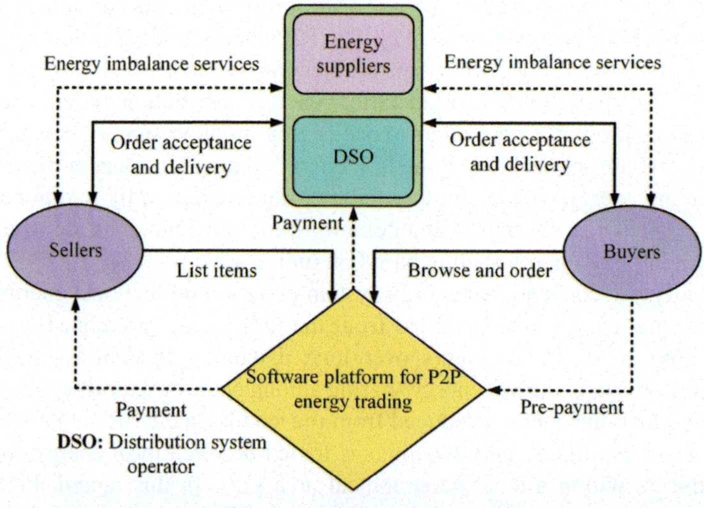

Fig. 5.2 Energy interactions of the different players in the P2P energy trading platform [12]

Given Fig. 5.2, each energy seller lists its excess energy over half an hour for selling. On the other side, each energy buyer browses all the listed items and regulates its orders. The orders consist of the information including the price of the energy to be traded, the amount of energy to be shared, the time period for the energy trade, and the details about the buyer and seller such as the ratings of energy supply reliability, the identities. Alternatively, each energy seller browses the orders of the buyers, that is, the amount of required energy at each time period, and then places its orders. After peers have placed their orders, they are analyzed by the distribution system operator (DSO), P2P energy trading platform, and suppliers for either accepting or rejecting orders based on the energy network constraints, for example, energy balance, thermal overloading, and voltage excursion. After identifying the accepted or rejected orders, each peer becomes aware of the acceptance or rejection of its order for scheduling to act based on its commitments in the accepted orders. Finally, the energy is supplied based on the scheduled orders through the distribution network. In this regard, the smart meters are used throughout the energy network for recording the actual amount of energy consumed or generated by peers with the aim of enabling the DSO for full supervision of accepted commitments.

5.6 Blockchain Technology

Over recent decades, energy networks are undergoing rapid evolutions in accommodating the day-by-day increment of energy consumption as well as the proliferation of the mushrooming of the clean energy production units. RERs have recently been experienced massive developments and are enabled by privatization, boosted by energy policy initiatives and financial incentives, and unbundling of the energy sector [15]. RERs are pretty variable, do not have specific behavior to predict, and their output directly depends on the climate changes, hence raise new concerns and challenges in optimal operation and management of the energy networks as well as flexibility and stability of the system for ensuring safe operation [14]. By rapidly increasing the transformational changes of the DERs particularly RERs, the energy sector is on the brink of entering the digital era through massive deploying of the sophisticated technologies for effectively handling the intelligent systems throughout the energy network. Increasing the penetration of the decentralized energy generation systems along with the smart devices has vastly enhanced the volume of the generated data, which has created a great need for the advanced communication protocols as well as the comprehensive energy management and control schemes. In order to effectively accommodate these digitalization and decentralization trends, capable local distributed management and control techniques should be adopted for the wider group of intelligent systems [16]. For this new energy environment, blockchains or distributed ledger technologies are proposed aiming to enable the system for distributed transactions by eliminating central management [15].

5.6.1 Blockchain Definition and Fundamental Principles

A blockchain is a distributed and shared data structure or ledger and database that comprises a continuously expanding log of transactions and their chronological order [15]. Indeed, blockchains are the digital data structures that make the storing of digital transactions in the system possible without adopting a central point of authority [17]. The digital transactions are congregated into larger formations and build the blocks that are not only time-stamped but also cryptographically linked to previous blocks. Linking the created blocks to each other forms a chain of records that identifies the sequencing order of events or the blockchain. In blockchain technology, the records can be simultaneously changed by multiple users in the ledger that results in several new blocks and then multiple chain versions. The blockchain technology dismantles the authority-based system by removing the managing of the ledger based on a single trusted center. Alternatively, blockchain technology provides a copy of the records' chain for each of the network members that uses consensus for reaching an agreement on the valid state of the ledger. The way that how consensus can be effectively reached is the key issue in the blockchain technology that is created an ongoing area of research and is going toward a wide range of application domains.

Blockchain technology uses cryptographies for securely linking the new transactions with the previous ones with the aim of making the networks more resilient and safe. Every network user is permitted to check the transactions if they are valid in the blockchain platform. This issue allows the system to provide trustable and transparency, tamper-proof records. For this aim, blockchains need to run on digital networks. In digital networks, data transmission refers to copying data from one certain place to the other. For example, in the cryptocurrency domain, data transmission is equivalent to copying digital coins from the electronic wallet of one user to another one. One of the primary challenges here is how to reliably ensure that coins are only spent once for avoiding double-spending [15]. Heretofore, using a central point of authority is taken into account as a traditional solution that is targeted to act as the trusted intermediary among the transacting parties. Indeed, a central point of authority is responsible for storing and safeguarding the valid state of the ledger and remain the records up to date [15]. In this regard, the central authority has to implement significant control and consolidates different changes in the ledger when the multiple parties require to write in the ledger [15]. Because central authority management offers intermediary costs and develops the system that prompted the network uses to trust a third party for operating of the system, it may not be desirable or feasible on several occasions [18]. Moreover, a single point of failure mechanism has caused substantial disadvantages for the centralized system that render it more vulnerable to both malicious attacks and technical failures [19].

The blockchain technologies are targeted to effectively remove the authority-based intermediaries and convert the system from a centralized network to the distributed system of digital users who are responsible for verifying transactions and act in partnership for safeguarding the integrity of the ledger. In the blockchain-

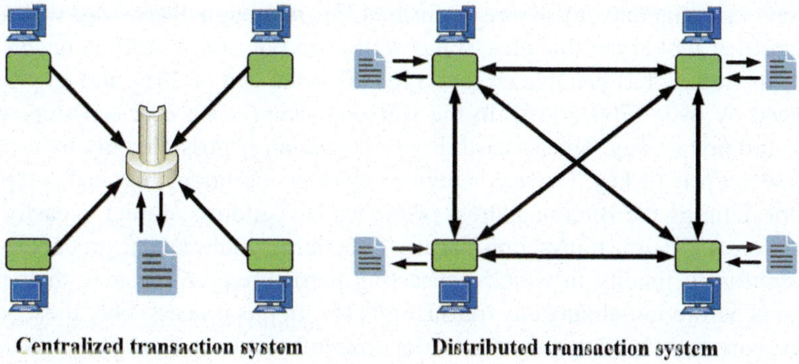

Fig. 5.3 Centralized and distributed transactional platforms in the blockchain technology [15]

based paradigm, unlike centralized systems, the accessibility of members of the blockchain network to their copy of ledgers is provided in the open cloud or every member holds his/her own copy of the ledger as illustrated in Fig. 5.3. Thereby, in order to create a high level of transparency, anyone has access to the historical log of the system in the network for verifying their validity. By removing the central management system and moving toward the distributed network of digital users, the challenge of adopting an efficient solution for consolidating and synchronizing multiple copies of the ledger remains a significant challenge. The ledger consolidation process as well as the exact validation procedure varies for various types of blockchains. This is while, in principle, the distributed network members are enabled to evaluate their versions of the ledger by developing a process intuitively akin to distributed voting with the aim of validating the state of the ledger [20]. These are the blockchain mechanisms for validating the ledgers that are recognized as distributed consensus algorithms. In this respect, game-theoretic rewards or incentives are commonly used for establishing the collaboration and honest behaviors of the distributed nodes [21]. In fact, without the substantial part of the network colluding, the chance of tampering in the blockchain technology is very low. Therefore, blockchain technology can provide a sufficient level of security and can be tamper-resistant (Fig. 5.3).

5.6.2 Blockchain Cryptocurrencies: Bitcoin and Ethereum as Two Important Paradigms

In the blockchain area, Bitcoin was established in 2009 as the world's first cryptocurrency by Nakamoto [22]. He proposed a distributed electronic cash payment platform that is developed based on the P2P communication system with unknown and anonymous Internet users [15]. In this process, there is not a central bank to control or issue the digital cash transacted among end-users but a group of computers creates a network of computers that is targeted to be used in collaborations and develops cryptography to assure security [23]. Although there was widespread skepticism at first about this idea, it has well gone out of obscurity in recent years so that its trading has become around $20,000 at the end of 2017 and its price has enhanced by over 1700% [15]. In the Bitcoin system, each user can store with a public and private key. The accessibility to the wallet is possible only by using the user's secret private key. The user's public key offers pseudonymity and is typically used for defining the Bitcoin address or the wallet's address aiming to easily identify a user. A Bitcoin transaction usually experiences subsequent procedure from initialization to finality in which transacting parties require to know their public addresses before initializing any transaction [15]. In this process, only those groups of users can create an outgoing transaction that sufficient coins are already stored in their wallet. A simple overview of a blockchain transaction is depicted in Fig. 5.4.

Each transaction in the blockchain technology is encrypted with the receiver's public key and it contains two prominent information, including addresses of transacting parties and the amount of coin trade [15]. In this respect, before transmitting transactions to the Bitcoin network, they need to be digitally signed by the sender. All outgoing transactions in the last 10 minutes are aggregated by the special nodes into a single block for fine-tuning the validation process. In this process, on average, approximately 10 minutes are required for validating one block and including in the blockchain [24]. Afterward, validator nodes that are broadly known as miners attempt to participate in the competition with other miners for solving the cryptographic puzzle to acquire the right of adding the block in the existing ledger with the aim of receiving a financial reward. This financial reward has consisted of two parts. The first part refers to the amount of reward agreed by all network members that currently stands at 12.5 Bitcoins every time a miner is successful while the second part resides on the transaction fees that are proposed by transaction parties [15]. A random selection process is developed based on the computational work required for selecting the successful miner that is broadly known as proof of work [25].

At the time that a miner succeeds in solving the available block, the relevant solution is transferred to the network and then the next block will be made available

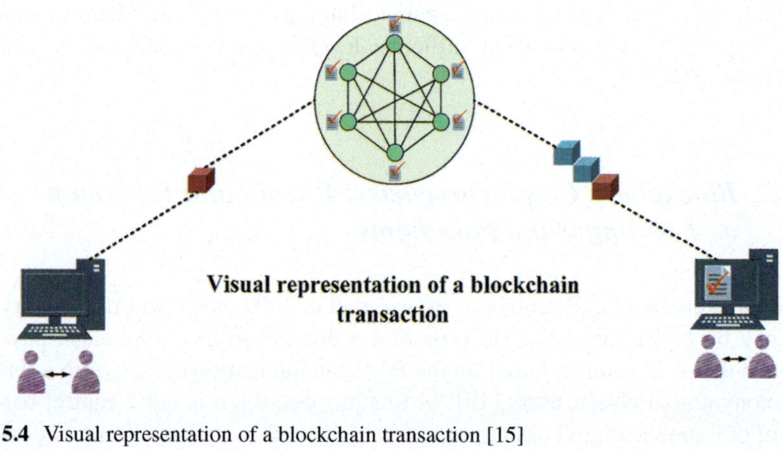

Fig. 5.4 Visual representation of a blockchain transaction [15]

to other miners to start to work on it. Succeeding blocks are recognized by two characteristics, including the hash output of the new validated block and the related transactions. Because the newly validated block is linked with previous blocks and is computationally expensive to have been produced, it can provide enough assurance for the users about its validation. In the running process of block validation, many miners are involved in a parallel process, which can be led to several blocks creating the block of chains that should be consolidated. Thereby, the blockchain structure is a tree of blocks with a valid path from the genesis block (root) to the blocks comprising the most recent transactions (leaf) [26]. In this regard, among the tracked of multiple chains that are kept by the network, the network members intend the chain with the most computationally expensive or longest chain formed in the system to indicate the valid state of the ledger. The renewed computational efforts will be required when any changes are applied in a single block with the aim of proofing work for all succeeding blocks. A computational minority of the Bitcoin has sufficiently empowered it in front of the computational power of all other truthful miners [21] that have made it enormous resilient to malicious attacks.

Bitcoin is designed in a way to majorly contribute to developing a significant solution for establishing trust between two mutually unrelated and unknown parties to such extent that secure and sensitive transactions can be conducted in a confidence way over an open environment like Internet [23]. Bitcoin is the first cryptocurrency in the blockchain world that establishes the largest blockchain application up to date. Cryptocurrency is a special type of digital currency that makes secure and reliable transactions possible while it contains all the properties of the conventional physical currencies for establishing instantaneous transactions [17]. As the Bitcoin is the first cryptocurrency in the blockchain world that creates the largest blockchain application up to date, Ethereum [27] has emerged as another cryptocurrency that is developed in a way to effectively dominate blockchain applications besides cryptocurrencies. Ethereum is an innovative version of the cryptocurrencies that runs on top of the blockchain architectures and enables users to create their own applications [15]. One of the considerable capabilities of the Ethereum is to enable the users to create smart contracts that allow the system to develop a technology platform on which different applications may be built based on the diverse transaction concepts [26]. Decentralized applications, as well as smart contracts, are taken into account as two major applications of the Ethereum. Decentralized applications are opensource systems that provide the autonomous operation of the system without human intervention and are executed by using tokens or cryptocurrencies in the network of computers for storing outputs in public ledgers [15]. The smart attributes of Ethereum have made it as one of the beneficial tools for redesigning services proportional to the energy market requirements.

5.6.3 Blockchain Potential Impacts on the Energy Sector

The widespread applications of blockchain technology have been led to the expansion of it even on the energy sector. The utility companies and decision-makers have asserted that this technology has enough potential to offer practical solutions for effectively addressing the challenges in the energy sector. According to the German Energy Agency's report [28], the blockchain technology can be widely used for accelerating the development of digital applications and IoT platforms, improving the efficiency of current energy processes and practices, and providing innovations in decentralized energy generation and P2P energy trading. Moreover, this technology resides in improving customer services and costs as well as internal processes with the aim of making the current practices of utility companies and energy enterprises better than ever before [28].

Recently, the energy industry has faced substantial evolutions triggered by the developments and advancements in information and communication technologies and DERs. In this sector, recent trends in digitalization and decentralization have created new requirements for the operation and adoption of advanced paradigms and innovative technologies. Due to the sufficient potential of the blockchain technology in decentralization and digitalization of the systems, it is recognized as a promising way for dealing with the challenge of optimal control and management of the microgrids with decentralized complex energy systems [29]. In the ongoing structure of the energy networks, integrating distributed generation, clean energy production systems, flexibility services, sophisticated information technologies, and consumer involvement in the energy market interactions is known as a demanding task [15]. Active participation of the consumers can be facilitated by applying the blockchain technology for securely recording their interactions into transparent, immutable, and tamper-proof smart contracts. Developing such automated information exchanging platforms enables energy networks to safely deliver price signals and other energy-based information to consumers altogether by providing them for demand-side energy management purposes [30]. Blockchain technology can make the feasibility of establishing the local energy trading in the microgrids possible with the aim of supporting continuous energy supply in the renewable-based systems. Deferring expensive energy grid upgrades and reducing the transmission losses can be two valuable achievements of this approach. The blockchain technology offers a widespread platform for securely recording ownership and origins of the energy supplied or consumed. Therefore, developing community microgrids, sharing of resources, smart charging arrangements, data storing in the smart grids, and cybersecurity objectives can be effectively realized by using the potential of the blockchain technology in the energy sector [31].

5.6.4 Blockchain-Based Energy Trading

Blockchain technology develops a secure structure for actualizing the automated execution of the smart contracts in the P2P energy trading network. In the P2P energy trading category, we can witness the presence of business-to-business energy trading, bilateral transactions between consumers and prosumers, and decentralized energy trading in microgrids [15]. From the energy system stakeholders' point of view [28], blockchain technology can effectively develop solutions for the following:

- Coordination of RERs portfolio
- Community energy projects
- Control of decentralized energy systems
- Management of energy storage systems
- Energy grid management and control
- Coordination of virtual power plants
- Demand response services

In this regard, P2P energy trading has emerged as a truly decentralized form of energy markets. This is while this contrasts with those applications such as imbalance settlement that still largely pursue the current structure of energy markets. The P2P energy trading enabled by the blockchain-based systems is one of the significant domains of a blockchain application that allows for direct energy trading among energy consumers and energy prosumers/producers, who can involve in this platform for effectively controlling of their energy production and demand. Although this can be realized in the small level of the grid such as the microgrid level, a key concern here is how it can fit with existing distribution network operation and management. Therefore, the operation of the system has mostly relied on network-based companies to manage the grid infrastructure and control the system for reliable energy supply. Consequently, blockchain technology is expected to play a key role in the truly decentralized energy markets for potentially improving the operation of the system by developing P2P energy trading mechanisms.

5.6.5 Edge Computing-Based Energy Trading

Nowadays, energy system technologies are upgrading day by day to effectively address the ever-increasing concerns over the energy crisis in a reliable manner. Creating energy trading possibility is recognized as one of the appropriate choices for dealing with the energy demand in the renewable-based grids. As mentioned in the previous section, Blockchain-based energy trading can accelerate the establishment of the widespread and secure energy trading environment in the future modern MCENs. As security and privacy are taken into account as two prominent factors in the blockchain technology, the edge computing mechanism is developed to address them for enabling efficient and secure energy sharing in the system. Indeed, the energy trading framework designed by leveraging the edge computing and blockchain technology can provide a suitable platform for secure and efficient energy exchange [32]. The edge computing is incorporated to effectively address the high computational cost challenge in solving proof-of-work puzzles associated with the blockchain creation [32]. Solving this issue is key because it causes a severe and unacceptable delay in the process of energy trading, which can be properly addressed by applying the edge computing-based energy trading that is aimed to improve the success probability of block creation [32]. Therefore, incorporating edge computing with blockchain technology is necessary for mitigating the challenges of blockchain technology in designing an efficient energy trading framework for the future modern MCENs. The mathematical implementation of the edge computing energy trading mechanism can be found in [32].

5.6.6 Blockchain Challenges

As the blockchain technology is recognized as a good idea for facilitating the secure promotion of decentralized energy systems throughout the energy network, it brings several challenges that need to be completely addressed before achieving larger adoption. For the future modern MCENs, using the maximum potential of the environment for 100% clean energy production is intended as one of the main principles for the hybrid networks. Because this issue requires for adopting full clean energy generation systems, the volume of RERs penetration in the presence of the blockchain technology has created the challenge of maintaining the security of energy supply at the permissible level as well as improving the network resilience. In order to effectively address this issue, the decentralization and digitalization should be promoted in the vast area of the hybrid energy network that seriously needs the exploitation of the intelligent systems and adopting innovative technologies. The scalability and cost are two other critical challenges when blockchain technology is considered to be employed on this network, which is expected to maintain the desired properties of security and decentralization. Other concerns are mostly related to user privacy and anonymity, along with the governance of blockchain systems that are often going against conventional practices adopted by industry and governments.

These are only some of the most important challenges of the blockchain technology that has increased the risk of applying it in the future modern MCENs. All of the aforementioned items inspire the fact that the blockchain technology needs to be developed in a way to effectively fill the mentioned gaps to be eligible enough for widespread presence in the future modern MCENs structure. Therefore, although blockchain technology can provide appropriate conditions for securely developing the decentralization in the system and improve the P2P energy trading mechanisms, especially for the renewable-based systems, it needs to be better refined to fit with the future modern MCENs, particularly in terms of security, scalability, economically, privacy, and timely aspects.

5.7 Bilateral Energy Trading

Energy networks mainly comprise the energy generation, transmission, and distribution companies that are united with other market entities and independent system operator (ISO) for optimally operating of the system [33]. Generation companies (GENCOs) are responsible for selling energy to the distribution companies (DISCOs) via negotiations or auctions under the supervision of the market operator in a pool market. In the energy networks, the DISCOs typically own the transmission lines and are responsible for maintaining them and occurring any interruptions and fault in the system. The pattern of energy consumption is commonly monitored in the system to track and forecast energy demand due to dynamically dispatching it during a day. One of the active energy networks' entities is the retailer who is responsible for buying energy from DISCOs at wholesale prices and then selling it to the end consumers at the time of use rates [34] or fixed prices [35]. Here, a centralized system is developed as the energy trading via the pool market in which the energy producers play the main role in determining the pool energy price and are usually operated on an hourly basis [36]. In addition, the pool market upgrades the energy trading scope by enabling those groups of energy generators to participate in the energy interactions through selling their energy to specific customers that they had not been able to do before. On the contrary, bilateral energy contracts are suggested as a decentralized model to establish a more flexible platform for energy trading between the energy generators and consumers that realize the energy exchanging up to 15 years. The effective energy control and management of the hybrid energy networks with increasingly growing energy demand and stochastic producers is a vital issue that can be realized by developing the energy trading markets. In this context, the bilateral energy trading mechanism can allow energy consumers and producers to effectively participate in transparent energy interactions. Therefore, it can be effectively applied to the new structure of the future modern MCENs for improving the energy exchanging platforms.

5.8 Game Theoretic-Based Energy Trading

In the hybrid energy networks with a high penetration of RERs, the deployment of the different energy trading platforms is inevitable. This issue has prompted the research area in the context of energy to find appropriate solutions for expanding energy sharing among various energy market players. Thereby, diverse forms of energy trading platforms have appeared to make the energy sector more flexible and sustainable. In this regard, the state of art game theoretic-based models is developed to enable participants in the energy trading market to share energy with each other by pursuing special goals. For example, a multileader multifollower game allows some players to act as the leaders for deciding about the amount of energy traded in the system. On the other hand, some other players work as a follower in deciding about the price per unit energy [37]. Therefore, the game-theoretic energy trading models are suitable for improving the quality of the multi-carrier energy trading environment in terms of the competition, economic, and reliability in the future modern MCENs.

5.9 Bargaining-Based Cooperative Energy Trading

For the cooperative interaction problems, the Nash bargaining theory is proposed as one branch of the cooperative game applicable to the energy trading models [38]. This theory is extensively used for the communication networks [39] with the aim of improving the social optimal and Pareto-optima of the interactive system. The Nash bargaining theory is recently applied in the energy engineering filed for improving the performance of the group interaction [40]. The Nash bargaining cooperative game is structured in a way to yield a fair outcome and Pareto-efficient as well as improve the economic benefits of players in comparison with the Stackelberg game or frequently used noncooperative game [38]. Indeed, the Nash bargaining-based framework can provide suitable conditions for achieving the socially optimal solution without jeopardizing the desired objectives of the players in a noncooperative state [38]. Thereby, it provides a proper opportunity for self-interested entities to cooperate in the energy trading market at the social optimum. As a special result, developing energy trading mechanisms based on the bargaining-based cooperative model enables the DISCOs and demand response aggregators to collaboratively make optimal decisions about the amounts of energy sharing and the associated payments. The cooperative interaction upgrades the capability of the system in reducing the aggregated peak demand and substantially increasing the economic benefits that are desired for both DISCOs and demand response aggregators. Therefore, a bargaining-based cooperative energy trading mechanism can be implemented as significant energy exchanging model with substantial achievements in the future modern MCENs aiming to improve the hybrid energy network's condition both economically and technically.

5.10 Bidirectional Energy Trading

As the hybrid energy networks underwent extensive changes over the recent years, the end-user part of the system has become a more active sector in the energy networks interactions. In this regard, smart buildings and smart homes have commenced a new round by moving from the passive elements toward the active systems in the prosumer-side. Given smart buildings and homes are targeted to be equipped with a high level of RERs, creating the possibility of energy trading between them is essential for improving the flexibility of the demand-side and subsequently overall energy network. The bidirectional energy trading mechanism is introduced to enable the customer-side elements to exit a unidirectional-based system and become active players in the energy market interactions [41]. This energy trading mechanism allows smart buildings/homes to trade energy with each other and with the main grid in the bidirectional platform of the hybrid energy networks [42]. By implementing this mechanism, end-users can not only involve in the energy networks interactions and maximize their economic benefits but also this mechanism significantly supports the system by improving the flexibility and reliability of the energy supply in the presence of a high level of RERs. Therefore, employing the bidirectional energy trading mechanism can make the prosumer-side of the future modern MCENs more flexible than before by giving the opportunity of energy trading for the smart buildings/homes in the vast area.

5.11 Cross-Border Energy Trading

Generally, modern energy networks' idea has accelerated the promotion of the transition necessity from the islanding systems to hybrid interconnected energy networks. The community-based energy networks provide the possibility of implementing the network-to-network support in the energy world. The idea of network-to-network support enables the multi-carrier energy networks to effectively connect to each other and interact with each other to build a sustainable energy environment with the aim of maintaining the reliability, resiliency, and stability of each of them. Here, adopting appropriate optimization strategies is essential for reaching the optimal situations in the complex structure of the hybrid energy networks. Cross-border energy trading is one of the effective ways for optimizing energy resources that is developed based on the energy sharing protocols in the interconnected energy networks [43]. The stochastic clean energy production units are scheduled to be a main part of the energy generation sector in various energy networks that yield a high level of fluctuations in the energy production process. In this condition, the probability of facing the renewable-based system with the imbalances between the scheduled energy production and its real-time production is a very high and undeniable issue. For these high-reality scenarios, the future modern MCENs need to design strong strategies and solutions for ensuring the continuous energy supply during a day. Herein, the cross-border energy trading mechanism can extend the energy trading among the multi-carrier energy grids aiming to improve the reliability, resiliency, flexibility, and stability of the energy network.

5.12 Emission-Aware Energy Trading

Nowadays, the advent of innovative technologies along with the developments in the energy systems has faced the energy industry with revolutionary changes. In this regard, almost all of the energy market players and decision-makers are united with each other in emphasizing on high adoption of the pollutant-free and cost-effective energy production units due to negative environmental impacts of fossil fuel-based units. These efforts are largely made at pursuing specific goals including mitigation of certain risks, reduction of emissions, operating cost reduction, and profit maximization either individually or collectively [44]. In order to alleviate the greenhouse gas emissions produced by energy sector, the European Union is proposed an innovative emission trading mechanism and implemented it by developing an emissions trading scheme [44]. This trading mechanism is a generalized scheme that is targeted to involve all industries like the GENCOs [45] in the different environmentally friendly plans such as decarbonization. The emissions trading mechanism is a comprehensive scheme that limits industries' destructive environmental activities by allowing them to generate only a specified amount of emissions [44]. This scheme enables each industry to trade its emission allowance percentage for maximizing its economic benefits when its emission generation is less than the emission limit assigned to it. This is while each industry will suffer heavy penalties or will be prompted to buy more allowance from the market if the amount of its emission production exceeds the allowable limit. Therefore, the emission-aware energy trading mechanism can be extensively used in the future modern MCENs infrastructure with the aim of supporting the different decarbonization activities.

5.13 Event-Driven Energy Trading Mechanism

An event-driven energy trading system is developed to facilitate the participation of the microgrids in the energy trading market either as a provider or a consumer based on their energy production or consumption [46]. According to this energy trading mechanism, energy exchanging is immediately conducted when a consumer requests a certain amount of energy from the energy trading market [46]. The consumer introduces a special amount of reward and determines the amount of posted reward for procuring energy depending on its energy consumption [46]. On the other side of the energy trading market, providers are notified about this consumer's request and react to this proposed reward by submitting their energy bid. Eventually, the eligible providers are considered for delivering the required energy of the consumer in return for the posted reward. In other words, in the event-driven energy trading system, a consumer triggers energy sharing to the distributor by determining aperiodic energy exchanging epoch when its energy falls below a certain level. One of the distinguishing features of this energy trading mechanism is that no expected waiting time for energy sharing exists under this approach [46]. This is because each consumer can have an energy exchanging interaction whenever its energy for consumption is insufficient. Hence, the event-driven energy trading mechanism is a promising way for the diverse structures of the hybrid energy networks like microgrids to immediately start energy exchanging in the energy trading market with the aim of dynamically establishing energy balance in the future modern MCENs with a full share of RERs.

5.14 Auction Mechanisms in Energy Trading

Auction mechanisms are one of the transparent ways for developing energy trading models that enable the district manager to sell multi-carrier energy to the users and/or trade energy with outer energy networks [47]. These mechanisms can be designed based on the various energy markets' conditions such as day-ahead and real-time energy markets. Furthermore, they can also be structured with the aim of pursuing some specific objectives such as social welfare maximization and economic efficiency improvement. Such special features of the auction mechanisms incentivize users to effectively participate in the auctions with fairness and make the energy trading market more active than ever before. In other words, the implementation of the auction mechanisms for establishing energy trading possibility in the system can maximize the economic benefits for the market participants and bring technical benefits through guaranteeing the optimality of energy allocation accompanying with satisfying all physical system constraints. Moreover, it motivates users to participate in the energy market interactions by guaranteeing this issue that no one cheats in auctions and they never have any monetary loss [47]. Thus, auctions mechanisms are satisfactory tools for developing the energy trading models and make the widespread participation of the users in the energy market interactions possible by reaching the significant economic and technical benefits for both users and energy networks. Thereby, they can be effectively used for enhancing the quality of the interoperability in the hybrid structure of the future modern MCENs.

5.15 Machine Learning in Energy Trading: Reinforcement Learning Mechanism

In recent years, the smart grid has become a common word in a set of efforts for making the energy networks more autonomous, reliable, and sustainable. A significant attribute of the smart grids is to optimally manage the energy supply- and demand-side that regularly face technical challenges like dynamic operation, adaptive decision-making, interactions, and pricing [48]. All of such challenges are ripe to be effectively addressed by employing game theory-based techniques considering the need for the combination of the energy market players with the decision-making and pricing [48]. In this regard, designing the energy management scheme by implementing game equilibria technique as well as developing energy trading games have attracted extensive attention as a central problem [49]. Another problem is to enable the different energy market players to obtain a certain desired game outcome using various learning algorithms that are needed to be significantly developed. Over recent years, there has insured several research endeavors in developing learning techniques to seek game equilibria, which one of the effective of them is the reinforcement learning algorithm. This algorithm can be used for different objectives such as estimation of pay off under incomplete information and computation of game equilibria [48]. In other words, this algorithm can be effectively applied for developing energy trading frameworks based on the game theory when adequate information is not available in the system. Therefore, employing the reinforcement learning-based energy trading mechanism allows the future modern MCENs to seek the Nash equilibria for accommodating the variety of situations in the presence of full RERs. The mathematical modeling of this algorithm can be accessed in [48].

5.16 Zero-Energy Trading Mechanism

As the advantages of clean energy production units have become obvious particularly in terms of economic and environmental aspects, their popularity is substantially enhanced day by day that has made them one of the main and key players in special movements such as decarbonization. This issue has been led to the significant proliferation of mushrooming them throughout the energy networks with the aim of maximum usage of the environmental potential for clean energy production. Such evolutions have been strengthened to more support the efforts for equipping energy networks with high or full level of RERs. In this condition, although several faults may be occurred during the operation of the system and led to various interruptions, intermittency in the RERs outputs is another prominent concern that can upset the energy balance in the system and cause a break in the system. For example, a severe windstorm can be led to stop the wind generation in the system. In such contingency conditions, different structures of the energy network such as microgrids attempt to minimize the consequence of the extreme events through rescheduling of their resources for meeting energy demand. In renewable-based systems, energy deficiency can occur several times during a day that needs effective actions to be properly addressed and prevent the system from load shedding. In this regard, microgrids faced with such events can utilize the capabilities of other microgrids for effectively accommodating the energy deficiency. Herein, the zero-energy trading mechanism is proposed to make the utilization of the capabilities of some microgrids by other ones possible in the energy networks [50]. The zero-energy trading mechanism is an efficient strategy that develops a cooperative energy trading environment among the microgrids for dynamic balancing energy in the deregulated environment. By assuming fixed energy generation costs for each microgrid during all time intervals in this strategy, the surplus energy production of the healthy microgrids is exported to the faulty ones and the same amount of energy is targeted to be imported to the healthy ones in the critical time intervals. Therefore, applying this strategy can effectively develop the local energy trading between the renewable-based microgrids by providing suitable conditions for using the capabilities of microgrids aiming to create a dynamic energy balance in the system. Consequently, a zero-energy trading strategy can be employed in the future modern MCENs with the aim of maintaining the system reliability and resiliency as well as providing significant economic benefits for the market participants in a sustainable manner.

5.17 Transactive Energy

As the demand for multi-carrier energy is drastically growing in subsequent to the world's population growth, sustainability has become a common word in the ongoing activities of the research area in the field of energy. In this condition, developing the energy networks for unremitting energy generation is the necessary step in reliable energy supply for the consumers. The current energy networks mostly rely on conventional energy generation units for accommodating the need for different dimensions of energy. Because of undeniable concerns of the traditional systems both environmentally and economically, the transition from the fossil fuel-based system to the pollutant-free energy networks has become a great need of today's world. The promotion of the operation of DERs in the high level has been typically approved as a satisfactory solution to solve the aforementioned challenges. In this respect, the remarkable advantages of clean energy production units have driven the development trends of the energy networks toward the 100% RERs in the future modern grids. This means that, most likely, we will witness the modern energy landscape with full RERs in the near future. Despite the decentralized energy production units particularly renewable ones promise significant technically, economically, and environmentally benefits, but all together will pose substantial challenges regarding the optimal management of the system in terms of social, technological, and policy-related aspects [51]. Therefore, in order to harness the great potential of the energy as a key player in the most effective transitions toward sustainability, the current infrastructure of energy networks requires to be rigorously re-engineered into an interconnected and smart system that is most suitable for addressing the great societal challenges of our time in accessing efficient, clean, and secure multi-carrier energy. In this light, transactive energy technology has emerged as a promising way of orchestrating the coordinated exploitation of intelligent devices [52]. In other words, transactive energy is an innovative vision of sophisticated technologies that bring professional techniques and implement reliable strategies for achieving sustainable modern grids.

5.17.1 *Transactive Energy Concept and Definition*

As the idea of modern grids with 100% RERs is appeared along with the increment in the pressure of the demand for secure, reliable, and clean multi-carrier energy, transactive energy has become an essential tool for developing sustainable modern MCENs. In other words, transactive energy has emerged as a strong contender for solving the challenges ahead of renewable-based energy networks in a reliable and sustainable manner. The GridWise Architecture Council (GWAC) defines transactive energy as "a set of economic and control mechanisms that allows the dynamic balance of supply and demand across the entire electrical infrastructure using value as a key operational parameter [53]." Four key parts can be identified in this definition as below.

- The first is "a set of economic and control mechanisms," which means that transactive energy not only provides economic benefits for the market's participants but also adopts appropriate mechanisms for controlling the system in a reliable manner.
- The second is "allows the dynamic balance of supply and demand" that refers to establish a dynamic energy balance in the system.
- The third is "across the entire electrical infrastructure," which declares this technology engages the entire electricity system from the transmission level to the end-user's part.
- The fourth is "using value as a key operational parameter," which identifies this technology enables the system for making operational decisions based on exchanging value-based information captured in transactions among transactive energy market participants.

The GWAC was given a mission from the U.S. Department of Energy (DOE) to develop transactive energy technology with the aim of properly addressing the challenges of the energy networks with ever-increasing the renewable systems. In response to this need, the GWAC held several workshops and gather the experts to finalize some prominent achievements for altering the energy landscape by employing transactive energy technology. Thereby, a series of efforts were made to transform transitive energy technology into a paradigm that is supposed to play a central role in the energy market interactions. Some of the substantial features of the transactive energy technology that made it an essential need for the system are listed in the following [52].

- It sets up a new system that allows for integrating all energy suppliers, buyers, and consumers in a unique environment to promote interoperability using sophisticated information and communication technologies.
- It is a market-based technology that designs a new architecture of energy with widespread synergies across all platform providers and market participants.
- It develops a transparent energy market that welcomes all participants by eliminating all intermediaries.
- It builds an energy environment that provides personal services for end-users on the prosumer-side.
- It enables the system for incorporating different capable technologies for making hybrid energy networks more sustainable, flexible, reliable, secure, resilient, and stable than ever before.

5.17.2 Transactive Energy Missions and Principles

Nowadays, there is a general belief that the integration of information and communication technology with smart systems can be useful for e-commercing with distributed intelligent devices and networks while makes tomorrow's energy infrastructure more secure and efficient. Herein, a great need is felt to integrate technologies and intelligent systems in the smartization activities since their lack can easily postpone the realization of the modern grids. In this regard, transactive energy technology is introduced to develop automated energy architectures to transform energy system by focusing on completely intending the new requirements of the future modern MCENs. Due to this, DOE has supported the GWAC to extend the scope of transactive energy-centric activities as follows [54]:

- Energy production and delivery sector
- Building automation
- Industrial systems control
- Economic and regulatory policy
- Information technology and telecommunications

In a community of practice, transactive energy needs to develop a holistic framework for effectively facing with the challenges of modern grids. Therefore, transactive energy systems are constructed with the aim of facilitating the implementation of the transactive energy goals in the modern area of energy networks. Thereby, the transactive energy systems are incorporated in the transactive energy framework with special features as shown in Fig. 5.5 [52].

One of the main missions of the GWAC is to significantly improve the intelligent systems' performance in the transactive energy market. To this end, the GWAC has defined some special principles for the transactive energy as below [52, 54].

- It needs to develop a self-optimization system that allows for the highly coordinating of the system.
- It needs to become auditable and visible at the interface.
- It needs to design a fair energy environment for the participation of eligible players in the energy market interactions.
- It needs to construct the energy trading market in a way to be scalable, adaptable, extensible, and sustainable in all parts of the system.
- All participants in the transactive energy market need to act based on the standards set in the system.
- It needs to adopt significant strategies for making hybrid energy networks more reliable, flexible, stable, and resilient than ever before.

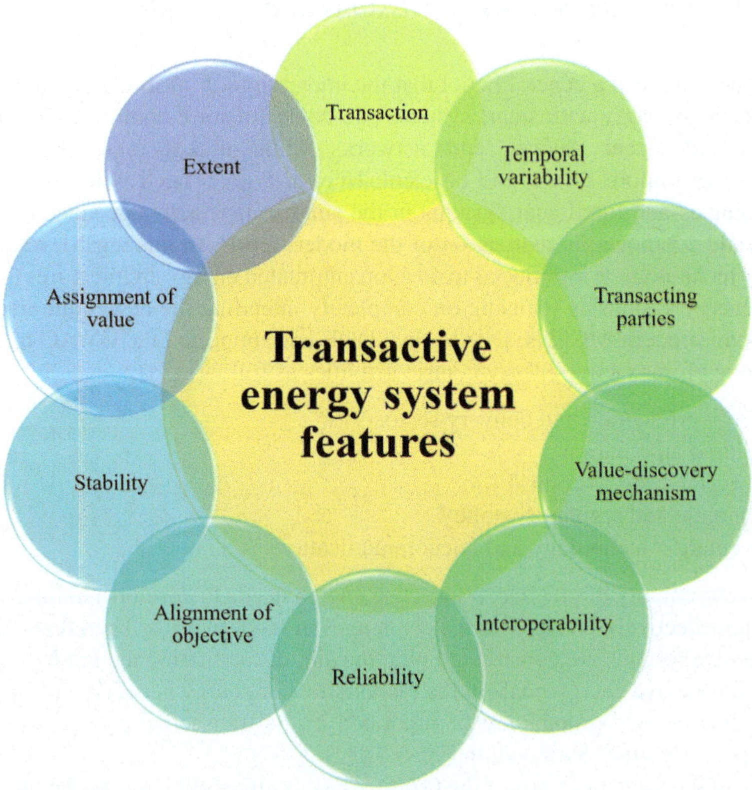

Fig. 5.5 Some special features of the transactive energy systems [52]

5.17.3 A New Definition of Transactive Energy Technology for the Future Modern MCENs

As mentioned earlier, the GWAC defines transactive energy as “a set of economic and control mechanisms that allows the dynamic balance of supply and demand across the entire electrical infrastructure using value as a key operational parameter [53].” Given this definition, four key parts and questionable items can be identified that are given as:

- First, “a set of economic and control mechanisms” that only limits transactive energy to the economic and control mechanisms.
- Second, “allows the dynamic balance of supply and demand” that lacks information about the structure of the system to make a dynamic energy balance.
- Third, “the entire electrical infrastructure” that only limits transactive energy interactions for electrical energy and encourages the one-dimension energy trading market.

- Fourth, “using value as a key operational parameter” that provides incomplete information as the value and key operational parameters remained vague.

The aforementioned shortages in the transactive energy definition have limited the area of its usage in the energy networks. Furthermore, the aforementioned definition cannot completely reflect all features and capabilities of the transactive energy technology. This is while the application of transactive energy is far beyond the current descriptions of this technology. The future MCENs is targeted to be modern and adopt sophisticated technologies and 100% RERs for clean energy generation. In such an environment, transactive energy technology can be widely employed to create a dynamic energy balance by developing appropriate strategies. In order to effectively use the potential of the transactive energy technology in the future modern MCENs, we need to develop its definition in accordance with its possible applications in the future modern MCENs. Hence, for the future modern MCENs, the new version of transactive energy technology as the “transactive multi-carrier energy (TMCE) technology”, which is *a set of economic, social, environmental, and control mechanisms that is responsible to establish a dynamic balance for multi-dimensional energy across the entire multi-carrier energy infrastructure with a full/high share of clean energy resources using economical, technical, social, and environmental targets as the key operational parameters for making the hybrid energy networks more reliable, resilient, flexible, efficient, sustainable, and stable than ever before* can be defined.

The new definition for the TMCE technology consists of the five key parts that are as follows:

- The first part is “*a set of economic, social, environmental, and control mechanisms,*” which refers that transactive multi-carrier energy develops systematic plans in a way to effectively improves the economic benefits of the participants in the transactive multi-carrier energy environment, adopts suitable strategies to substantially enhance social welfare in the modern energy area, develops the special local multi-carrier energy trading market to deal with the uncertain behaviors of the stochastic producers for establishing a dynamic energy balance, and reliably controls the systems for ensuring the system resiliency, flexibility, stability, and sustainability in the high penetration of RERs.
- The second part is “*establish a dynamic balance for multidimensional energy*” that indicates the main goal of the transactive multi-carrier energy technology to establish a dynamic balance in the multi-carrier energy system instead of focusing only on the one-dimension of energy.
- The third part is “*across the entire multi-carrier energy infrastructure with a full/ high share of clean energy resources*” indicating that the application scope of the transactive multi-carrier energy technology not only is limited to a special energy network but also can effectively support the energy interactions simultaneously in the multi-carrier energy networks. Moreover, this technology can be used as a promising way for providing a reliable energy supply in the systems with a high or full share of RERs.

- The fourth part is "*using economical, technical, social, and environmental targets as the key operational parameters*" describing the objectives that can be considered in the transactive multi-carrier energy system. The main and overall scope of objectives in the transactive multi-carrier energy system is in the area of economical, technical, social, and environmental, but they are not limited to them and other goals can also be intended in line with the future modern MCENs.
- The fifth part is "*making the hybrid energy networks more reliable, resilient, flexible, efficient, sustainable, and stable than ever before*" that indicates the significant effects of the transactive multi-carrier energy technology on the future modern MCENs.

Given the aforementioned details regarding the TMCE technology, it can be broadly intended in the future modern MCENs landscape with the aim of effectively developing sustainable hybrid energy networks. Therefore, the widespread presence of TMCE technology is highly expected in the energy interactions of the future modern MCENs.

5.18 Grid Modernization (GM)-Based Modern Multi-Carrier Energy Networks

Over recent decades, as sophisticated technologies have appeared along with the advanced intelligent systems, the addressing of the ever-increasing concerns over carbon emission and energy deficiency has become a central topic. For this aim, accelerating the transition from the conventional energy grids with centralized energy systems to the modern energy networks with decentralized energy production units is recognized as one of the effective steps in decarbonization plans and overcoming the energy crisis. In this regard, emerging the different hybrid systems indicates the key fact of the interdependency of the energy networks with each other. This is while their proliferating throughout the energy networks highlights this fact that the dependently analyzing of the energy networks is inevitable for achieving the near results due to their direct impact on each other's performance. This issue has driven the research trends to mostly focus on multi-carrier energy networks rather than scrutinizing the special type of energy network independently. On the other hand, due to advances in the technologies of smart devices, developments in the information and communication technologies, emerging new version of the hybrid energy systems, ever-increasing the penetration of the DERs especially renewable types of them, the modernization of the MCENs has become a common effort in the energy sector. Indeed, grid modernization is known as the necessary step for ensuring from continuous energy supply in the system with 100% RERs, providing an acceptable level of safety and privacy for different users, increasing the quality of the energy networks infrastructure to make the hybrid energy networks enough reliable, flexible, resilient, stable, secure, adaptable, scalable, and sustainable. Therefore, the research efforts in the field of energy are rapidly growing for realizing the future modern MCENs. Since this trend has not become mature enough at this moment, there are not clear and comprehensive models for it. This is while the success of a special trend and objective directly depends on defining the transparent roadmap as well as enacting new appropriate guidelines. Despite proposing new models for modernizing the future MCENs is necessary, but a holistic model is not proposed for this aim yet. The lack of such models in the grid modernization process has prompted us to use our best of knowledge and experience to propose a holistic model for the future modern MCENs that can be helpful for directing the research trends in this filed. Therefore, the new model is proposed for the future modern MCENs with the aim of facilitating the modernization process of hybrid energy networks. This model is proposed in this book for the first time by authors, and it is named the GM (Grid Modernization) model. This model is structured to first determine the targeted objectives in the grid modernization process and then enact some necessary guidelines for the future modern MCENs. In fact, these guidelines open up the role of side guards for the grid modernization route so that their adherence helps the set of efforts not to deviate from the main path of modernizing future MCENs. Because of the ever-developing technologies in all fields of energy, so completing the previously proposed models will be essential to effectively cover the new structure of the modern grids. Due to this, the characteristics of the first version of the GM model are described for the future modern MCENs that is called a GM-V1 model. The next version of the GM model will be proposed in accordance with the advancements in the field of grid modernization. According to the GM-V1 model for the future modern MCENs, the objectives of the grid modernization at first identified and then the required guidelines for reaching the determined objectives are discussed. Therefore, the objectives, guidelines, and standards are respectively described in the following subsection based on the GM-V1 model for the future modern MCENs.

5.18.1 GM-V1 Model Goals for the Future Modern Multi-Carrier Energy Networks

As a rule, each path is created to reach a specific destination and goals. If destination and objectives are not set correctly, it will not be possible to follow the right path with a set of efforts to achieve the targeted goals. This rule is true not only for every path but also for energy schemes. The main goals of modernizing the future MCENs based on the GM-V1 model are described below. Indeed, according to the GM-V1 model, the future modern MCENs needs to:

1. Equip with 100% RERs as the main source of energy generation for producing fully clean energy in the system.
2. Supply cost-effective and environmentally friendly energy to all consumers in a reliable manner.

3. Increase customer choices, improve their convenience by adopting smart devices, provide suitable conditions for their participation in the MCENs' interactions.
4. Build a sustainable and flexible energy environment for effectively integrating the high level of clean energy production units.
5. Demonstrate and develop novel control systems aiming to improve system observability and control it at a very high-resolution level.
6. Utilize the information, measurement, and sensing technologies for optimal operation and planning of the modern system for improving the grid reliability, resiliency, efficiency, and stability as well as reducing outages.
7. Improve the physical security of energy production, the privacy of all participants, and other grid-related assets.
8. Adopt the expert controlling technologies for monitoring the overall MCENs interactions.
9. Develop the system platform for flexible adaptation to the emerging cycles such as food–energy–water nexus.
10. Implement automated maintenance and operation systems as well as the self-healing paradigm.
11. Develop a comprehensive architecture with several control centers aiming to simultaneously monitor the status of the electric power system, natural gas network, the district heating network (DHN), and water distribution system to adopt the optimal strategies.
12. Develop a holistic reconfigurable system capable of dealing with any changes in modern MCENs.
13. Only adopt the large number of those self-action systems that their selfing feature is brought through analyzing the deterministic data not those ones with stochastic resources.
14. Develop a multi-layer energy system aiming for layer-to-layer controlling of the system to cover all sensitive layers of the hybrid energy networks such as physical, cyber, and social.
15. Develop a decentralized sense-to-sense reactive system for enabling the modern network to be a highly fast response system in facing numerous changes at each moment.
16. Develop a widespread level of the multiple-energy process instead of the dual-energy process.

5.18.2 GM-V1 Model Guidelines for the Future Modern Multi-Carrier Energy Networks

For facilitating the path of reaching to the described objectives in the previous subsection, defining several effective guidelines would be very efficient. Therefore, the different groups of systems, technologies, models, paradigms, architectures, patents, and other schemes that we call their collection as the modernization movement in this book, which are be ranked at different levels of the GM-V1 model, depending on how many of them satisfy the GM-V1 model guidelines. The GM-V1 model is classified into three levels of 1–3. The schematic for this leveling is depicted in Fig. 5.6. Accordingly, if any modernization movement complies with only between 1 and 35% of the GM-V1 model guidelines, it is classified at the level 1, meaning that it has performed the weakest performance in modernizing MCENs. Modernization movements that meet between 36 and 80% of the proposed model guidelines are classified as level 2, indicating their moderate performance in realizing MCENs modernization. Finally, those modernization movements that can comply with 81–100% of the guidelines are at level 1 of the model, which is the best way to modernize MCENs. Each modernization movement at different levels can be upgraded to higher levels by improving its conditions and following more guidelines. Leveling the performance of each modernization movement in the path of MCENs modernization not only identifies the important role of each of them in this process but also can be an appropriate way to consider a variety of incentives for manufacturers, developers, or other relevant sectors of the modernization movements.

Each modernization movement should satisfy at least one of the following GM-V1 guidelines to be considered in the path of modernizing the future MCENs. It should be noted that satisfying one feature among the several features in each of the following items indicates that the related modernization movement satisfies that item.

1. Contains cost-effective and environmentally friendly features in its application throughout the modern network
2. Increases the reliability of multi-carrier energy supply in the complex hybrid energy networks
3. Improves flexibility, resiliency, stability, safety, sustainability, security, adaptability, scalability, auditable, and extensibility of the hybrid energy networks
4. Facilitates the adoption of the different innovative technologies in the system

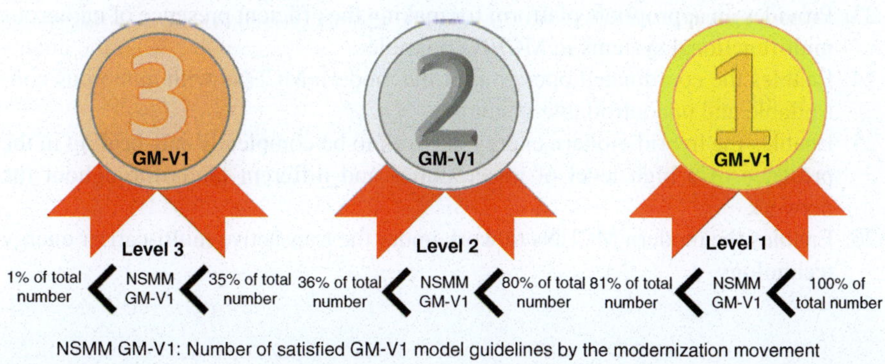

Fig. 5.6 Different levels of the GM-V1 model in the modernization process of future MCENs

5. Increases the interactions between different energy networks
6. Reduces the energy losses throughout the hybrid network
7. Increases the interoperability and synergies among all players of the energy networks
8. Improves the multi-carrier energy quality in both supply- and prosumer-sides
9. Increases customer choices, improves their convenience, and provides suitable conditions for their participation in the MCENs' interactions
10. Improves network's conditions for optimal integrating decentralized energy systems
11. Facilitates development of hybrid energy networks with 100% RERs
12. Increases the energy control and management systems with the aim of improving the observability of the system and controlling it at a very high resolution
13. Increases the capability of the system for big data processing in the system with a large number of intelligent devices
14. Allows for jointing each smart system into the energy network's interactions
15. Improves the status of physical, cyber, social, and other layers of modern networks
16. Increases the capability of the system to fast monitoring of the system
17. Enables the system platform for adopting different effective emerging cycles such as food–energy–water nexus
18. Enables modern energy infrastructure for implementing a fully automated system
19. Enables the MCENs structure to develop a holistic architecture for simultaneously controlling several energy networks
20. Enables the MCENs infrastructure to be a fast reconfigurable system in facing with various events
21. Enables the modern MCENs to rapidly foster and efficiently recover the large disruptions
22. Enables the modern MCENs to effectively handle the complexity of the large-scale hybrid system, the computational burden of the expert computing and control systems, and different kinds of computational language related to the various intelligent systems
23. Provides an appropriate platform for making the efficient presence of numerous multifunctional systems in MCENs possible
24. Enables the coordinated operation of the modern MCENs with numerous controllable and uncontrollable systems
25. Enables the hybrid modern energy network to be completely self-healing in the presence of a high level of uncertainties and different faults throughout the network
26. Enables the modern MCENs to work under the transactive multi-carrier energy technology

5.19 Summary

Nowadays, increasing the tendency of research communities to the modern version of the hybrid energy networks has prompted us to effectively evaluate one of the critical issues of the future modern MCENs and address its challenges as much as possible in this chapter. This key issue is an energy trading possibility that is recognized as one of the essential tools for making renewable-based energy networks more sustainable than ever before. Indeed, creating an appropriate energy trading environment is necessary to reliably support the modern MCENs with 100% RERs and to ultimately provide continuous energy supply during a day. Therefore, the benefits of energy trading and its potential are deeply investigated for the future modern MCENs. The different energy exchanging techniques are discussed and their usage chance in the future modern MCENs interactions is also assessed from various aspects. Afterward, the capability of transactive energy technology for creating a dynamic energy balance in modern energy networks is examined by focusing on its features, missions, and principles. In order to effectively address the shortages of transactive energy technology, the new version of it is proposed, which is called the transactive multi-carrier energy technology. The new definition is proposed for the transactive multi-carrier energy technology and its potential is scrutinized for widely present in the future modern MCENs interactions as the sustainable and reliable technology for establishing a dynamic energy balance in the system with a full share of RERs. Finally, the new GM-V1 model is proposed for the future modern MCENs with the aim of facilitating the success of the grid modernization process by providing transparent objectives and guidelines for the future modern MCENs.

5.20 Questions

After careful reading and evaluating this chapter, we are expected to be able to answer the following questions:

1. Why creating the energy trading possibility is essential for the renewable-based hybrid energy networks?
2. Why are RERs, energy storage systems, and energy sharing possibility three sides of the triangle needed for future modern MCENs?
3. What are the energy trading benefits for energy networks?
4. Why do we need to create a strong energy trading environment for the future modern MCENs?
5. What is the real-time energy trading mechanism and how it can be used in the future modern MCENs?
6. What is the peer-to-peer energy trading mechanism and how it can be used in the future modern MCENs?
7. What are the features of the peer-to-peer energy trading platform?

8. What is the blockchain technology and for which goals it can be used in energy networks?
9. What are the blockchain definition and fundamental principles in the energy industry?
10. Why Bitcoin and Ethereum are taken into account as two important paradigms of blockchain technology?
11. What are the blockchain potential impacts on the energy sector?
12. How blockchain technology can develop an energy trading possibility in the future modern MCENs?
13. How edge computing-based energy trading can be used in the energy network interactions?
14. What are the blockchain challenges in the energy sector?
15. What is the bilateral energy trading mechanism and how it can be used in the future modern MCENs?
16. What is the game theoretic-based energy trading mechanism and how it can be used in the future modern MCENs?
17. What is the bargaining-based cooperative energy trading mechanism and how it can be used in the future modern MCENs?
18. What is the bidirectional energy trading mechanism and how it can be used in the future modern MCENs?
19. What is the cross-border energy trading mechanism and how it can be used in the future modern MCENs?
20. What is the emission-aware energy trading mechanism and how it can be used in the future modern MCENs?
21. What is the event-driven energy trading mechanism and how it can be used in the future modern MCENs?
22. What are the auction mechanisms in energy trading and how they can be used in the future modern MCENs?
23. How the machine learning techniques can be used for developing energy trading possibility in the future modern MCENs?
24. What is the reinforcement learning mechanism and how it can be used in the future modern MCENs?
25. What is the zero-energy trading mechanism and how it can be used in the future modern MCENs?
26. What are the transactive energy technology concept and definition?
27. What are the distinguishing features of the transactive energy technology in the future modern MCENs?
28. What are the transactive energy missions and principles in hybrid energy networks?
29. What are the special features of the transactive energy systems?
30. What is the transactive multi-carrier energy technology and why it is proposed for the future modern MCENs?
31. What are the key parts of the transactive multi-carrier energy technology and how they can support the special parts of the future modern MCENs?

32. What is the GM-based modern multi-carrier energy network and for which motivations is proposed?
33. Why the GM-V1 model is essential for the modern structure of the future MCENs?
34. What are the goals of the GM-V1 model in the future modern MCENs?
35. What are the guidelines of the GM-V1 model in the future modern MCENs?

5.21 Suggestions

In this section, some suggestions are provided related to this chapter that can give a good overview of some substantial issues in the future modern MCENs.

1. Creating an energy trading possibility is useful for reliable supporting the continuous energy supply in the future modern MCENs with 100% RERs.
2. Developing a strong energy trading environment is essential for establishing an effective multi-carrier energy sharing in the future modern MCENs.
3. Developing various energy trading mechanisms can significantly increase the flexibility of the future modern MCENs for energy management and control.
4. Employing transactive multi-carrier energy technology is essential for developing the local multi-carrier energy trading environment aiming to establish a dynamic energy balance in the future modern MCENs with a full share of stochastic producers.
5. The operation of the future modern MCENs based on the proposed GM-V1 model can be an effective way of accelerating the transition from the traditional systems to the modern hybrid energy networks.

5.22 Future Trends and Discussion Topics

Here, future trends and some discussion topics are proposed for the interested readers regarding this chapter that can be useful for developing the MCENs modernization process.

1. Developing new energy trading mechanisms for increasing the capability of the future modern MCENs in energy management and control
2. Upgrading the energy control systems for developing a strong energy trading environment in the renewable-based modern MCENs
3. Improving the MCENs infrastructure for easily adopting various energy trading mechanisms
4. Developing effective energy sharing environments for facilitating the management of energy interactions in hybrid energy networks

References

1. M. Daneshvar, B. Mohammadi-ivatloo, S. Asadi, K. Zare, A. Anvari-Moghaddam, Optimal day-ahead scheduling of the renewable based energy hubs considering demand side energy management, in *2019 International Conference on Smart Energy Systems and Technologies (SEST)*, (2019), pp. 1–6
2. M. Daneshvar, M. Abapour, B. Mohammadi-ivatloo, S. Asadi, Impact of optimal DG placement and sizing on power reliability and voltage profile of radial distribution networks. Majlesi J. Electr. Eng **13**(2), 91–102 (2019)
3. M. Daneshvar, B.M. Ivatloo, M. Abapour, S. Asadi, R. Khanjani, Distributionally robust chance constrained transactive energy framework for coupled electrical and gas microgrids. IEEE Trans. Ind. Electron., **68**(1), 1–1 (2021)
4. M. Daneshvar, B. Mohammadi-Ivatloo, A. Anvari-Moghaddam, Optimal energy trading strategy for proactive DISCO considering demand response programs in the distribution networks, in *CIGRE Symposium Aalborg 2019*, (CIGRE (International Council on Large Electric Systems), 2019)
5. M. Daneshvar, B. Mohammadi-ivatloo, S. Asadi, M. Abapour, A. Anvari-Moghaddam, A Transactive energy management framework for regional network of microgrids, in *2019 International Conference on Smart Energy Systems and Technologies (SEST)*, (2019), pp. 1–6
6. M. Daneshvar, B. Mohammadi-Ivatloo, M. Abapour, S. Asadi, Energy exchange control in multiple microgrids with transactive energy management. J. Modern Power Syst. Clean Energy, **8**(4), 1–8 (2020)
7. M. Daneshvar, B. Mohammadi-ivatloo, S. Asadi, S. Galvani, Short term optimal hydro-thermal scheduling of the transmission system equipped with pumped storage in the competitive environment. Majlesi J. Electr. Eng **14**(1), 77–84 (2020)
8. M. Daneshvar, B. Mohammadi-Ivatloo, K. Zare, Two-stage optimal robust scheduling of hybrid energy system considering the demand response programs. J. Clean. Prod. **248**, 119267 (2020)
9. M. Daneshvar et al., The chance-constrained models for transactive energy management of interconnected microgrids clusters. J. Clean. Prod. **271**, 122177 (2020)
10. I.S. Bayram, M.Z. Shakir, M. Abdallah, K. Qaraqe, A survey on energy trading in smart grid, in *2014 IEEE Global Conference on Signal and Information Processing (GlobalSIP)*, (IEEE, 2014), pp. 258–262
11. S. Bahrami, M. H. Amini, A decentralized framework for real-time energy trading in distribution networks with load and generation uncertainty, arXiv preprint arXiv:1705.02575, 2017
12. C. Zhang, J. Wu, Y. Zhou, M. Cheng, C. Long, Peer-to-peer energy trading in a microgrid. Appl. Energy **220**, 1–12 (2018)
13. J. Hamari, M. Sjöklint, A. Ukkonen, The sharing economy: Why people participate in collaborative consumption. J. Assoc. Inf. Sci. Technol. **67**(9), 2047–2059 (2016)
14. C. Eid, P. Codani, Y. Perez, J. Reneses, R. Hakvoort, Managing electric flexibility from distributed energy resources: A review of incentives for market design. Renew. Sust. Energ. Rev. **64**, 237–247 (2016)
15. M. Andoni et al., Blockchain technology in the energy sector: A systematic review of challenges and opportunities. Renew. Sust. Energ. Rev. **100**, 143–174 (2019)
16. U. Ahsan, A. Bais, Distributed big data management in smart grid, in *2017 26th Wireless and Optical Communication Conference (WOCC)*, (IEEE, 2017), pp. 1–6
17. M. Daneshvar, B. Mohammadi-Ivatloo, M. Abapour, The possibility of using blockchain based cryptocurrency in transactive energy markets: Ongoing activities and opportunities ahead, presented at the the fourth international energy management and technology conference, Tehran, 2018. Available: https://www.civilica.com/Paper-IEAC04-IEAC04_005.html
18. J. Mattila, C. Naucler, R. Stahl, M. Tikkanen, A. BÂdenlid, Industrial blockchain platforms: An exercise in use case development in the energy industry, ETLA Working Papers 2016
19. M. Armbrust et al., A view of cloud computing. Commun. ACM **53**(4), 50–58 (2010)

20. J. Mattila, The blockchain phenomenon–the disruptive potential of distributed consensus architectures, ETLA working papers 2016
21. A. Back et al., Enabling blockchain innovations with pegged sidechains, http://www.opensciencereview.com/papers/123/enablingblockchain-innovations-with-pegged-sidechains, vol. 72, 2014
22. S. Nakamoto, Bitcoin: A peer-to-peer electronic cash system [Electronic resource], Bitcoin [Official website]. https://bitcoin.org/bitcoin.pdf (дата обращения: 22.10. 2017), 2018
23. S. Muftic, I. Sanchez, E. JRC, L. Beslay, Overview and analysis of the concept and applications of virtual currencies, Joint Research Centre, Italy Raspoloživo na: http://publications.jrc.ec.europa.eu/repository/handle/JRC105207, 2016
24. G.P. Dwyer, The economics of Bitcoin and similar private digital currencies. J. Financ. Stab. **17**, 81–91 (2015)
25. S. Nakamoto, Bitcoin: A peer-to-peer electronic cash system, Manubot 2019
26. G. Wood, Ethereum: A secure decentralised generalised transaction ledger. Ethereum Project Yellow Paper **151**(2014), 1–32 (2014)
27. Ethereum Foundation. Ethereum blockchain app platform. (9 June, 2020). Available: https://ethereum.org/
28. C. Burger, A. Kuhlmann, P. Richard, J. Weinmann, *Blockchain in the Energy Transition. A Survey Among Decision-Makers in the German Energy Industry*, vol 60 (DENA German Energy Agency, Berlin, 2016)
29. M. Mylrea, S.N.G. Gourisetti, Blockchain for smart grid resilience: Exchanging distributed energy at speed, scale and security, in *2017 Resilience Week (RWS)*, (IEEE, Piscataway, 2017), pp. 18–23
30. E. Mengelkamp, J. Gärttner, K. Rock, S. Kessler, L. Orsini, C. Weinhardt, Designing microgrid energy markets: A case study: The Brooklyn microgrid. Appl. Energy **210**, 870–880 (2018)
31. M. Mylrea, S.N.G. Gourisetti, Cybersecurity and optimization in smart "autonomous" buildings, in *Autonomy and Artificial Intelligence: A Threat or Savior?* (Springer, Cham, 2017), pp. 263–294
32. Z. Zhou, L. Tan, G. Xu, Blockchain and edge computing based vehicle-to-grid energy trading in energy internet, in *2018 2nd IEEE Conference on Energy Internet and Energy System Integration (EI2)*, (IEEE, 2018), pp. 1–5
33. J.J. Roberts, A.M. Cassula, J.L. Silveira, E. da Costa Bortoni, A.Z. Mendiburu, Robust multi-objective optimization of a renewable based hybrid power system. Appl. Energy **223**, 52–68 (2018)
34. A. Anees, Y.-P.P. Chen, True real time pricing and combined power scheduling of electric appliances in residential energy management system. Appl. Energy **165**, 592–600 (2016)
35. F. Farzan, M.A. Jafari, J. Gong, F. Farzan, A. Stryker, A multi-scale adaptive model of residential energy demand. Appl. Energy **150**, 258–273 (2015)
36. A. Anees, T. Dillon, Y.-P.P. Chen, A novel decision strategy for a bilateral energy contract. Appl. Energy **253**, 113571 (2019)
37. A. Mondal, S. Misra, Game-theoretic energy trading network topology control for electric vehicles in mobile smart grid. IET Netw **4**(4), 220–228 (2015)
38. S. Fan, Q. Ai, L. Piao, Bargaining-based cooperative energy trading for distribution company and demand response. Appl. Energy **226**, 469–482 (2018)
39. C. Yang, J. Li, A. Anpalagan, M. Guizani, Joint power coordination for spectral-and-energy efficiency in heterogeneous small cell networks: A bargaining game-theoretic perspective. IEEE Trans. Wirel. Commun. **15**(2), 1364–1376 (2015)
40. K. Dehghanpour, H. Nehrir, An agent-based hierarchical bargaining framework for power management of multiple cooperative microgrids. IEEE Trans. Smart Grid **10**(1), 514–522 (2017)
41. A. Al Hasib, N. Nikitin, L. Natvig, Load scheduling in smart buildings with bidirectional energy trading, in *2014 IEEE 33rd International Performance Computing and Communications Conference (IPCCC)*, (IEEE, 2014), pp. 1–2

42. B.-G. Kim, S. Ren, M. Van Der Schaar, J.-W. Lee, Bidirectional energy trading and residential load scheduling with electric vehicles in the smart grid. IEEE J. Sel. Areas Commun **31**(7), 1219–1234 (2013)
43. S.K. Zimba, M. Chudý, F. Janíček, Imbalance energy in Southern African power pool, in *2017 IEEE PES PowerAfrica*, (IEEE, Piscataway, 2017), pp. 407–412
44. G.A. Akhtar, A.T. Al-Awami, M.W. Khalid, Coordinating emission-aware energy trading with V2G services, in *Eurocon 2013*, (IEEE, 2013), pp. 1293–1298
45. I. Kockar, A.J. Conejo, J.R. McDonald, Influence of emissions trading scheme on market clearing and prices, in *2009 IEEE Power & Energy Society General Meeting*, (IEEE, 2009), pp. 1–5
46. S. Park, J. Lee, G. Hwang, J.K. Choi, Event-driven energy trading system in microgrids: Aperiodic market model analysis with a game theoretic approach. IEEE Access **5**, 26291–26302 (2017)
47. W. Zhong, K. Xie, Y. Liu, C. Yang, S. Xie, Auction mechanisms for energy trading in multi-energy systems. IEEE Trans. Ind. Inf **14**(4), 1511–1521 (2017)
48. H. Wang, T. Huang, X. Liao, H. Abu-Rub, G. Chen, Reinforcement learning in energy trading game among smart microgrids. IEEE Trans. Ind. Electron. **63**(8), 5109–5119 (2016)
49. J. Lee, J. Guo, J.K. Choi, M. Zukerman, Distributed energy trading in microgrids: A game-theoretic model and its equilibrium analysis. IEEE Trans. Ind. Electron. **62**(6), 3524–3533 (2015)
50. M.M. Arsoon, S.M.M. Tafreshi, Zero energy trading strategy for enhancing resilience of networked microgrids, in *2019 7th International Istanbul Smart Grids and Cities Congress and Fair (ICSG)*, (IEEE, 2019), pp. 86–90
51. J. Qiu, J. Zhao, H. Yang, Z.Y. Dong, Optimal scheduling for prosumers in coupled transactive power and gas systems. IEEE Trans. Power Syst. **33**(2), 1970–1980 (2017)
52. M. Daneshvar, M. Pesaran, B. Mohammadi-ivatloo, Transactive energy in future smart homes, in *The Energy Internet*, (Elsevier, Amsterdam, 2019), pp. 153–179
53. R. Ambrosio, Transactive energy systems. IEEE Electrific. Mag **4**(4), 4–7 (2016)
54. D. Forfia, M. Knight, R. Melton, The view from the top of the mountain: Building a community of practice with the gridwise transactive energy framework. IEEE Power Energy Mag **14**(3), 25–33 (2016)

Chapter 6
Mathematical Modeling and Uncertainty Management of the Modern Multi-Carrier Energy Networks

Abstract In recent years, renewable energy resources (RERs) are taken into account as the popular generation units due to economic advantages as well as trouncing the environmental emissions. Indeed, increasing the level of RERs in the energy generation cycle has made the prediction of the amount of energy production more difficult and intermittent. Therefore, in spite of the existence of a strong motivation for high usage of RERs in the energy generation process, some basic challenges as well as uncertainty modeling of the system should be considered in optimal operation of the renewable-based multi-carrier energy networks (MCENs). After analyzing the MCENs from different aspects in the previous chapters, this chapter is targeted to provide the mathematical models for MCENs and their uncertainty quantification with the aim of facilitating their implementation in real hybrid energy networks. Given the provided mathematical models, the optimal scheduling of the interconnected microgrids with a full share of RERs is considered as a case study for examining a sample MCENs with real-recorded data. Indeed, each microgrid is only equipped with the RERs with the aim of the complete production of clean energy. In the studied system, the probabilistic modeling of the problem has been performed using the information gap decision theory (IGDT) method, in which both the robustness and opportunistic states of the RERs are respectively modeled as the risk-averse and risk-seeker strategies. In order to reliably meet the energy demand in the presence of 100% RERs, transactive multi-carrier energy technology is exerted for building a local energy trading environment considering both the economic and control mechanisms. The result analysis indicated the effectiveness of the proposed model for the MCENs with 100% RERs.

Keywords Modern multi-carrier energy networks · Hybrid energy systems · Renewable energy resources · Mathematical energy models, demand-side energy management · Uncertainty quantification · Realistic system modelign

© Springer Nature Switzerland AG 2021

M. Daneshvar et al., *Grid Modernization – Future Energy Network Infrastructure*, Power Systems, https://doi.org/10.1007/978-3-030-64099-6_6

6.1 Introduction

Over recent decades, the energy crisis has been widely felt across the globe by increasing the pressure on the energy demand due to emerging new consumers with diverse energy consumption patterns as well as enhancing the people's life dependency mostly on the multi-carrier energy for effectively meeting the social welfare and well-being standards. This ever-increasing trend in the multi-carrier energy consumption indicates the great need to strengthen the energy production sector along with effectively managing the energy consumption side. For this aim, different energy production units and energy management schemes are proposed for upgrading the current status of the energy networks for properly facing the energy changes. Typically, every device or scheme, from the proposal stage to the implementation stage in practice, requires passing different steps. In other words, after providing the proposal for the new device or scheme, the complete evaluation should be carried out for analyzing its various dimensions in energy networks. In the next step, the theoretical concepts of the related device or scheme need to be implemented in the mathematical format to make detailed analysis of its impact on the energy network possible. Afterward, simulating the related device or scheme using the proposed mathematical models needs to be conducted in real case studies to assess its feasibility. A similar process needs to be considered for all sectors of the future modern MCENs. Therefore, in Chapters 1, 2, 3, 4, and 5, we have analyzed the conceptual requirements for the future modern MCENs and in this chapter, we will indicate how the MCENs can be mathematically modeled. For this aim, the mathematical models are completely provided for different energy networks, including electric power systems, natural gas grid, district heating network, and water distribution system along with their various devices. Future modern MCENs are expected to be faced with a high level of uncertainties, especially in the energy production sector due to equipping them to the 100% RERs. Furthermore, energy prices in the energy market and multi-carrier energy demand are other typical uncertainties in the MCENs that directly affect the optimal operation of the system. Because uncertainties are an integral part of the future modern MCENs and have direct impacts on the application of the different parts of the system, appropriately modeling them is necessary for achieving the near reality results for reliable operation of hybrid energy networks. To this end, different mathematical models are proposed for effectively capturing the fluctuations in various sectors of energy networks. Each of the proposed methods is developed for modeling special uncertain parameters and certain objectives. Given the key role of the uncertainties in the reliable operation of the future modern MCENs, we also have motivated to fully examine different uncertainty modeling approaches with their mathematical models in this chapter. Indeed, various uncertainty quantification techniques are completely investigated in this chapter along with their distinguishing features to transparently identify which uncertainty modeling approaches are appropriate for which types of uncertain parameters with special behaviors. On the other hand, simulating the real problem is an effective way to examine how the proposed models are applicable in real energy networks. Thus, to prove the effectiveness of the proposed model for the future modern MCENs, the sample problem is analyzed in the case study with several interconnected microgrids with 100% RERs. The scrutinizing of the simulation results is also highlighted the valuable achievements of applying the proposed model in the MCENs with real data.

6.2 Mathematical Modeling of the Modern Multi-Carrier Energy Networks

To make any idea implementable in practice, developing its mathematical form is one of the inevitable steps. This section covers the mathematical models for most common devices in the MCENs. Additionally, new mathematical models are proposed for some effective network-based technologies along with the different energy networks. All mathematical models are formulated for multi-microgrids using the index *m* in the equations. This does not mean their application is limited only to the interconnected microgrids but also all the presented models can be effectively applied for other structures of the MCENs such as energy hubs.

6.2.1 Electric Power System

The electric power system is responsible for reliably meeting the electrical energy demand at every time of a day for various consumers. The interconnecting of this grid with other energy networks in the future modern MCENs has made the need for some hybrid systems necessary. This subsection is focused on presenting the mathematical models for electricity-based systems, hybrid energy systems and technologies, and electricity network model.

6.2.1.1 Clean Energy Production Units

Clean energy production units are intended to play a central role in the future modern MCENs for producing 100% energy using the RERs. Because the main burden of fully clean energy generation is mostly on the RERs especially wind turbines and PV panels, appropriately modeling them in the future modern MCENs structure is necessary. Therefore, their mathematical models are provided in this section to make their deployment in hybrid energy networks easier.

6.2.1.1.1 Wind Turbine Model

Wind power is taken into account as one of the RERs that can provide a high amount of clean energy in the system. The generated power by the wind turbines can be formulated as follows [1].

$$P_{m,t}^{Wind} = \begin{cases} 0 & 0 \le \omega_t \le \omega_m^{Cut-In} \\ \left(\tau_1 + \tau_2\omega_t + \tau_3\omega_t^2\right) P_m^{W,Rated} & \omega_m^{Cut-In} \le \omega_t \le \omega_m^{Rated} \\ P_m^{W,Rated} & \omega_m^{Rated} \le \omega_t \le \omega_m^{Cut-Out} \\ 0 & \omega_m^{Cut-Out} \le \omega_t \end{cases} \tag{6.1}$$

$$P_{m,t}^{Wind} / \sqrt{\left(P_{m,t}^{Wind}\right)^2 + \left(Q_{m,t}^{Wind}\right)^2} = \text{Constant} \tag{6.2}$$

where $P_{m,t}^{Wind}$ presents the amount of electricity generation by the wind turbine in microgrid m at time t. $P_m^{W,Rated}$ represents the rated power of the wind turbine in microgrid m. ω_t is the forecasted wind speed at time t. ω_m^{Cut-In} and $\omega_m^{Cut-Out}$ are the amount of cut-in and cut-out wind speeds, respectively. ω_m^{Rated} is the amount of rated wind speed. $Q_{m,t}^{Wind}$ is the amount of reactive power production by the wind turbine. τ_1, τ_2, and τ_3 are the coefficients in modeling wind power production.

Equation (6.1) is used for modeling the wind power generation, while Eq. (6.2) presents the possibility of reactive power generation by wind turbines.

6.2.1.1.2 PV Panel Model

For the effective usage of solar radiation to produce clean energy in the studied region, each microgrid can be equipped with several PV panels, which can be modeled as follows [2].

$$P_{m,t}^{PV} \le \eta_m^{PV} . Sz_m^{PV} . SR_t^{PV} \quad \forall m \in 1 : N_m, \forall t \in 1 : N_t \tag{6.3}$$

where $P_{m,t}^{PV}$ is the amount of electricity generation by the PV panel in microgrid m at time t. η_m^{PV} and Sz_m^{PV} are the efficiency and size of the PV panel, respectively. SR_t^{PV} is solar radiation. Equation (6.3) indicates how the amount of power production by the PV panels can be computed and its amount depends on which factors.

6.2.1.2 Hybrid Energy Systems

Hybrid energy systems are the integral parts of the MCENs, which usually deal with more than one type of energy carrier. These systems play a key role in interconnecting the different energy networks to each other that are worked based on energy conversion techniques. The number of famous hybrid devices along with the related technologies is focused here. The complete descriptions regarding the hybrid energy systems are already provided in Chap. 3 of this book.

6.2.1.2.1 Diesel Generator (DGs) Unit Model

The DG units are one of the uni-generation systems that typically burn the natural gas fuel for generating only electrical energy. Due to achieving a higher amount of efficiency, DGs are recognized as one of the popular devices for a controlled amount of energy, especially in the remote area. The DG unit operation can be mathematically modeled as follows.

$$\chi_{m,t}^{DG}.\overline{P}_{m,t}^{DG} \leq P_{m,t}^{DG} \leq \chi_{m,t}^{DG}.\underline{P}_{m,t}^{DG} \quad \forall t, \forall m \tag{6.4}$$

$$\overline{P}_{m,t}^{DG} \leq P_{m,t}^{GD,dg}.HR^{Gas}.\Delta t \quad \forall t, \forall m \tag{6.5}$$

$$P_{m,t}^{DG} - P_{m,t-1}^{DG} \leq \overline{Ramp}_{m}^{DG}, \text{if } P_{m,t}^{DG} \geq P_{m,t-1}^{DG} \quad \forall t, \forall m \tag{6.6}$$

$$P_{m,t-1}^{DG} - P_{m,t}^{DG} \leq \underline{Ramp}_{m}^{DG}, \text{if } P_{m,t-1}^{DG} \geq P_{m,t}^{DG} \quad \forall t, \forall m \tag{6.7}$$

$$\left(\overline{T}_{m,t-1}^{DG} - MUT_{m}^{DG}\right).\left(\chi_{m,t-1}^{DG} - \chi_{m,t}^{DG}\right) \geq 0 \quad \forall t, \forall m \tag{6.8}$$

$$\left(\underline{T}_{m,t-1}^{DG} - MDT_{m}^{DG}\right).\left(\chi_{m,t-1}^{DG} - \chi_{m,t}^{DG}\right) \geq 0 \quad \forall t, \forall m \tag{6.9}$$

$$\chi_{m,t}^{DG} - \chi_{m,t-1}^{DG} \leq \chi_{m,t}^{su} \quad \forall t, \forall m \tag{6.10}$$

$$\chi_{m,t-1}^{DG} - \chi_{m,t}^{DG} \leq \chi_{m,t}^{sd} \quad \forall t, \forall m \tag{6.11}$$

$$\chi_{m,t}^{DG} - \chi_{m,t-1}^{DG} \leq \chi_{m,t}^{su} - \chi_{m,t}^{sd} \quad \forall t, \forall m \tag{6.12}$$

where $P_{m,t}^{DG}$ is the amount of power production by the DG unit in microgrid m at time t. $\chi_{m,t}^{DG}$ is the binary variable presenting the ON/OFF status of the DG unit. $\overline{P}_{m,t}^{DG}$ and $\underline{P}_{m,t}^{DG}$ are the up and down limitations for the DG unit's output. $P_{m,t}^{GD,dg}$ is the amount of gas consumption by the DG unit. HR^{Gas} presents the amount of heat rate gas. $\overline{Ramp}_m^{DG}$ and $\underline{Ramp}_m^{DG}$ are the amounts of ramp up and down for the DG unit. $\overline{T}_{m,t-1}^{DG}$ and $\underline{T}_{m,t-1}^{DG}$ are the number of hours that the DG unit is respectively ON and OFF. MUT_m^{DG} and MDT_m^{DG} are the minimum up and down times for the DG unit. $\chi_{m,t}^{su}$ and $\chi_{m,t}^{sd}$ state the binary variables that respectively presenting the status of the startup and shut down of the DG unit.

Equation (6.4) is used for considering the allowable range of the DG unit's power production. Equation (6.5) models the limitation of the natural gas fuel burned by the DG unit. Equations (6.6) and (6.7) are used for modeling the ramp up and down limitation for the DG unit. Equations (6.8) and (6.9) formulate the minimum up and downtime limitations for the DG unit. Equations (6.10), (6.11), and (6.12) intend the regulation of the ON/OFF status of the DG unit.

6.2.1.2.2 Combined Heating and Power (CHP) Unit Model

In hybrid energy networks, the CHP unit is one of the bi-generation units that is used as a reliable system not only for supporting RERs in meeting the electrical energy demand but also for supplying heating energy to the consumers. The CHP unit can be modeled using the following formulations.

$$\begin{aligned} C_{m,t}^{CHP} &= \gamma_1^{CHP} + \gamma_2^{CHP} P_{m,t}^{CHP} + \gamma_3^{CHP} H_{m,t}^{CHP} \\ &+ \gamma_4^{CHP} \left(P_{m,t}^{CHP} H_{m,t}^{CHP}\right) + \gamma_5^{CHP} \left(P_{m,t}^{CHP}\right)^2 + \gamma_6^{CHP} \left(H_{m,t}^{CHP}\right)^2 \end{aligned} \tag{6.13}$$

$$\chi_{m,t}^{CHP} \underline{P}_m^{CHP} \le P_{m,t}^{CHP} \le \chi_{m,t}^{CHP} \overline{P}_m^{CHP} \tag{6.14}$$

$$P_{m,t}^{CHP} - P_{m,t-1}^{CHP} \le \overline{Ramp}_m^{CHP} \tag{6.15}$$

$$P_{m,t-1}^{CHP} - P_{m,t}^{CHP} \le \underline{Ramp}_m^{CHP} \tag{6.16}$$

$$\left(\overline{T}_{m,t-1}^{CHP} - MUT_m^{CHP}\right).\left(\chi_{m,t-1}^{CHP} - \chi_{m,t}^{CHP}\right) \ge 0 \tag{6.17}$$

$$\left(\underline{T}_{m,t-1}^{CHP} - MDT_m^{CHP}\right).\left(\chi_{m,t-1}^{CHP} - \chi_{m,t}^{CHP}\right) \ge 0 \tag{6.18}$$

$$\left(P_{m,t}^{CHP} - P_M^{CHP}\right) - \left(\frac{P_M^{CHP} - P_N^{CHP}}{H_M^{CHP} - H_N^{CHP}}\right).\left(H_{m,t}^{CHP} - H_M^{CHP}\right) \le 0 \tag{6.19}$$

$$-\left(1 - U_{m,t}^{CHP}\right).\Omega \le \left(P_{m,t}^{CHP} - P_N^{CHP}\right) - \left(\frac{P_N^{CHP} - P_E^{CHP}}{H_N^{CHP} - H_E^{CHP}}\right).\left(H_{m,t}^{CHP} - H_N^{CHP}\right) \tag{6.20}$$

$$-\left(1-U_{m,t}^{CHP}\right).\Omega \le \left(P_{m,t}^{CHP}-P_{E}^{CHP}\right)-\left(\frac{P_{E}^{CHP}-P_{F}^{CHP}}{H_{E}^{CHP}-H_{F}^{CHP}}\right).\left(H_{m,t}^{CHP}-H_{E}^{CHP}\right) \quad (6.21)$$

where $C_{m,t}^{CHP}$ is the energy cost of the CHP unit m at time t, which is modeled by Eq. (6.4). $P_{m,t}^{CHP}$ indicates the amount of electricity generation by the CHP unit. $\gamma_{1,\ldots,6}^{CHP}$ are the cost coefficients of the CHP unit. $\overline{P}_{m}^{CHP}$ and $\underline{P}_{m}^{CHP}$ are the maximum and minimum amounts of energy production by the CHP unit. $\overline{Ramp}_{m}^{CHP}$ and $\underline{Ramp}_{m}^{CHP}$ are the amounts of ramp up and down for the CHP unit. MUT_{m}^{CHP} and MDT_{m}^{CHP} are the minimum up and down times for the CHP unit. Ω indicates a sufficient big number. $H_{m,t}^{CHP}$ is the amount of heating energy produced by the CHP unit. $\chi_{m,t}^{CHP}$ presents the ON/OFF status of the CHP unit. $\overline{T}_{m,t-1}^{CHP}$ and $\underline{T}_{m,t-1}^{CHP}$ represent the number of ON and OFF hours for the CHP unit. $U_{m,t}^{CHP}$ states the operation state of the CHP unit in the feasible region.

Equation (6.14) presents the limitation of the CHP unit's output. Equations (6.15) and (6.16) represent the ramping up and down limits for the CHP units. Equations (6.17) and (6.18) state the limitations for the number of hours that the CHP units have been ON or OFF. Equations (6.19), (6.20), and (6.21) are used for modeling the feasible operation region (FOR) of the CHP unit m at time t. The FOR of the CHP unit is demonstrated in Fig. 6.1.

6.2.1.2.3 Combined Cooling, Heating, and Power (CCHP) Unit Model

The CCHP unit is taken into account as one of the main members of tri-generation systems that can support the electrical, heating, and cooling energy production in the system. It contains both the PGU and boiler units. The PGU unit has a gas tur-

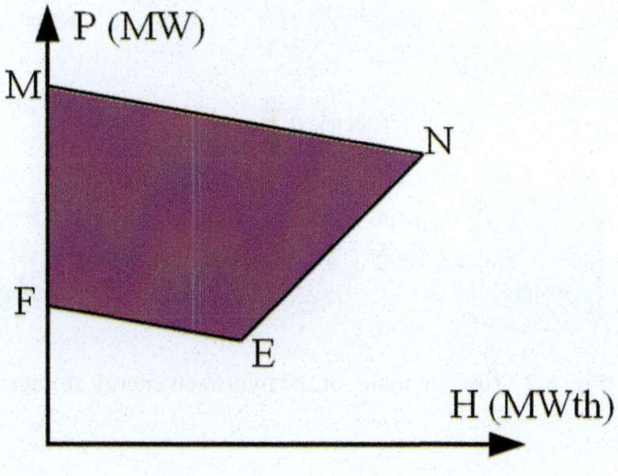

Fig. 6.1 Feasible operation region of the CHP unit

bine for generating electrical energy while the boiler is used for meeting the thermal energy. The related constraints for them are as follows.

$$\cos t_{m,t}^{CCHP} = F_{m,t}^{PGU}.P_{m,t}^{PGU} + F_{m,t}^{Bo}.P_{m,t}^{BO} \quad \forall t, \forall m \tag{6.22}$$

$$F_{m,t}^{PGU} \leq \chi_{m,t}^{PGU}.S_{m}^{PGU} \quad \forall t, \forall m \tag{6.23}$$

$$F_{m,t}^{Bo} \leq S_{m}^{BO} \quad \forall t, \forall m \tag{6.24}$$

$$P_{m,t}^{PGU} \leq \left(F_{m,t}^{PGU} - \chi_{h}^{PGU}.\Upsilon_{1}^{PGU}\right) / \Upsilon_{2}^{PGU} \quad \forall t, \forall m \tag{6.25}$$

where $F_{m,t}^{PGU}$ and $F_{m,t}^{Bo}$ present the amount of fuel consumed by the PGU unit and boiler. $P_{m,t}^{Bo}$ represents the amount of energy produced by the boiler. $\chi_{m,t}^{PGU}$ denotes the ON or OFF state of PGU. S_{m}^{PGU} (S_{m}^{BO}) is the size of PGU (boiler) in energy hub *h*. Υ_{1}^{PGU} and Υ_{2}^{PGU} state the coefficients related to the fuel to electricity conversion in the PGU unit.

6.2.1.2.4 Hydrogen Storage System Model

Hydrogen storage is one of the key systems for developing decarbonization plans in hybrid energy networks. This system consists of an electrolyzer system, hydrogen storage tank, and fuel cell for structuring the clean energy production process. This process is illustrated in Fig. 6.2.

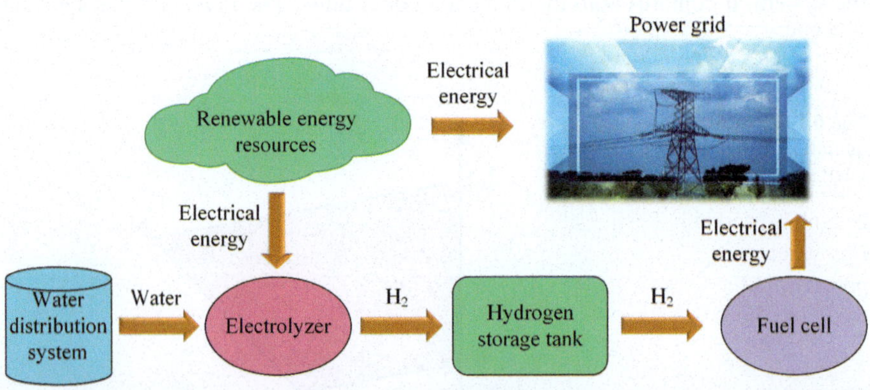

Fig. 6.2 The schematic of the hydrogen energy storage system

6.2.1.2.4.1 Electrolyzer Model

The electrolyzer is a hydrogen-based device that receives water and electrical energy for decomposing the water molecule and producing hydrogen energy. In the hybrid energy network, when the amount of clean energy produced by the RERs is greater than the energy consumption, the excess energy can be used for producing hydrogen energy by the electrolyzer system. The mathematical model for the electrolyzer system is as follows.

$$\chi^{EL}_{m,t}.\underline{P}^{EL}_{m} \leq P^{EL}_{m,t} \leq \chi^{EL}_{m,t}.\overline{P}^{EL}_{m} \quad \forall m, \forall t \tag{6.26}$$

$$He^{EL}_{m,t} \leq \chi^{EL}_{m,t}.\overline{He}^{EL}_{m} \quad \forall m, \forall t \tag{6.27}$$

$$He^{EL}_{m,t} = \frac{P^{EL}_{m,t}.\eta^{EL}_{m}}{LHV^{\mathrm{H2}}_{m}} \quad \forall m, \forall t \tag{6.28}$$

where $P^{EL}_{m,t}$ is the amount of power consumed by the electrolyzer for hydrogen production in microgrid m at time t. $\chi^{EL}_{m,t}$ presents the binary variable that indicates the ON/OFF status of the electrolyzer system in microgrid m at time t. $\overline{P}^{EL}_{m,t}$ and $\underline{P}^{EL}_{m,t}$ represent the upper and lower bounds for the power consumption by the electrolyzer. $He^{EL}_{m,t}$ states the amount of hydrogen energy produced by the electrolyzer in microgrid m at time t. $\overline{He}^{EL}_{m,t}$ denotes the maximum amount of hydrogen energy that can be produced by the electrolyzer. η^{EL}_{m} is the efficiency of the electrolyzer system. LHV^{H2}_{m} denotes the lower heating value of hydrogen energy in microgrid m.

Equation (6.26) is used for considering the allowable range of power consumption for the electrolyzer system. Equation (6.27) intends the maximum possible amount of producing hydrogen energy by the electrolyzer. Finally, Eq. (6.28) is used to consider the produced hydrogen energy is a function of the consumed electrical energy by the electrolyzer.

6.2.1.2.4.2 Hydrogen Storage Tank Model

After generating the hydrogen energy by the electrolyzer, the produced hydrogen needs to be stored in the hydrogen tank for later use. Due to this, the hydrogen storage tank is used here for storing the hydrogen energy to utilize it for various applications when it is required in the hybrid network. The required constraints for the operation of the hydrogen storage tank are given as:

$$\underline{P}^{P,\mathrm{H2}}_{m} \leq P^{P,\mathrm{H2}}_{m,t} \leq \overline{P}^{P,\mathrm{H2}}_{m} \quad \forall m, \forall t \tag{6.29}$$

$$P^{P,H2}_{m,t} = P^{P,H2}_{m,Initial} \quad \forall m, \forall t = 0 \tag{6.30}$$

where $P^{P,H2}_{m,t}$ presents the amount of hydrogen tank pressure in microgrid m at time t. $\overline{P}^{P,H2}_{m}$ and $\underline{P}^{P,H2}_{m}$ represent the maximum and minimum limitations for the pressure of hydrogen in the tank. $P^{P,H2}_{m,Initial}$ states the amount of initial hydrogen pressure in the tank.

Equation (6.29) formulates the permissible range of hydrogen pressure in the tank while Eq. (6.30) considers the amount of initial pressure of the hydrogen in the tank.

6.2.1.2.4.3 Fuel Cell Model

In the hybrid energy network, when the amount of energy demand is greater than energy production particularly at peak times, the fuel cell device can be used for generating clean energy aiming to support the system for continuous energy supply. In this process, the fuel cell receives the hydrogen energy stored in the hydrogen storage tank for simultaneously producing electricity, heating energy, and water. The hydrogen energy is decomposed in the anode part of the fuel cell system and release the hydrogen and electron ions. Then, the electrons are extracted through a wire for using in the electric part and then the combination of the hydrogen and electron ions along with the oxygen is used for producing water and heating energy. The fuel cell operation can be modeled using the following equations in the hybrid network.

$$\chi^{FC}_{m,t}.\underline{P}^{FC}_{m} \leq P^{FC}_{m,t} \leq \chi^{FC}_{m,t}.\overline{P}^{FC}_{m} \quad \forall m, \forall t \tag{6.31}$$

$$He^{FC}_{m,t} \leq \chi^{FC}_{m,t}.\overline{He}^{FC}_{m} \quad \forall m, \forall t \tag{6.32}$$

$$He^{FC}_{m,t} = \frac{P^{FC}_{m,t}}{\eta^{FC}_{m}.LHV^{H2}_{m}} \quad \forall m, \forall t \tag{6.33}$$

where $P^{FC}_{m,t}$ presents the amount of electrical energy produced by the fuel cell in microgrid m at time t. $\chi^{FC}_{m,t}$ denotes the ON/OFF status of the fuel cell. $\overline{P}^{FC}_{m}$ and $\underline{P}^{FC}_{m}$ are used for considering the permissible range of power production by the fuel cell system. $He^{FC}_{m,t}$ states the amount of hydrogen energy consumed by the fuel cell system in microgrid m at time t. $\overline{He}^{FC}_{m}$ denotes the maximum amount of hydrogen energy that can be consumed by the fuel cell. η^{FC}_{m} is the efficiency of the fuel cell.

Equation (6.31) is used to limit the amount of power production by the fuel cell in the standard range. Equation (6.32) considers the allowable upper level for fuel cell hydrogen consumption. Equation (6.33) indicates the relationship between the consumed hydrogen energy and power generation by the fuel cell.

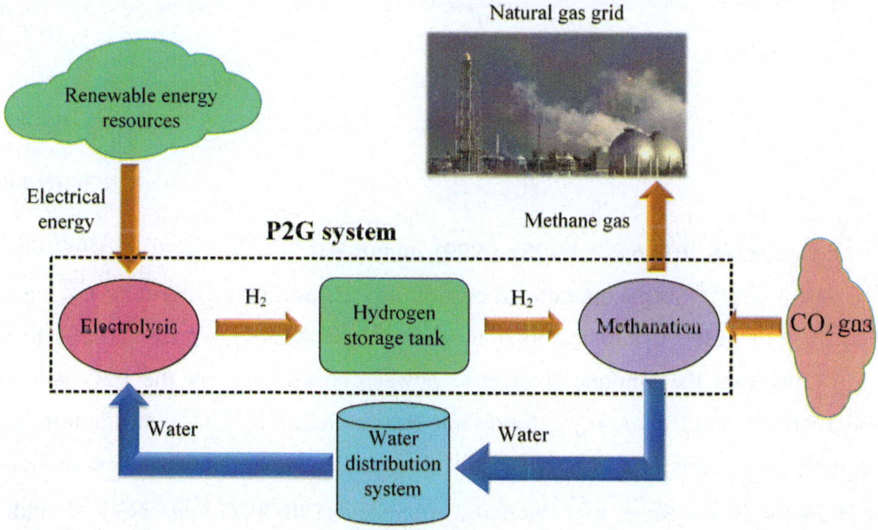

Fig. 6.3 The schematic of the P2G system

6.2.1.2.5 Power to Gas Energy Storage System

Power to gas (P2G) system is one of the innovative technologies that plays a central role in coupling different energy networks to each other. The P2G system consumes the surplus of clean energy produced in the system with the aim of producing the natural gas (CH_4) using the electrolysis and methanation processes. The schematic of the P2G technology is depicted in Fig. 6.3.

According to Fig. 6.3, the P2G system receives the excess amount of energy produced in the system to produce hydrogen energy by the electrolyzer system. Afterward, the produced hydrogen can be stored in the hydrogen store tank for future use. The process of the electrolyzer and hydrogen storage tank and their mathematical models are the same as described in the hydrogen storage system section. However, in the P2G system, the hydrogen energy received from the hydrogen storage tank continues the methanation process for generating the methane gas. In the P2G system process, water and electrical energy are the inputs for the electrolyzer system while the hydrogen energy along with the CO_2 gas is the requirement for the methanation process for CH_4 generation. Assuming that the cost of water is negligible, the costs of electricity and CO_2 gas are the two main factors that make up the operation cost of the P2G system. The mathematical model for P2G technology is as follows [3].

$$\mathrm{Cost}_m^{P2G} = \sum_{t=1}^{N_t} \left[\left(C_{P2G}^{El} . P_{m,t}^{P2G} \right) + \left(\alpha_m^{P2G} . \eta^{P2G} . C_{P2G}^{CO_2} . P_{m,t}^{P2G} \right) \right] \quad \forall m \tag{6.34}$$

$$G_{m,t}^{P2G} = \frac{\eta^{P2G}}{HHV_m^{P2G}} P_{m,t}^{P2G} \quad \forall m, \forall t \tag{6.35}$$

$$0 \leq G_{m,t}^{P2G} \leq \overline{G}_m^{P2G} \quad \forall m, \forall t \tag{6.36}$$

$$P_{m,t}^{EL} = P_{m,t}^{P2G} \quad \forall m, \forall t \tag{6.37}$$

Equations (6.26), (6.27), (6.28), (6.29), and (6.30).

where $\cos t_m^{P2G}$ is the operational cost of the P2G unit in microgrid m. C_{P2G}^{El} and $C_{P2G}^{CO_2}$ are the respective indicators of the electricity and CO_2 gas price in the system. $P_{m,t}^{P2G}$ presents the amount of electric power consumption by the P2G unit in microgrid *m* and time *t*. α_m^{P2G} represents the coefficient of CO_2 consumption per unit gas generation in the P2G unit. $G_{m,t}^{P2G}$ denotes the amount of natural gas production by the P2G system. η^{P2G} states the efficiency of the P2G unit. HHV_m^{P2G} indicates the higher heating value of synthetic natural gas (SNG) in microgrid *m*. $\overline{G}_m^{P2G}$ is the maximum amount of CH_4 production by the P2G unit.

Equation (6.34) models the amount of operational cost of the P2G system. Equation (6.35) formulates the amount of methane gas production considering the amount of power consumption by the P2G system. Equation (6.36) intends the permissible range of natural gas generation by the P2G unit. Equation (6.37) indicates that the amount of electricity consumption by the electrolyzer system is equal to the power consumption of the P2G unit.

6.2.1.3 Electrical Energy Storage Systems

Electrical energy storage systems will be the inevitable devices of the future modern MCENs with 100% RERs. The features of the different types of electrical energy storage systems are fully described in Chap. 3. Herein, we want to show how we can mathematically model the most popular energy storage systems.

6.2.1.3.1 Battery Energy Storage System (BESS) Model

The BESS is introduced as one of the appropriate ways for alleviating the fluctuation behaviors of the stochastic producers by storing energy at high energy production times for later use with lower energy production. Since each microgrid is targeted to be equipped with a large number of RERs or full share of them in the modern grids, the effective potential of the BESS can be used in the microgrids to provide a proper condition for a high/full-level operation of the clean energy resources in the power grid. In order to exploit the BESS in the microgrids, the following constraints should be met during the optimization process [4].

$$\mathrm{C}^{BESS}_{m,t} = \left(\kappa^{BESS}_{m} .P^{BESS}_{m,t} .\Delta t\right) + \left(\kappa^{BESS}_{m} .En^{BESS}_{m,t} .\eta^{LLF} .\Delta t\right) \tag{6.38}$$

$$\kappa^{BESS}_{m} = \frac{IC^{BESS}}{En^{BESS}_{RC} .LCN^{BESS}} \tag{6.39}$$

$$\begin{aligned} En^{BESS}_{m,t+1} &= En^{BESS}_{m,t} - P^{BESS}_{m,t} .\Delta t - \left|P^{BESS}_{m,t}\right| .\eta^{LF} .\Delta t \\ &\quad - En^{BESS}_{m,t} .\eta^{LLF} .\Delta t \quad \forall m \in 1:N_m, \forall t \in 1:N_t \end{aligned} \tag{6.40}$$

$$SOC^{BESS}_{m,t} = En^{BESS}_{m,t} \,/\, En^{BESS}_{RC} \tag{6.41}$$

$$SOC^{Down}_{m,t} \text{ " } SOC^{BESS}_{m,t} \text{ " } SOC^{Up}_{m,t} \tag{6.42}$$

$$-P^{Cha,Up}_{m,t} \leq P^{BESS}_{m,t} \leq P^{Dis,Up}_{m,t} \tag{6.43}$$

$$P^{Cha,Up}_{m,t} \geq 0 \ , \ P^{Dis,Up}_{m,t} \geq 0 \tag{6.44}$$

$$En^{BESS}_{m,Initial} = En^{BESS}_{m,0} \ ; \ En^{BESS}_{m,End} \leq En^{BESS}_{m,N_t} \tag{6.45}$$

where $P^{BESS}_{m,t}$ is the amount of energy charging and discharging in the BESS system in microgrid m at time t. $\mathrm{C}^{BESS}_{m,t}$ presents the amount of BESS energy cost. $En^{BESS}_{m,t}$ represents the amount of energy stored in the BESS. $SOC^{BESS}_{m,t}$ states the BESS's state of charge. η^{LF} denotes the charging and discharging loss factor. κ^{BESS}_{m} indicates the lifetime degradation coefficient of the BESS. η^{LLF} is the leakage loss factor of the BESS. IC^{BESS} is the investment cost of the BESS. En^{BESS}_{RC} presents the rated energy capacity of the BESS. LCN^{BESS} represents the life cycle number of the BESS. $SOC^{Up}_{m,t}$ and $SOC^{Down}_{m,t}$ are the maximum and minimum amount of state of charge for the BESS.

Equations (6.38) and (6.39) are used for computing the energy cost of the BESS. The energy balance constraint for the BESS is modeled by Eq. (6.40). Equations (6.41) and (6.42) are used for modeling the state of charge for the BESS in microgrid m at time t. The amounts of charging and discharging energy in the BESS are described in Eq. (6.43). Equation (6.45) presents the initial and final amounts of the stored energy in the BESS.

In the optimization problems, the linear models can achieve the optimal solutions in comparison with the non-linear ones. If all equations in our optimization problem are linear and we have a BESS in our model, using linear equations for modeling the BESS would be very effective in achieving optimal solutions. The usage of Eqs. (6.38), (6.39), (6.40), (6.41), (6.42), (6.43), (6.44), and (6.45) will convert our problem from the mixed-integer linear problem (MILP) to the mixed-integer non-linear programming (MINLP) one. Therefore, using the following linear equations can maintain the problem in the MILP format with optimal solutions.

$$\chi_{m,t}^{Ba,Ch} + \chi_{m,t}^{Ba,Dis} \le 1 \quad \forall m, \forall t \tag{6.46}$$

$$S_m^{BESS} . \sigma_{\min}^{BESS} \le P_{m,t}^{BESS} \le S_m^{BESS} \quad \forall m, \forall t \tag{6.47}$$

$$P_{m,t}^{BESS} = \left(P_{m,t}^{Ba,Ch} - P_{m,t}^{Ba,Dis}\right).\Delta t + E_{Ini}^{BESS} \quad \forall m, \forall t = 1 \tag{6.48}$$

$$P_{m,t}^{BESS} - P_{m,t-1}^{BESS} = \left(P_{m,t}^{Ba,Ch} - P_{m,t}^{Ba,Dis}\right).\Delta t \quad \forall m, \forall t \ge 2 \tag{6.49}$$

$$\chi_{m,t}^{Ba,Ch} . S_m^{BESS} . \sigma_{\min}^{Ba,Ch} \le P_{m,t}^{Ba,Ch} \le \chi_{m,t}^{Ba,Ch} . S_m^{BESS} . \sigma_{\max}^{Ba,Ch} \quad \forall m, \forall t \tag{6.50}$$

$$\chi_{m,t}^{Ba,Dis} . S_m^{BESS} . \sigma_{\min}^{Ba,Dis} \le P_{m,t}^{Ba,Dis} \le \chi_{m,t}^{Ba,Dis} . S_m^{BESS} . \sigma_{\max}^{Ba,Dis} \quad \forall m, \forall t \tag{6.51}$$

where $\chi_{m,t}^{Ba,Ch}$ and $\chi_{m,t}^{Ba,Dis}$ are the respective indicators of binary variables for charging and discharging states of the BESS in microgrid m at time t. S_m^{BESS} and $\sigma_{\min}^{BESS}$ present the size of BESS and coefficient for the minimum storage limit of the BESS. $P_{m,t}^{BESS}$ represents the amount of electricity stored in the BESS. $P_{m,t}^{Ba,Ch}$ and $P_{m,t}^{Ba,Dis}$ are the respective indicators of charging and discharging of electrical energy in the BESS. E_{Ini}^{BESS} is the initial amount of electricity in the BESS. $\sigma_{\min}^{Ba,Ch}$ and $\sigma_{\max}^{Ba,Ch}$ state the minimum and maximum charging coefficients in the BESS while $\sigma_{\min}^{Ba,Dis}$ and $\sigma_{\max}^{Ba,Dis}$ play the same role for the discharging mode of the BESS.

Equation (6.46) indicates that the BESS cannot be charged and discharged at the same time. Equation (6.47) presents the allowable range of the BESS changes during a day. Equations (6.48) and (6.49) are used for balancing the charging and discharging energy in the BESS during a day. Equations (6.50) and (6.51) model the permissible range of the charging and discharging of the BESS.

6.2.1.3.2 Pumped Storage Model

Pumped storage is one of the large-scale energy storage systems that can be used in the frequency control due to its capability for charging and discharging a large amount of energy. As the pumped storage system is very beneficial for developing the clean energy production units in the system, its operation directly depends on the geographical conditions and it cannot be structured everywhere. The mathematical model for the pumped storage system is as follows [5].

$$P_{m,t}^{Pu,T} = \gamma_m^{Pu,T} . Q_{m,t}^{Pu,T} \quad \forall m, \forall t \tag{6.52}$$

$$P_{m,t}^{Pu,P} = \gamma_m^{Pu,P} . Q_{m,t}^{Pu,P} \quad \forall m, \forall t \tag{6.53}$$

$$V_{m,t+1}^{Pu,Ur} = V_{m,t}^{Pu,Ur} + Q_{m,t}^{Pu,P} - Q_{m,t}^{Pu,T} + S_{m,t}^{Pu,Inflow} \quad \forall m, \forall t \tag{6.54}$$

$$V_{m,t+1}^{Pu,Lr} = V_{m,t}^{Pu,Lr} + Q_{m,t}^{Pu,T} - Q_{m,t}^{Pu,P} \quad \forall m, \forall t \tag{6.55}$$

$$\underline{V}_{m}^{Pu,Ur} \leq V_{m,t}^{Pu,Ur} \leq \overline{V}_{m}^{Pu,Ur} \quad \forall m, \forall t \tag{6.56}$$

$$\underline{V}_{m}^{Pu,Lr} \leq V_{m,t}^{Pu,Lr} \leq \overline{V}_{m}^{Pu,Lr} \quad \forall m, \forall t \tag{6.57}$$

$$V_{Ini,m}^{Pu,Ur} \leq V_{m,t}^{Pu,Ur} \quad \forall m, t = T \tag{6.58}$$

$$V_{Ini,m}^{Pu,Lr} \leq V_{m,t}^{Pu,Lr} \quad \forall m, t = T \tag{6.59}$$

$$0 \leq Q_{m,t}^{Pu,T} \leq \overline{Q}_{m}^{Pu,T} \quad \forall m, \forall t \tag{6.60}$$

$$0 \leq Q_{m,t}^{Pu,P} \leq \overline{Q}_{m}^{Pu,P} \quad \forall m, \forall t \tag{6.61}$$

where $P_{m,t}^{Pu,T}$ and $P_{m,t}^{Pu,P}$ are the power production and consumption of the pumped storage system in microgrid m at time t, respectively. $Q_{m,t}^{Pu,T}$ and $Q_{m,t}^{Pu,P}$ are the turbining and pumping flow in the pumped storage system. $\gamma_{m}^{Pu,T}$ and $\gamma_{m}^{Pu,P}$ present the conversion factors of power production and consumption in the pumped storage system. $V_{m,t}^{Pu,Ur}$ and $V_{m,t}^{Pu,Lr}$ represent the volume of water in the upper and lower reservoirs of the pumped storage system. $S_{m,t}^{Pu,Inflow}$ states the amount of natural inflow of water into the upper reservoir. $\overline{V}_{m}^{Pu,Ur}$ and $\underline{V}_{m}^{Pu,Ur}$ are the upper and lower levels for the volume of water in the upper reservoir while the $\overline{V}_{m}^{Pu,Lr}$ and $\underline{V}_{m}^{Pu,Lr}$ play the same roles for the lower reservoir in the pumped storage system. $V_{Ini,m}^{Pu,Ur}$ and $V_{Ini,m}^{Pu,Lr}$ denote the amount of initial volume of water in the upper and lower reservoirs, respectively. $\overline{Q}_{m}^{Pu,T}$ and $\overline{Q}_{m}^{Pu,P}$ are the respective indicators of the maximum amount of turbining and pumping flow.

Equations (6.52) and (6.53) model the amount of power production and consumption by the pumped storage system in turbining and pumping modes. Equations (6.54) and (6.55) indicate the balance equations for the volume of water in the upper and lower reservoirs. Equations (6.56) and (6.57) consider the allowable range of the changes for the volume of water in the upper and lower reservoirs. Equations (6.58) and (6.59) indicate the initial amounts of water in the upper and lower reservoirs. Equations (6.60) and (6.61) formulate the permissible range of changes for the turbining and pumping flows in the pumped storage system.

6.2.1.3.3 Compressed Air Energy Storage (CAES) Model

The compressed air energy storage (CAES) system is one of the large-scale storage technologies that uses the effective potential of the caverns for storing energy. The CASE technology is described in Chap. 3 of this book. The CASE system can be formulated as follows.

$$Q_{m,t}^{Ca,inj} = \kappa_m^{Ca,inj}.P_{m,t}^{Ca,Ch} \quad \forall m, \forall t \tag{6.62}$$

$$P_{m,t}^{Ca,Dis} = \kappa_m^{Ca,p}.Q_{m,t}^{Ca,p} \forall m, \forall t \tag{6.63}$$

$$\underline{Q}_m^{Ca,inj}.\chi_{m,t}^{Ca,inj} \le Q_{m,t}^{Ca,inj} \le \overline{Q}_m^{Ca,inj}.\chi_{m,t}^{Ca,inj} \quad \forall m, \forall t \tag{6.64}$$

$$\underline{Q}_m^{Ca,p}.\chi_{m,t}^{Ca,p} \le Q_{m,t}^{Ca,p} \le \overline{Q}_m^{Ca,p}.\chi_{m,t}^{Ca,p} \quad \forall m, \forall t \tag{6.65}$$

$$S_{m,t+1}^{Ca} = S_{m,t}^{Ca} + Q_{m,t}^{Ca,inj} - Q_{m,t}^{Ca,p} \quad \forall m, \forall t \tag{6.66}$$

$$\underline{S}_m^{Ca} \le S_{m,t}^{Ca} \le \overline{S}_m^{Ca} \quad \forall m, \forall t \tag{6.67}$$

$$\chi_{m,t}^{Ca,inj} + \chi_{m,t}^{Ca,p} \le 1 \quad \forall m, \forall t \tag{6.68}$$

where $Q_{m,t}^{Ca,inj}$ is the amount of injected energy to the CASE system in microgrid m at time t. $Q_{m,t}^{Ca,p}$ is the energy consumed by the CAES system for pumping air to the combustion chamber in microgrid m at time t. $\kappa_m^{Ca,inj}$ and $\kappa_m^{Ca,p}$ are the respective indicators of the efficiency factors for received and produced power by the CASE system. $P_{m,t}^{Ca,Ch}$ and $P_{m,t}^{Ca,Dis}$ present the charging and discharging energy of the CASE system. $\chi_{m,t}^{Ca,p}$ and $\chi_{m,t}^{Ca,inj}$ denote the air pumping and injection states of the CASE system. $\overline{Q}_m^{Ca,p}$ and $\underline{Q}_m^{Ca,p}$ are the upper and lower limits for the energy consumption by the CASE system for pumping air to the combustion chamber. $\overline{Q}_m^{Ca,inj}$ and $\underline{Q}_m^{Ca,inj}$ are the upper and lower limits for the amount of injected energy to the CASE system. $S_{m,t}^{Ca}$ states the amount of stored energy in the CASE system. $\overline{S}_m^{Ca}$ and $\underline{S}_m^{Ca}$ indicate the maximum and minimum amounts of the stored energy in the CASE system.

Equations (6.62) and (6.63) model the amount of energy consumption and production by the CASE system. Equations (6.64) and (6.65) consider the allowable range of energy injecting and pumping by the CASE system. Equation (6.66) formulates the stored energy in the CASE system and Eq. (6.67) limits it in the standard range. Equation (6.68) indicates that the CASE system cannot be simultaneously charged and discharged.

6.2.1.4 Electricity Network Model

6.2.1.4.1 AC Power Flow Model

In order to make the proposed model feasible and practical from the network's point of view, the AC power flow should be considered in the studied system. The following formulations are applied to model the AC power flow with the related constraints.

$$P_{i,t}^{F} + P_{i,t}^{D} = P_{i,t}^{Ge} \quad \forall i \in 1:N_b, \forall t \in 1:N_t \tag{6.69}$$

$$Q_{i,t}^{F} + Q_{i,t}^{D} = Q_{i,t}^{Ge} \quad \forall i \in 1:N_b, \forall t \in 1:N_t \tag{6.70}$$

$$P_{i,t}^{F} = \sum_{j=1}^{N_b} V_{i,t}.V_{j,t}.Y_{i,j}.\cos\left(\varpi_{i,j} + \theta_{j,t} - \theta_{i,t}\right) \ \forall i \in 1:N_b, \forall t \in 1:N_t \tag{6.71}$$

$$Q_{i,t}^{F} = -\sum_{j=1}^{N_b} V_{i,t}.V_{j,t}.Y_{i,j}.\sin\left(\varpi_{i,j} + \theta_{j,t} - \theta_{i,t}\right) \ \forall i \in 1:N_b, \forall t \in 1:N_t \tag{6.72}$$

$$\underline{S}_{i,j} \le S_{i,j,t} \le \overline{S}_{i,j} \quad \forall i,j \in 1:N_b, \forall t \in 1:N_t \tag{6.73}$$

$$\underline{V}_{i} \le V_{i,t} \le \overline{V}_{i} \quad \forall i \in 1:N_b, \forall t \in 1:N_t \tag{6.74}$$

$$\underline{\theta}_{i} \le \theta_{i,t} \le \overline{\theta}_{i} \quad \forall i \in 1:N_b, \forall t \in 1:N_t \tag{6.75}$$

where $P_{i,t}^{Ge}$ and $Q_{i,t}^{Ge}$ are the active and reactive power generation in bus i at time t. $P_{i,t}^{F}$ and $Q_{i,t}^{F}$ are the amount of active and reactive power flow. $P_{i,t}^{D}$ and $Q_{i,t}^{D}$ present the amount active and reactive power load in the system. $V_{i,\,t}$ and $\theta_{i,\,t}$ represent the voltage magnitude and its phase angle. $S_{i,\,j,\,t}$ is the amount of the complex power between line i-j. $Y_{i,\,j}$ and $\varpi_{i,\,j}$denote the amount of admittance and its phase angle.

Equations (6.69) and (6.70) present the active and reactive power balance constraints at bus i and time t. Equations (6.71) and (6.72) are used for modeling the active and reactive power flow in the system. Equations (6.73), (6.74), and (6.75) state the allowable range of the complex power, voltage magnitude, and voltage phase angle, respectively.

6.2.1.4.2 DC Power Flow Model

DC power flow model allows for easily modeling of the optimization problems by using the linear equations. Although the DC power flow is one of the simple ways of modeling electricity networks, their results are not reliable to be considered for practical application. By assuming $X_{i,j}^{DC} >> R_{i,j}^{DC}$, the DC power flow model can be formulated as follows [6].

$$P_{i,j,t}^{F,DC} = \frac{\theta_{i,t} - \theta_{j,t}}{X_{i,j}^{DC}} \quad \forall i,j \in 1:N_b, \forall t \in 1:N_t \tag{6.76}$$

$$\underline{P}_{i,j}^{F,DC} \le P_{i,j,t}^{F,DC} \le \overline{P}_{i,j}^{F,DC} \quad \forall i \in 1:N_b, \forall t \in 1:N_t \tag{6.77}$$

$$P_{i,t}^{Ge} - P_{i,t}^{D} = \sum_{j=1}^{N_b} B_{i,j}^{DC} \theta_{i,t} \quad \forall i \in 1:N_b, \forall t \in 1:N_t \tag{6.78}$$

where $X_{i,j}^{DC}$ and $R_{i,j}^{DC}$ are the respective indicators of reactance and resistance of line *i*-*j*. $P_{i,j,t}^{F,DC}$ presents the amount of DC power flow in line *i*-*j*. $\overline{P}_{i,j}^{F,DC}$ and $\underline{P}_{i,j}^{F,DC}$ represent the upper and lower bounds for the DC power flow. $B_{i,j}^{DC}$ states the amount of admittance of line *i*-*j*.

Equation (6.76) indicates how the DC power flow can be computed for line *i*-*j*. Equation (6.77) considers the allowable range for the DC power flow changes in line *i*-*j*. Equation (6.78) models the power balance at bus *i*.

6.2.2 Natural Gas Network

As the number of hybrid gas-fired energy systems is drastically increased in the system, the role of the natural gas networks has become central in developing sustainable hybrid energy networks. Natural gas networks, as the main entities, are responsible for reliably accommodating the gas energy demand in hybrid energy networks. Due to the ever-increasing penetration of the gas-fired hybrid energy systems in MCENs, the interconnecting of the natural gas networks with other energy structures is inevitable for the future modern MCENs. Therefore, developing appropriate mathematical models for the natural gas networks is necessary for achieving the near reality results.

6.2.2.1 Linepack Model

In the natural gas networks, Linepack (LP) refers to the pressurized gas energy stored in the pipelines [7]. The amount of LP directly depends on the average pressure of a pipe and can be described by four variables illustrated in Fig. 6.4 [8].

The relationships between gas variables can be expressed as follows in accordance with Boyle's law [9].

$$\frac{P_0^{LP}}{D_0^{LP}.T_0^{LP}} = \frac{P^{LP}}{D^{LP}.T^{LP}} = Q_{Gas}^{LP}.Y_{Gas}^{LP} = \text{Constant} \tag{6.79}$$

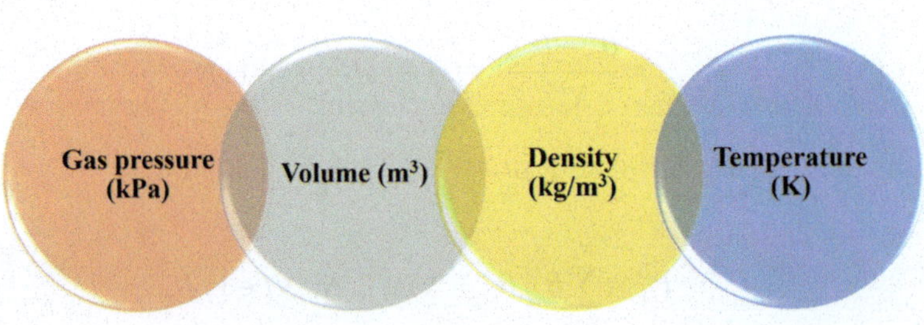

Fig. 6.4 Effective variables for the amount of gas in the pipelines [8]

$$P_{i,j}^{Ave}.V_{i,j}^{LP} = P_0^{Ave}.V_0^{LP} \tag{6.80}$$

$$P_{i,j}^{Ave} = \sqrt{\frac{1}{2\left(P_i^2 - P_j^2\right)}} \tag{6.81}$$

$$V_{i,j}^{LP} = \pi d_{i,j}^2 L_{i,j}^{Line} / 4 \tag{6.82}$$

$$LP_{i,j}^{Initial} = \frac{\sigma^{LP}.P_{i,j}^{Ave}.V_{i,j}^{LP}}{D_0^{LP}.T_0^{LP}.Q_{Gas}^{LP}.Y_{Gas}^{LP}} \tag{6.83}$$

$$LP_{i,j,t+1} = LP_{i,j,t} + \Delta P_{i,j,t}^{Gas}.\Delta t \tag{6.84}$$

$$\Delta P_{i,j,t}^{Gas} = P_{i,j,t}^{Ge,Gas} - P_{i,j,t}^{D,Gas} - P_{i,j,t}^{Com,Gas} \tag{6.85}$$

where P^{LP}, T^{LP}, D^{LP}, and $V_{i,j}^{LP}$ are the respective indicators of gas pressure, temperature, density, and volume in the pipeline. Q_{Gas}^{LP} and Y_{Gas}^{LP} presents the gas compressibility factor and gas constant. $L_{i,j}^{Line}$ and $d_{i,j}$ represent the length of the pipeline and internal diameter of pipe *i-j*. P_i denotes the amount of gas pressure at bus *i*. $LP_{i,j,t}$ and $LP_{i,j}^{Initial}$ state the real and initial amounts of the LP in line *i-j*. σ^{LP} is the conversion factor between gas volume and energy (GJ/m³). $P_{i,j,t}^{Ge,Gas}$ and $P_{i,j,t}^{D,Gas}$ are the gas supplied and consumed at time *t*. $P_{i,j,t}^{Com,Gas}$ denotes the amount of gas consumption by the compressor.

Equations (6.79), (6.80), (6.81), and (6.82) are Boyle's law for the natural gas network. Equation (6.83) indicates how the initial amount of LP can be computed in the system. Equation (6.84) models the balance conditions of the LP in the pipelines.

6.2.2.2 Gas Flow Equations

In the steady state, Weymouth's formula can be used for modeling the gas flows [10]. In the pipelines, the nodal pressure differences can affect the gas flows and their directions. The gas flow equations in the stead-state are as follows [10].

$$\begin{aligned} P_{i,j,t}^{F,Gas} &= Sgn_{i,j}.\xi_{Gas}^{F}.\frac{T_0^{LP}}{P_0^{LP}}.\sqrt{Sgn_{i,j}.\frac{\left(p_{i,t}^2 - p_{j,t}^2\right).d_{i,j}^5}{F_{i,j}^{Di}.G_{Gas}^{F}.L_{i,j}^{Line}.T^{LP}.Y_{Gas}^{LP}}} \\ &= Sgn_{ij}.W_{ij}.\sqrt{\left(p_{i,t}^2 - p_{j,t}^2\right)} \end{aligned} \tag{6.86}$$

$$Sgn_{i,j} = \begin{cases} +1 & \text{if } \ \mathrm{p}_i - p_j > 0 \\ -1 & \text{if } \ \mathrm{p}_i - p_j < 0 \end{cases} \tag{6.87}$$

$$\xi_{Gas}^{F} = \sqrt{\pi^{2}.\vartheta_{Air}^{Gas,F} / 64} \tag{6.88}$$

$$G_{Gas}^{F} = \frac{\vartheta_{Air}^{Gas,F}}{Q_{Gas}^{LP}} = \frac{1}{0.6} = 1.6667 \tag{6.89}$$

$$F_{i,j}^{Di} = \frac{0.032}{d_{i,j}^{(1/3)}} \tag{6.90}$$

where $P_{i,j,t}^{F,Gas}$ presents the amount of gas flow in line *i-j* at time *t*. $Sgn_{i,j}$ represents the direction of the gas flow in line *i-j*. ξ_{Gas}^{F} and G_{Gas}^{F} are the respective indicators of air constant and specific gravity ration. $F_{i,j}^{Di}$ denotes the dimensionless friction factor that its amount directly depends on the $d_{i,j}$ according to Eq. (6.90). $\vartheta_{Air}^{Gas,F}$ is the coefficient related to the specific gravity ration.

Equation (6.86) models the amount of natural gas flow in pipeline *i-j*. Equation (6.87) denotes how the direction of the gas flow can be determined in the pipeline *i-j*. Equations (6.88) and (6.89) compute some constants related to the air constant and specific gravity ration.

6.2.2.3 Compressor Station Equation

The compressor is one of the key devices in the natural gas networks that is essential for maintaining the gas pressure in the standard range. The amount of gas consumption by the compressor directly depends on the difference between inlet and outlet gas pressures and gas flows. The related empirical equations are as follows [11].

$$P_{i,j,t}^{Com,Gas} = Sgn\left(p_i,p_j\right).\frac{HP_{i,j,t}^{Com}}{\tau_{i,j}^{Com1} - \tau_{i,j}^{Com2}.\left[\frac{\max\left(p_i,p_j\right)}{\min\left(p_i,p_j\right)}\right]^{\tau_{i,j}^{Com3}}} \tag{6.91}$$

$$P_{i,j,t}^{Com,Gas}\left(HP_{i,j,t}^{Com}\right) = \beta_{i,j}^{Com1} + \beta_{i,j}^{Com2}.HP_{i,j,t}^{Com} + \beta_{i,j}^{Com3}.\left(HP_{i,j,t}^{Com}\right)^{2} \tag{6.92}$$

$$\underline{PR}_{i,j,t}^{Gas,Com} \le \frac{\max\left(p_i,p_j\right)}{\min\left(p_i,p_j\right)} \le \overline{PR}_{i,j,t}^{Gas,Com} \tag{6.93}$$

where $P_{i,j,t}^{Com,Gas}$ is the amount of gas flow in the compressor. $HP_{i,j,t}^{Com}$ states the amount of compressor's horse power in line *i-j* at time *t*. $\tau_{i,j}^{Com1}$, $\tau_{i,j}^{Com2}$, and $\tau_{i,j}^{Com3}$ are the coefficients for computing the amount of gas flow in the compressor. $\beta_{i,j}^{Com1}$, $\beta_{i,j}^{Com2}$, and $\beta_{i,j}^{Com3}$ are the coefficients for calculating the amount of gas consumption by the gas compressor. $\overline{PR}_{i,j,t}^{Gas,Com}$ denotes the amount of compressor pressure ratio.

Equation (6.91) models the amount of gas flow in pipeline *i-j* at time *t*. Equation (6.92) formulates the amount of gas consumption by the compressor. Equation (6.93) considers the allowable range of changes for the compressor pressure ratio.

6.2.2.4 Natural Gas Storage Model

The natural gas storage system is one of the necessary devices for ensuring the continuous gas supply in hybrid energy networks. It makes the storing of natural gas energy at sometimes with a large amount of gas production and using it later possible. Thus, natural gas storage can increase the overall reliability and flexibility of the hybrid networks at the desirable range. The following equations can be used for making the possibility of examining the implementation of the natural gas storage system in the real-world feasible.

$$NG_{m,t}^{S,Gas} = NG_{m,t-1}^{S,Gas} + G_{m,t}^{S,Ch}.\eta_m^{GS,Ch} - \frac{G_{m,t}^{S,Dis}}{\eta_m^{GS,Dis}} \quad \forall m, \forall t \tag{6.94}$$

$$\underline{NG}_m^{S,Gas} \le NG_{m,t}^{S,Gas} \le \overline{NG}_m^{S,Gas} \quad \forall m, \forall t \tag{6.95}$$

$$0 \le G_{m,t}^{S,Ch} \le \overline{G}_m^{S,Ch} \quad \forall m, \forall t \tag{6.96}$$

$$0 \le G_{m,t}^{S,Dis} \le \overline{G}_m^{S,Dis} \quad \forall m, \forall t \tag{6.97}$$

$$NG_{m,t=0}^{S,Gas} = NG_{m,t=N_T}^{S,Gas} \quad \forall m \tag{6.98}$$

where $NG_{m,t}^{S,Gas}$ presents the amount of gas energy stored in the natural gas storage in microgrid *m* at time *t*. $G_{m,t}^{S,Ch}$ and $G_{m,t}^{S,Dis}$ represent the amount of gas charging and discharging in the gas storage system. $\eta_m^{GS,Ch}$ and $\eta_m^{GS,Dis}$ state the amounts of charging and discharging efficiency of the gas storage system. $\overline{NG}_m^{S,Gas}$ and $\underline{NG}_m^{S,Gas}$ denote the upper and lower bounds for the stored gas energy in the natural gas storage. $\overline{G}_m^{S,Ch}$ ($\overline{G}_m^{S,Dis}$) is the maximum amount of charging (discharging) gas energy in the gas storage system. $NG_{m,t=0}^{S,Gas}$ and $NG_{m,t=N_T}^{S,Gas}$ are the amounts of the initial and final amount of gas energy stored in the gas storage system.

Equation (6.94) models the gas energy balance in the gas storage system. Equation (6.95) intends the allowable range of stored gas energy in the natural gas storage. Equations (6.96) and (6.97) formulate the permissible bounds for the amount of natural gas charging and discharging in the storage system. Equation (6.98) indicates that the initial and final amounts of gas stored in the storage system should be equal.

6.2.2.5 Natural Gas Network Model

The natural gas network model is one of the primary steps for analyzing the implementability of the proposed models in natural gas networks. This model covers different gas energy balance equations in the system. The natural gas network model can be formulated using the following equations [12].

$$P_{i,j,t}^{Ge,Gas} + \sum_{j=1}^{L} P_{i,j,t}^{F,Gas} = \sum_{j=1}^{L} P_{j,i,t}^{F,Gas} + P_{i,j,t}^{D,Gas} + P_{i,j,t}^{Com,Gas} \quad \forall i \in 1 : N_b \tag{6.99}$$

$$P_{i,j,t}^{D,Gas} = P_{i,j,t}^{D,DGas} + P_{i,t}^{D,Heat} \quad \forall i, \forall j, \forall t \tag{6.100}$$

$$\underline{LP}_{i,j} \le LP_{i,j,t} \le \overline{LP}_{i,j} \quad \forall i, \forall j \tag{6.101}$$

$$LP_{i,j,t=0} = LP_{i,j,Initial}; LP_{i,j,t=N_T} \ge LP_{i,j,End} \tag{6.102}$$

where $P_{i,j,t}^{D,DGas}$ is the amount of consumed gas by the gas-fired systems. $P_{i,t}^{D,Heat}$ states the consumed gas energy for producing heating energy. $\overline{LP}_{i,j}$ and $\underline{LP}_{i,j}$ are the upper and lower bounds of the LP. $LP_{i,j,Initial}$ and $LP_{i,j,End}$ are the initial and final amounts of LP in the system.

Equation (6.99) models the nodal balance of natural gas energy in the gas network. Equation (6.100) presents the different types of the gas energy demand. Equation (6.101) intends the permissible range for the LP changes in the system. Equation (6.102) indicates the initial and final amounts of gas energy.

6.2.3 *District Heating Network (DHN)*

The DHNs are developed to easily meet the heating energy demand of the consumers. The DHNs typically rely on the co- and tri-generation systems for generating heating energy, but central solar heating, heat pumps, geothermal heating, and heat-only boiler stations are also used for supporting DHNs. Herein, we present some of the common mathematical models for DHN.

6.2.3.1 District Heating Network Model

The district heating network (DHN) can be operated in four modes; the two most useful of them are variable-flow variable-temperature (VF-VT) and constant-flow variable-temperature (CF-VT) [13]. The mathematical models for both of them are presented in the following subsections.

6.2.3.1.1 Variable-Flow Variable-Temperature (VF-VT) Mode

A DHN mainly resides on return pipelines and symmetric supply network modes [14]. At each source node, heating energy either is injected into the network or withdrawn from the network via a heat exchanger among the return and supply sides. Hydraulic and thermal conditions are two common network models for the DHN [15]. Hydraulic conditions include the fluid pressure and mass flow in pipelines and comprise the following formula [14].

(a) *Continuity of mass flow:* The amount of mass flows in both entering and leaving sides of each node in the DHN are equal according to the following equation.

$$\sum_{h\in H_i^{Pe}} m_{h,t}^{H,Ps} + \sum_{s\in H_i^{S}} m_{s,t}^{H,S} = \sum_{h\in H_i^{Ps}} m_{h,t}^{H,Ps} + \sum_{\ell\in H_i^{D}} m_{\ell,t}^{H,D} \tag{6.103}$$

$$\sum_{h\in H_i^{Ps}} m_{h,t}^{H,\mathrm{Pr}} + \sum_{\ell\in H_i^{D}} m_{\ell,t}^{H,D} = \sum_{h\in H_i^{Pe}} m_{h,t}^{H,\mathrm{Pr}} + \sum_{s\in H_i^{S}} m_{s,t}^{H,S} \tag{6.104}$$

where $m_{h,t}^{H,Ps}$ and $m_{h,t}^{H,\mathrm{Pr}}$ are the respective indicators of mass flow rate in the supply and return pipelines. H_i^{Pe} and H_i^{Ps} present the set of pipelines ending and starting at node i. $m_{s,t}^{H,S}$ and $m_{\ell,t}^{H,D}$ represent the mass flow rate of heat source and load, respectively. H_i^{S} and H_i^{D} denote the set of heat sources and loads in the DHN.

(b) *Fluid pressure balance:* The friction of pipelines causes a decrease in fluid pressure along the pipe. This loss in fluid pressure can be expressed as follows.

$$P_{h,t}^{HL,s} = \phi_h^{Heat,L} . \left(m_{h,t}^{H,Ps}\right)^2 \tag{6.105}$$

$$P_{h,t}^{HL,r} = \phi_h^{Heat,L} . \left(m_{h,t}^{H,\mathrm{Pr}}\right)^2 \tag{6.106}$$

where $P_{h,t}^{HL,s}$ and $P_{h,t}^{HL,r}$ state the pressure losses in the supply and return pipelines. $\phi_h^{Heat,L}$ is the coefficient of heat pressure loss in the pipeline. The amount of pressure loss in the closed-loop is equal to zero given the principle of pressure balance as shown in the following equations [14]:

$$\sum_{b\in H^{loop}} \psi_b^{DHN} . P_{h,t}^{HL,s} = 0 \tag{6.105}$$

$$\sum_{b\in H^{loop}} \psi_b^{DHN} . P_{h,t}^{HL,r} = 0 \tag{6.106}$$

$$\psi_b^{DHN} = \begin{cases} +1 \Rightarrow \text{The direction of mass fow in pipeline } b \text{ is consistent} \\ \text{with the loop direction.} \\ -1 \Rightarrow \text{The direction of mass fow in pipeline } b \text{ is opposite} \\ \text{to the loop direction.} \end{cases}$$

where ψ_b^{DHN} presents the mass flow direction in pipeline b. H^{loop} represents the set of loops in the DHN.

(c) *Heat source model*: The amount of heat consumption at a load node can be modeled as follows.

$$H_{i,t}^{S} = \upsilon_H^W . m_{i,t}^{H,S} . \left(T_{i,t}^{H,SS} - T_{i,t}^{H,RS} \right) \tag{6.107}$$

where $H_{i,t}^{S}$ presents the generation of heat source in bus i at time t. υ_H^W denotes the specific heat capacity of water. $T_{i,t}^{H,SS}$ and $T_{i,t}^{H,RS}$ are the supply and return temperatures of the heat source.

(d) *Heat load model*: The amount of heat consumption at a load node can be modeled as follows.

$$H_{i,t}^{D} = \upsilon_H^W . m_{i,t}^{H,D} . \left(T_{i,t}^{H,SD} - T_{i,t}^{H,RD} \right) \tag{6.108}$$

where $H_{i,t}^{D}$ is the heat energy demand in bus i at time t. $T_{i,t}^{H,SD}$ and $T_{i,t}^{H,RD}$ are the supply and return temperatures of heat load.

(e) *Pipeline model:* The temperature of fluid typically decreases in the pipelines due to inevitable heat losses. The inlet and outlet temperatures of the pipelines can be formulated as follows.

$$T_{h,t}^{H,So} = \left(T_{h,t}^{H,Si} - T_t^{Am} \right) e^{-\frac{\lambda_h^{DHN} L_h^{DHN}}{\upsilon_H^W . m_{h,t}^{H,Ps}}} + T_t^{Am} \quad \forall h, \forall t \tag{6.109}$$

$$T_{h,t}^{H,Ro} = \left(T_{h,t}^{H,Ri} - T_t^{Am} \right) e^{-\frac{\lambda_h^{DHN} L_h^{DHN}}{\upsilon_H^W . m_{h,t}^{H,Pr}}} + T_t^{Am} \quad \forall h, \forall t \tag{6.110}$$

where $T_{h,t}^{H,So}$ and $T_{h,t}^{H,Si}$ ($T_{h,t}^{H,Ro}$ and $T_{h,t}^{H,Ri}$) are the respective indicators of temperatures at outlet and inlet temperatures of supply (return) pipeline. T_t^{Am} is the ambient temperature at period t. λ_h^{DHN} and L_h^{DHN} present the coefficient of heat transfer and length of the pipeline.

(f) *Temperature at confluence nodes:* At confluence nodes, when the fluids with various temperatures come across a node, the temperature of the mixture fluid can be calculated as follows.

$$\sum_{h \in H_i^{Pe}} \left(T_{h,t}^{H,So} . m_{h,t}^{H,Ps} \right) + \sum_{s \in H_i^S} \left(T_{s,t}^{H,S} . m_{s,t}^{H,S} \right) = T_{i,t}^{H,Sm} . \left(\sum_{h \in H_i^{Pe}} m_{h,t}^{H,Ps} + \sum_{s \in H_i^S} m_{s,t}^{H,S} \right) \quad \forall t \tag{6.111}$$

$$\sum_{h\in H_i^{Ps}}\left(T_{h,t}^{H,Ro}.m_{h,t}^{H,\Pr}\right)+\sum_{\ell\in H_i^{D}}\left(T_{\ell,t}^{H,R}.m_{\ell,t}^{H,D}\right)$$
$$=T_{\ell,t}^{H,Rm}.\left(\sum_{h\in H_i^{Ps}}m_{h,t}^{H,\Pr}+\sum_{\ell\in H_i^{D}}m_{\ell,t}^{H,D}\right)\quad\forall t \tag{6.112}$$

where $T_{s,t}^{H,S}$ and $T_{\ell,t}^{H,R}$ are the supply temperature of the heat source and return temperature of heat load, respectively. $T_{i,t}^{H,Sm}$ and $T_{\ell,t}^{H,Rm}$ denote the mixture temperature at the supply and return node. In order to consider the equality of the temperatures of leaving mass flows with the temperature during mixing at each node, the following equations should be intended.

$$T_{h,t}^{H,Sh}=T_{i,t}^{H,Sm}\quad,\quad T_{\ell,t}^{H,SD}=T_{i,t}^{H,Sm} \tag{6.113}$$

$$T_{h,t}^{H,Rh}=T_{\ell,t}^{H,Rm}\quad,\quad T_{s,t}^{H,RS}=T_{\ell,t}^{H,Rm} \tag{6.114}$$

6.2.3.1.2 Constant-Flow Variable-Temperature (CF-VT) Mode

The CF-VT mode of DHN operation not only is one of the useful modes for examining the DHN but also allows for developing a linear optimization problem. In the CF-VT mode, the hydraulic conditions are fixed while the output of the heat source and the nodal temperatures are variable. The mathematical model for the CF-VT mode is as follows [16].

$$T_{h,t}^{H,So}=\left(T_{h,t}^{H,Si}-T_{t}^{Am}\right)e^{-\frac{\lambda_h^{DHN}L_h^{DHN}}{\upsilon_H^{W}.m_h^{H,Ps}}}+T_t^{Am}\quad\forall h,\forall t \tag{6.115}$$

$$\left(\rho_t^{H,E}.P_{s,t}^{H,E}\right)+H_{s,t}^{H,Co}=\upsilon_H^{W}.\dot{m}_s.\left(T_{s,t}^{H,SN}-T_{s,t}^{H,RN}\right)\ \forall t,\forall s\in\Omega_s \tag{6.116}$$

$$H_{\ell,t}^{D}=\upsilon_H^{W}.m_\ell^{H,D}.\left(T_{\ell,t}^{H,SN}-T_{\ell,t}^{H,RN}\right)\ \forall t,\forall\ell\in\Omega_L \tag{6.117}$$

$$T_{k,t}^{H,SN}=\frac{\sum_{v\in k}\dot{m}_v T_{v,t}^{H,SN}}{\sum_{v\in k}\dot{m}_v}\quad\forall t,\forall k\in\Omega_k \tag{6.118}$$

where $P_{s,t}^{H,E}$ is the amount of purchased electrical energy from the electricity network for generating heat energy. $\rho_t^{H,E}$ denotes the price of purchased electricity. $H_{s,t}^{H,Co}$ states the amount of heat energy bought from the energy company. Ω_s and Ω_L indicate the set of heat energy suppliers and load nodes. Ω_k presents the set of nodes with several inlets.

The dropped temperature across the pipeline is modeled by Eq. (6.115). Equation (6.116) formulates the amount of thermal energy supplied by heat source while the amount of received thermal energy by heat load is formulated by Eq. (6.117). At confluence nodes, the temperature of the mixed fluid is modeled using Eq. (6.118).

6.2.3.2 Thermal Storage Model

Thermal storage systems are effectively used for increasing the reliability of the continuous thermal energy supply as well as the optimal operation of the system during a day. These systems support the hybrid energy networks for locally thermal energy supply with the aim of avoiding thermal energy losses. The mathematical formulations for the thermal storage system are given as [17]:

$$\chi_{m,t}^{TS,Ch} + \chi_{m,t}^{TS,Dis} \leq 1 \quad \forall m, \forall t \tag{6.119}$$

$$S_m^{TS}.\Omega_{\min}^{TS} \leq P_{m,t}^{TS} \leq S_m^{TS} \quad \forall m, \forall t \tag{6.120}$$

$$P_{m,1}^{TS} = TEI_m^{TS} + \left(P_{m,1}^{TS,Ch} - P_{m,1}^{TS,Dis}\right).\Delta t \ \forall m, t = 1 \tag{6.121}$$

$$P_{m,t}^{TS} - P_{m,t-1}^{TS} = \left(P_{m,t}^{TS,Ch} - P_{m,t}^{TS,Dis}\right).\Delta t \ \forall m, \forall t \geq 2 \tag{6.122}$$

$$C_{m,t}^{MS,C} + H_{m,t}^{MS,H} \leq S_m^{TS} \quad \forall m, \forall t \tag{6.123}$$

$$C_{m,t}^{SM,C} + H_{m,t}^{SM,H} \leq P_{m,t}^{TS,Dis}.\eta^{TS,Dis} \quad \forall m, \forall t \tag{6.124}$$

$$P_{m,t}^{TS,Ch} \leq \left(C_{m,t}^{MS,C} + H_{m,t}^{MS,H}\right).\eta^{TS,Ch} \quad \forall m, \forall t \tag{6.125}$$

$$S_m^{TS}.\Omega_{\min}^{TS,Ch}.\chi_{m,t}^{TS,Ch} \leq P_{m,t}^{TS,Ch} \leq S_m^{TS}.\Omega_{\max}^{TS,Ch}.\chi_{m,t}^{TS,Ch} \quad \forall m, \forall t \tag{6.126}$$

$$S_m^{TS}.\Omega_{\min}^{TS,Dis}.\chi_{m,t}^{TS,Dis} \leq P_{m,t}^{TS,Dis} \leq S_m^{TS}.\Omega_{\max}^{TS,Dis}.\chi_{m,t}^{TS,Dis} \quad \forall m, \forall t \tag{6.127}$$

where $\chi_{m,t}^{TS,Dis}$ and $\chi_{m,t}^{TS,Ch}$ are the discharging and charging states of thermal energy storage (TES); $|_{\min}^{TS}$ and S_m^{TS} are the coefficient of minimum storage limit and size of TES; $P_{m,t}^{TS}$ states the amount of thermal energy stored in TES; TEI_m^{TS} is the initial amount of thermal energy in TES; $P_{h,t}^{TS,Cha}$ and $P_{m,t}^{TS,Dis}$ are the respective indicators of the charging and discharging rate of TES; $\eta^{TS,\,Ch}$ and $\eta^{TS,\,Dis}$ present the charging and discharging efficiency of TES; $|_{\max}^{TS,Dis}$ and $|_{\max}^{TS,Ch}$ ($|_{\min}^{TS,Dis}$ and $|_{\min}^{TS,Ch}$) denote the upper (lower) coefficients for discharging and charging limit of TES. $C_{m,t}^{MS,C}$ and $C_{m,t}^{SM,C}$ ($H_{m,t}^{MS,H}$ and $H_{m,t}^{SM,H}$) are the respective indicators of charging and discharging cooling (heating) energy in the TES.

Equation (6.119) indicates the TES cannot be charged and discharged at the same time. Equation (6.120) intends the allowable range for the changes of the thermal energy in the storage system. Equations (6.121) and (6.122) model the balancing of thermal energy in the storage system during a day. Equations (6.123), (6.124), and (6.125) are used for considering the maximum possible amounts of thermal energy charging and discharging in the TES. Equations (6.126) and (6.127) consider the permissible range of thermal energy charging and discharging in the storage system.

6.2.3.3 Electric Water Boiler Model

The electric water boiler is one of the useful devices for generating heat energy in hybrid energy networks. This system receives electrical energy from the electric power system especially from the RERs to produce heat energy. The electric water boiler's behavior in generating heat energy can be formulated as follows [18].

$$H_{m,t}^{EWB} = \eta_m^{EWB}.\left(P_{m,t}^{EWB}\right) \quad \forall m, \forall t \tag{6.128}$$

where $H_{m,t}^{EWB}$ is the amount of generated heat energy by the electric water boiler system in microgrid m at time t. η_m^{EWB} is the efficiency of the electric water boiler system. $P_{m,t}^{EWB}$ is the amount of electricity consumed by the electric water boiler system.

6.2.3.4 Solar Water Boiler

The solar water boiler uses solar radiation for producing heat energy. The related mathematical model is as follows [18].

$$H_{m,t}^{SWB} = \eta_m^{SWB}.\left(SR_t^{SWB}.Nc_m^{SWB}.Ac_m^{SWB}\right) \quad \forall m, \forall t \tag{6.129}$$

where $H_{m,t}^{SWB}$ denotes the amount of heat energy produced by the solar water boiler in microgrid m at time t. η_m^{SWB} states the efficiency of the solar water boiler. SR_t^{SWB} is the solar radiation at time t. Nc_m^{SWB} presents the number of solar collectors in microgrid m. Ac_m^{SWB} represents the area of solar collectors.

6.2.3.5 Reciprocating Chiller

Reciprocating chiller consumes the electrical energy received from the power grid particularly RERs for producing cooling energy in the hybrid networks. The mathematical model for the reciprocating chiller is as follows [18].

$$C_{m,t}^{RC} = Co_m^{RC}.P_{m,t}^{RC} \quad \forall m, \forall t \tag{6.130}$$

where $C_{m,t}^{RC}$ is the amount of cooling energy generated by the reciprocating chiller in microgrid m at time t. Co_m^{RC} is the performance coefficient of the reciprocating chiller. $P_{m,t}^{RC}$ indicates the amount of electricity consumption by the reciprocating chiller.

6.2.3.6 Absorption Chiller

Absorption chiller typically receives the heat energy from the DHN for producing the cooling energy. It can be modeled using the following equation [18].

$$C_{m,t}^{AC} = Co_m^{AC}.H_{m,t}^{AC} \quad \forall m, \forall t \tag{6.131}$$

where $C_{m,t}^{AC}$ is the amount of cooling energy generated by the absorption chiller in microgrid m at time t. Co_m^{RC} is the performance coefficient of the absorption chiller. $H_{m,t}^{RC}$ indicates the amount of heat energy received by the reciprocating chiller from the DHN.

6.2.4 *Water Distribution Network (WDN) Model*

As the technology of the hybrid energy systems is significantly developed, the water distribution systems have become famous for their vital role in most of the hybrid process. By emerging the new water-based hybrid systems and advancements in the MCENs, water was no longer just a basic necessity for the human lives but became a key element for hybrid energy networks. Due to the necessity of effectively modeling the water distribution systems in the MCENs interactions, this section provides the required mathematical models.

In the future modern MCENs, the water distribution system is responsible not only for reliably meeting the potable water but also for supplying the required water of the hybrid energy networks. For this aim, this system consists of some devices including water storage, water well, and water desalination units. In this regard, water storage uses a water pump for pumping the water into the storage tank that consumes a certain amount of power in the charging mode to do this. This is while the water well is also equipped with the pump to transfer the water from the well to the water distribution system that consumes the electrical energy in discharging mode [19]. The water balance equation in the water distribution systems is as follows.

$$\sum_{t=1}^{N_t} \left(W_{m,t}^{Well} + W_{m,t}^{Desal} - W_{m,t}^{S,Ch} + W_{m,t}^{S,Dis} \right) = \sum_{t=1}^{N_t} W_{m,t}^{D} \quad \forall m, \forall t \tag{6.131}$$

where $W_{m,t}^{Well}$ is the extracted water from the water well in microgrid m at time t. $W_{m,t}^{Desal}$ presents the amount of produced water by the desalination unit. $W_{m,t}^{S,Ch}$ and $W_{m,t}^{S,Dis}$ represent the amounts of charging and discharging water in the water storage system. $W_{m,t}^{D}$ states the total water demand in the water distribution system. In the water distribution system, water storage is one of the key devices that plays a crucial role in providing continuous water supply in the hybrid network which can be mathematically modeled using the following equations.

$$W_{m,t}^{S,L} = W_{m,t-1}^{S,L} + \frac{Q_{m,t}^{WS,Ch}}{Cs_m^W} - \frac{Q_{m,t}^{WS,Dis}}{Cs_m^W} \quad \forall m, \forall t \tag{6.132}$$

$$0 \le W_{m,t}^{S,L} \le \overline{W}_m^{S,L} \quad \forall m, \forall t \tag{6.133}$$

$$0 \le Q_{m,t}^{WS,Ch} \le \chi_{m,t}^{WS,Ch} \overline{Q}_m^{WS,Ch} \quad \forall m, \forall t \tag{6.134}$$

$$0 \le Q_{m,t}^{WS,Dis} \le \chi_{m,t}^{WS,Dis} \overline{Q}_m^{WS,Dis} \quad \forall m, \forall t \tag{6.135}$$

$$0 \le \chi_{m,t}^{WS,Ch} + \chi_{m,t}^{WS,Dis} \le 1 \quad \forall m, \forall t \tag{6.136}$$

where $W_{m,t}^{S,L}$ states the level of stored water in the water storage system in microgrid m at time t. $Q_{m,t}^{WS,Ch}$ and $Q_{m,t}^{WS,Dis}$ are the respective indicators of the water charging and discharging in the water storage system. Cs_m^W denotes the cross section of the water storage system. $\overline{W}_m^{S,L}$ indicates the upper standard level for the stored water in the storage system. $\overline{Q}_m^{WS,Ch}$ and $\overline{Q}_m^{WS,Dis}$ present the maximum amount of water charging and discharging in the water storage system. $\chi_{m,t}^{WS,Ch}$ and $\chi_{m,t}^{WS,Dis}$ represent the binary variables that indicate the water charging and discharging status in the water storage system.

Equation (6.132) models the level of water in the water storage system. Equation (6.133) limits the amount of water stored in the water storage. Equations (6.134) and (6.135) consider the upper and lower bounds for changing the water in the charging and discharging modes of the water storage system. Equation (6.136) indicates that the water storage system cannot be charged and discharged at the same time.

The electrical energy consumption by the water pump is a function of the head of water and water flow [20]. The water head is the distance between the water pump placement and the water level. Also, the density of water and the gravity of the ground are two other effective factors in the amount of power consumption by the water pump. The amount of power consumption by the water pump can be formulated as follows [20].

$$P_{m,t}^{W,pump} = \frac{P_{m,t}^{Wp} . Q_{m,t}^{WP} . G^W . D_{m,t}^W}{\eta_m^{W,P}} \quad \forall m, \forall t \tag{6.137}$$

where $P_{m,t}^{W,pump}$ is the power consumption by the water pump in microgrid m at time t. $P_{m,t}^{Wp}$ states the amount of water pressure or head. $Q_{m,t}^{WP}$ denotes the flow of water through the water pipeline. G^W presents the gravity of the ground. $D_{m,t}^W$ represents the density of water. $\eta_m^{W,P}$ is the efficiency of the water pump. By simplifying Eq. (6.137) for the one-hour interval, we can reach to the following equations:

$$P_{m,t}^{W,Wpump} = \frac{L_{m,t}^{Wp}.Q_{m,t}^{WP}.G^{W}.D_{m,t}^{W}}{\left(3.6\times10^{6}\right).\eta_{m}^{W,P}} \quad \forall m,\forall t \tag{6.138}$$

$$P_{m,t}^{W,Spump} = Q_{m,t}^{WS,Ch}.\left(W_{m,t}^{S,L} + W_{m,t-1}^{S,L} + A_{m}^{WS,L}\right) .\frac{G^{W}.D_{m,t}^{W}}{2.\left(3.6\times10^{6}\right).\eta_{m}^{W,P}} \quad \forall m,\forall t \tag{6.139}$$

where $L_{m,t}^{Wp}$ is the water well level and $A_{m}^{WS,L}$ denotes the altitude of water storage location. $P_{m,t}^{W,Wpump}$ and $P_{m,t}^{W,Spump}$ are the respective indicators of the power consumed by the water well and water storage pumps in microgrid m at time t. The numerical coefficient 3.6 × 10 in the denominator of Eqs. (6.138) and (6.139) is used for converting the consumed electrical energy in W unit during one second to electricity consumption in the kW unit within a 1-hour interval.

Another key unit in the water distribution system is the desalination unit that works based on seawater reverse osmosis technology [19]. The power consumption by this unit can be modeled as follows.

$$P_{m,t}^{W,Desal} = \eta_{m}^{W,D}.Q_{m,t}^{W,Desal} \quad \forall m,\forall t \tag{6.140}$$

$$0 \leq Q_{m,t}^{W,Desal} \leq \overline{Q}_{m}^{W,Desal} \quad \forall m,\forall t \tag{6.141}$$

where $P_{m,t}^{W,Desal}$ presents the amount of power consumed by the desalination unit in microgrid m at time t. $\eta_{m}^{W,D}$ is the efficiency of the desalination unit. $Q_{m,t}^{W,Desal}$ denotes the amount of water produced by the desalination unit in microgrid m at time t. $\overline{Q}_{m}^{W,Desal}$ states the maximum amount of water that can be produced by the desalination unit.

Given the amount of power consumption of the different electric-based water distribution system devices, the total power consumption in the water distribution system can be formulated as follows.

$$P_{m,t}^{W,Tot} = P_{m,t}^{W,Wpump} + P_{m,t}^{W,Spump} + P_{m,t}^{W,Desal} \quad \forall m,\forall t \tag{6.142}$$

where $P_{m,t}^{W,Tot}$ is the total amount of power consumption in the water distribution system in microgrid m at time t.

6.2.5 *Demand-Side Energy Management Model*

In the systems with a high share of RERs, demand-side energy management (DSEM) programs are proposed as one of the effective strategies for increasing the flexibility of the power grid. In this regard, the energy load is classified into two categories (i.e., elastic and inelastic loads). The inelastic loads or indispensable demand should be met during a day while the effective potential of the elastic loads can be used for shifting loads (as the first priority) over the scheduling horizon or curtailing loads (as the second priority) in the emergency conditions. In this regard, price response and load response programs are considered with the aim of the demand-side energy management in the interconnected microgrids.

6.2.5.1 Price Response Programs

In this program, given the various energy prices during a day, consumers can shift their energy consumption from hours with a high energy price to the other times with the lower energy prices. This strategy can be used for maximizing the amount of consumers' energy cost savings. The price response program can be formulated as follows for the elastic loads.

$$P_{i,t}^{D} = P_{i,t}^{D,F} + Sh_{i,t}^{D} \quad \forall t \in 1:N_T,\ \forall i \in 1:N_b \tag{6.143}$$

$$\left|Sh_{i,t}^{D}\right| \le \lambda^{\max}.P_{i,t}^{D,F} \quad \forall t \in 1:N_T,\ \forall i \in 1:N_b \tag{6.144}$$

$$\sum_{t=1}^{N_t} Sh_{i,t}^{D} = 0 \quad \forall i \in 1:N_b \tag{6.145}$$

where $P_{i,t}^{D,F}$ is the amount of forecasted demand by bus i at time t. $Sh_{i,t}^{D}$ presents the amount of shifting load during a day. λ^{Max} is the coefficient that considers the maximum amount of shiftable load.

Equation (6.143) states the amount of energy demand by considering the shifted loads. The amount of shifting load during a day should be kept in the allowable range, which is considered in Eq. (6.144). The sum of the shifted loads should be zero during the day for each bus, which is modeled in Eq. (6.145).

6.2.5.2 Load Response Programs

Because of considering the high amount of cost for the load curtailment, this program is intended only for energy balancing in the emergency conditions. For example, it is assumed that up to 5% of the energy demand can be curtailed if necessary. The amount of interrupted load can be modeled as follows.

$$C_{i,t}^{IL} = \gamma_{1i}^{IL}\cdot\left(P_{i,t}^{IL}\right)^2 + \gamma_{2i}^{IL}.P_{i,t}^{IL} \tag{6.146}$$

$$P_{i,t}^{IL} \le \overline{P}_{i,t}^{IL} \quad \forall i \in 1:N_b, \forall t \in 1:N_T \tag{6.147}$$

$$\underline{P}_{i,t}^{D} \leq P_{i,t}^{D} - P_{i,t}^{IL} \leq \overline{P}_{i,t}^{D} \quad \forall i \in 1:N_b, \forall t \in 1:N_t \tag{6.148}$$

where $C_{i,t}^{IL}$ is the amount of interrupted load (IL) cost in bus i at time t. $P_{i,t}^{IL}$ is the amount of IL in bus i at time t. γ_{1i}^{IL} and γ_{2i}^{IL} are the coefficients for computing the IL cost. $\overline{P}_{i,t}^{IL}$ states the maximum amount of IL in bus i at time t. $\overline{P}_{i,t}^{D}$ and $\underline{P}_{i,t}^{D}$ denote the upper and lower amounts of energy demand after IL.

Equation (6.146) presents the load curtailment cost and Eq. (6.147) represents the maximum amount of IL at bus i and time t. Also, constraint (6.148) is used for ensuring that the amount of IL is within the permissible range.

6.2.6 *Transactive Multi-Carrier Energy Model*

In order to create another reliable way for establishing a dynamic energy balance between energy supply and demand in the systems with a high penetration of the RERs, the local transaction market (LTM) can be developed based on the transactive multi-carrier energy technology. In the LTM, microgrids can exchange energy with each other for free under a transactive energy management system. Indeed, when some microgrids in the specific geographic locations have more amount of clean energy production from their stochastic producers, they can deliver their surplus of energy production to the LTM. In the LTM, some other microgrids, which their energy demand is greater than their energy production can use the received energy from the LTM for free with the aim of balancing energy.

Based on Eq. (6.152), those microgrids that delivered their surplus of energy production for free to the LTM at sometimes they will receive the same amount of energy from the LTM for free at other times, which other microgrids generate more amount of energy due to their geographical location. In other words, microgrids can lend their surplus energy to the LTM for free in times that they have more energy production and they receive the same amount of energy for free from the LTM in other times that other microgrids have more energy production. Indeed, these microgrids can give back the same amount of energy that they already received from the LTM (these are microgrids that already received energy from the LTM in other hours of a day). This mechanism in the LTM not only can reduce the dependency of the microgrids to the main grid with the conventional power plants but also can establish a dynamic energy balance at the local level between microgrids equipped with a high level of RERs. The LTM between the interconnected microgrids can be modeled using the following formulations.

$$\chi e_{m,t}^{In} + \chi e_{m,t}^{Out} \leq 1 \quad \forall m \in 1:N_m, \forall t \in 1:N_T \tag{6.149}$$

$$\chi h_{m,t}^{In} + \chi h_{m,t}^{Out} \leq 1 \quad \forall m \in 1:N_m, \forall t \in 1:N_T \tag{6.150}$$

$$\chi c_{m,t}^{In} + \chi c_{m,t}^{Out} \leq 1 \quad \forall m \in 1:N_m, \forall t \in 1:N_T \tag{6.151}$$

$$P_{m,t}^{LtM} \leq \Omega.\chi e_{m,t}^{In} \quad \forall m \in 1:N_m, \forall t \in 1:N_T \tag{6.152}$$

$$P_{m,t}^{MtL} \leq \Omega.\chi e_{m,t}^{Out} \quad \forall m \in 1:N_m, \forall t \in 1:N_T \tag{6.153}$$

$$H_{m,t}^{LtM} \leq \Omega.\chi h_{m,t}^{In} \quad \forall m \in 1:N_m, \forall t \in 1:N_T \tag{6.154}$$

$$H_{m,t}^{MtL} \leq \Omega.\chi h_{m,t}^{Out} \quad \forall m \in 1:N_m, \forall t \in 1:N_T \tag{6.155}$$

$$C_{m,t}^{LtM} \leq \Omega.\chi c_{m,t}^{In} \quad \forall m \in 1:N_m, \forall t \in 1:N_T \tag{6.156}$$

$$C_{m,t}^{MtL} \leq \Omega.\chi c_{m,t}^{Out} \quad \forall m \in 1:N_m, \forall t \in 1:N_T \tag{6.157}$$

$$\sum_t P_{m,t}^{LtM} = \sum_t P_{m,t}^{MtL} \quad \forall m \in 1:N_m \tag{6.158}$$

$$\sum_m P_{m,t}^{LtM} = \sum_m P_{m,t}^{MtL} \quad \forall t \in 1:N_T \tag{6.159}$$

$$\sum_t H_{m,t}^{LtM} = \sum_t H_{m,t}^{MtL} \quad \forall m \in 1:N_m \tag{6.160}$$

$$\sum_m H_{m,t}^{LtM} = \sum_m H_{m,t}^{MtL} \quad \forall t \in 1:N_T \tag{6.161}$$

$$\sum_t C_{m,t}^{LtM} = \sum_t C_{m,t}^{MtL} \quad \forall m \in 1:N_m \tag{6.162}$$

$$\sum_m C_{m,t}^{LtM} = \sum_m C_{m,t}^{MtL} \quad \forall t \in 1:N_T \tag{6.163}$$

where $\chi e_{m,t}^{In}$ and $\chi e_{m,t}^{Out}$ (($\chi h_{m,t}^{In}$ and $\chi h_{m,t}^{Out}$)/($\chi c_{m,t}^{In}$ and $\chi c_{m,t}^{Out}$)) are the binary variables for electrical energy (heating energy/cooling energy) receiving and transmitting by microgrid *m* to the LTM at time *t*. $P_{m,t}^{LtM}$ and $P_{m,t}^{MtL}$ (($H_{m,t}^{LtM}$ and $H_{m,t}^{MtL}$)/($C_{m,t}^{LtM}$ and $C_{m,t}^{MtL}$)) are the respective indicators of the amount of electrical energy (heating energy/cooling energy) received by the microgrid from the LTM and the amount of electrical energy (heating energy/cooling energy) transmitted from the microgrid to the LTM. Ω is the big number used for limiting the energy trading between the interconnected microgrids.

Equations (6.149), (6.150), and (6.151) present that each microgrid cannot receive and inject electrical, heating, and cooling energy into the LTM simultaneously. Equations (6.152) and (6.153) represent the maximum amount of the received electrical energy from the LTM and delivered electrical energy to the LTM, respectively. Equations (6.154) and (6.155)/Eqs. (6.156) and (6.157) indicate the maximum amount of the received heating energy/cooling energy from the LTM and delivered heating energy/cooling energy to the LTM, respectively. Equations (6.158) and (6.159) are used for modeling an electrical energy balance between microgrids in the LTM. Equations (6.160) and (6.161)/Eqs. (6.162) and (6.163) are intended for modeling a heating energy/cooling energy balance between microgrids in the LTM.

6.3 Uncertainty Modeling of the Modern Multi-Carrier Energy Networks

An ever-increasing penetration of the stochastic clean energy production units in the energy production process highlights the fact that hybrid energy networks are rapidly going to become a more uncertain environment. This is while the future modern MCENs are planned to be equipped with the 100% RERs for energy generation. Thus, the uncertainties associated with the RERs from one-side and uncertainties in the users' behaviors, energy price, and other stochastic resources from another side are indicating that we need to effectively model the uncontrollable events. Indeed, effectively modeling of the intermittences in the future modern MCENs can be resulted in achieving the near reality results. Therefore, in this section, we want to evaluate the different uncertainty modeling methods and get acquainted with how they model the stochastic behaviors of the various uncertain parameters.

6.3.1 Stochastic Programming Model

Stochastic programming is one of the effective ways of capturing the unpredictable behaviors of uncertain parameters. This method is structured based on the scenario generation for the associated uncertain parameter. In other words, it intends numerous samples of occurrence state of the uncertain parameter to consider them with their occurrence probability in the optimization problem. The most important strength of this method is that it considers most of the possible occurrence states of uncertain parameters in a certain uncertainty set by producing numerous scenarios using various techniques. Indeed, the large number of possible occurrence states can be intended with their occurrence probability that allows the technique to effectively model the fluctuations of a special uncertain parameter in the predetermined and predicted uncertainty set. On the other hand, the stochastic programming method has some disadvantages that have created substantial concerns regarding its usage for practical problems. Generating a large number of scenarios for effectively considering various states of the uncertain parameters occurrence not only substantially increases the complexity and the computational burden of the problem but also is very time-consuming work that has made its application for practical problems as an unreasonable option. In other words, this method is very effective for those groups of uncertain parameters that have a large fluctuation during a day like the wind speed. This is because almost all the uncertain parameter fluctuations in the uncertainty set can be considered by taking a sample as a scenario from each period of uncertainty parameter's changes.

6.3.1.1 Scenario Generation Models

Given how taking a sample from the uncertainty set of uncertain parameters, several scenario generation techniques are proposed with different methodologies in the scenario production. Some of the commonly used techniques are described in the following subsections.

6.3.1.1.1 Monte Carlo Simulation Method

Monte Carlo (MC) simulation method is one of the stochastic programming approaches that its operation depends on the mean and standard deviation of the uncertain parameter. The MC technique has mostly relied on the generation of a large number of scenarios for the related uncertain parameter considering the amount of its mean and standard deviation. For systems with higher complexity, as the MC efficiency is independent of either the complexity or size of the system, it is preferred in comparison with analytical methods. However, this is not suitable for practical cases [21]. The realization of the MC approach can be examined using either sequential sampling or nonsequential sampling [21]. The component states can be modeled using the sequential sampling considering their transition probabilities and correlations among random variables by sampling in chronological order [22]. This is while it needs a longer time to reach convergence in comparison with nonsequential sampling due to differing of any two consecutive samples by only one state component. To make the convergence process easier, variance reduction methods can be applied such as stratified sampling, antithetic variates, conditional Monte Carlo, control variates, and importance sampling [23]. Although the MC method is simple, it cannot guarantee the selection of sample from each area of uncertainty set. Moreover, it takes a longer time for convergence and imposes a high computational burden for the system that are taken into account as the serious concerns in adopting it for the uncertainty quantification.

6.3.1.1.2 Latin Hyperbolic Sampling (LHS) Method

The Latin Hyperbolic Sampling (LHS) method is one of the stochastic programming techniques that focuses on the whole sample space in generating scenarios with different probabilities [24]. Indeed, the LHS method divides the deviational interval of the uncertain parameter to select a sample from all areas of the uncertainty set. To generate Ns scenarios, the LHS approach creates the number of Ns same-length non-overlapping intervals by dividing the scale of cumulative probability (0 to 1.0). In the LHS method, a midpoint of each created interval can be selected as a scenario or it can be randomly chosen as the representative of each interval [25]. The mathematical model of the LHS method for modeling the stochastic behaviors of the wind speed is as follows [17].

$$\omega_n^{LHS} = CDF^{-1}\left(\frac{n-0.5}{N_s}\right) = \sqrt{\ln\left(\frac{1}{\left(1-\frac{n-0.5}{N_s}\right)^{2\gamma_{LHS}^2}}\right)} \tag{6.164}$$

$$CDF(\omega) = \int_0^{\omega}\left(\frac{\omega}{\gamma_{LHS}^2}\right)\left(e^{-\left(\frac{\omega}{\sqrt{2}\gamma_{LHS}}\right)^2}\right) \tag{6.165}$$

where ω_n^{LHS} is the amount of wind speed in nth interval.

6.3.1.1.3 Autoregressive Integrated Moving Average (ARIMA) Method

Stochastic programming techniques are recognized as effective tools for uncertainty modeling by considering most of the occurrence states of the uncertain parameters. In this regard, an autoregressive moving average (ARMA) is one of the stochastic-based methods that is exerted for evaluating the time-series using the path-based process [26]. The stochastic process in the ARMA models is stationary and Gaussian distribution is used in this process, which is taken into account as two major concerns in these models. In the stochastic-based problems, an ARMA (x, y) technique z can be modeled as follows.

$$z_t = \sum_{\ell=1}^{x}\alpha_\ell z_{t-\ell} + \Gamma_t - \sum_{\ell=1}^{y}\beta_\ell \Gamma_{t-\ell} \tag{6.166}$$

In Eq. (6.166), the first term presents the autoregressive part while the reminder terms represent the moving average. Moreover, Γ_t states the error term with a normal stochastic process (mean zero and variance σ_Γ^2). In order to achieve the stationary for the mean, the ARMA model needs to be included with the differencing producer, which is appeared as a new version of ARMA and is called the ARIMA model [26]. This model can be mathematically formulated by three parameters (x, u, y), which is as follows.

$$\left(1-\sum_{\ell=1}^{x}\alpha_\ell B^\ell\right)(1-B)^u z_t = \left(1-\sum_{\ell=1}^{y}\beta_\ell B^\ell\right)\Gamma_t \tag{6.167}$$

where u indicates the differentiating order and B is the backshift operator.

6.3.1.2 Scenario Reduction Method

Scenario reduction methods are mainly proposed for reducing the number of scenarios using different algorithms. These techniques are widely considered for effectively addressing the disadvantages of the numerous generated scenarios in the system.

6.3.1.2.1 Fast Forward Selection (FFS) Method

High computational burden and complexity, as well as unreasonable running time, are the main disadvantages of the problems with a large number of scenarios [17]. These drawbacks have made the stochastic programming approach with numerous scenarios as an inappropriate option for practical problems. Due to this, scenario reduction methods are widely used to overcome the aforementioned challenges by reducing the number of scenarios to a reasonable amount [27]. In this regard, the FFS method is proposed as one of the effective techniques for the scenario reduction process, which operates based on minimizing the Kantorovich distance between the scenarios [28]. More information regarding the FFS approach can be found in ref. [29].

6.3.2 Chance-Constrained Programming (CCP) Method

The CCP approach is one of the significant techniques for probabilistic modeling of the systems that are involved with a large number of uncertainties [30]. This approach considers the realistic conditions of the power grid by allowing some constraints with uncertain parameters to be violated [31]. The general definition of the CCP method in the optimization problems is given as:

$$\min_{w} f\left(w,\psi\right) \tag{6.168}$$

$$\text{s.t } \Pr\left\{g_c\left(w,\psi\right) \geq 0,\ c = 1,\cdots,N_c\right\} \geq \vartheta \tag{6.169}$$

In Eq. (6.168), $f(w,\psi)$ is the objective function that includes the random variables. Equation (6.169) presents the set of joint probabilistic constraints. ϑ is the confidence level factor that is regulated by the decision-maker and is considered for the constraints with uncertain parameters. ψ is the vector of N_k random variables with the cumulative density functions (CDFs) $F_{\psi_k}\left(h\right) = \Pr\left(\psi_k \leq h\right)\left(k = 1,\cdots N_k\right)$. The probability of satisfying the constraints with uncertain parameters is targeted to be greater than ϑ%, in which only $(1 - \vartheta)$% of the candidate scenarios are allowed to violate the probabilistic constraints with the aim of considering the probabilistic nature of the problem based on the CCP method.

6.3.3 Robust Optimization Technique

As the uncertainties in the hybrid energy networks drastically raise, developing a sustainable infrastructure with a certain level of robustness for the energy networks has become a challenging duty for the MCENs. In this regard, although the stochastic programming considers various states of the uncertain parameter in its deviation interval or the CCP method allows some uncertain parameter for a bit reasonable deviation from its upper level, but both of them cannot guarantee a certain level of robustness for the system. Indeed, the stochastic programming and the CCP method do not intend the worst-case of the uncertain parameters for making a robust structure, especially in the renewable-based systems. Therefore, the robust optimization (RO) technique is proposed to make the development of the robust systems possible and was first introduced by Soyster [32]. The RO method is one of the suitable techniques that enable the optimization problem to deal with the intermittences even when the system faces with the lack of full information about the nature of relevant uncertainty [33]. Because the RO considers the worst-state of an uncertain parameter for a certain amount of budget of uncertainty, it is more suitable for those groups of uncertain parameters that fluctuate in the small interval such as energy prices [34]. The budget of uncertainty is a control parameter that allows the decision-maker to set the required degree of conservativeness [35]. In continuation, we want to mathematically describe the RO method using a simple example.

In the first step of applying RO mathematically, assume the simple optimization problem as follows.

$$\min_{X} R = U^T.X \tag{6.170}$$

Subject to:

$$AX \; '' \; B \tag{6.171}$$

$$\underline{U} \; '' \; U \; '' \; \overline{U} \tag{6.172}$$

where X presents the vector of decision variables. U is the uncertain parameter that RO is targeted to make the system robust by applying on it. We can rewrite Eq. (6.170) in the following format with the aim of making its structure simple for applying the RO technique.

$$\min_{X} R \tag{6.173}$$

Subject to:

$$U^{T}.X\ ''\ R \tag{6.174}$$

$$AX\ ''\ B \tag{6.175}$$

The actual amount of U (U^a) is equal to the sum of its predicted amount (U^p) and its deviations (ΔU), that is:

$$\underline{U} \le U^{a} \le \bar{U} \Rightarrow U^{a} = U^{p} + \Delta U \Rightarrow U^{a} = U^{p} + \left(U^{+} - U^{-}\right) \tag{6.176}$$

Now the question arises as to how the uncertainty parameter U needs to be changed to achieve the robust state of the system. Herein, note that those changes of the uncertain parameter lead to system robustness that simultaneously weakens the objective function. For example, here, because the problem is minimization, the worst-state of the problem is based on uncertainty parameter changes when its value is equal to:

$$U^{a} = U^{p} + U^{+} \tag{6.177}$$

By updating the optimization problem based on Eq. (6.177), we will have:

$$\min_{X} R$$

Subject to:

$$\left(U^{p} + U^{+}\right).X \le R \tag{6.178}$$

$$AX\ ''\ B$$

By considering several decision variables in the problem, Eq. (6.178) can be rewritten as follows.

$$\sum_{i}\left[\left(U_i^{p} + U_i^{+}\right).X_i\right] \le R \Rightarrow \sum_{i}\left(U_i^{p}.X_i\right) + \sum_{i}\left(U_i^{+}.X_i\right) \le R \tag{6.179}$$

The predicted amount of uncertain parameter is constant, but we define a new auxiliary variable δ_i^{RO} to control the amount of conservativeness degree in the problem. The updated version of Eq. (6.179) is:

$$\sum_{i}\left(U_i^{p}.X_i\right) + \sum_{i}\left(\delta_i^{RO}.U_i^{+}.X_i\right) \le R \tag{6.180}$$

Therefore, the updated optimization problem based on the RO method is as follows.

$$\min_{X} R \tag{6.181}$$

Subject to:

$$\sum_i \left(U_i^p . X_i\right) + \max_{\delta_i^{RO}} \sum_i \left(\delta_i^{RO} . U_i^+ . X_i\right) \leq R \tag{6.182}$$

$$0 \leq \delta_i^{RO} \leq 1 : \tau_i^{RO} \tag{6.183}$$

$$0 \leq \sum_i \delta_i^{RO} \leq \Theta^{RO} : \Phi^{RO} \tag{6.184}$$

$$AX \text{ ''} B \tag{6.185}$$

where Θ^{RO} is the budget of uncertainty. τ_i^{RO} and Φ^{RO} are the dual variables that will be used in the dual problem. In this RO-based optimization problem (robust counterpart), we want to minimize the objective function R from one side and maximize the robustness degree from another side given in Eq. (6.182). Indeed, by applying the RO method, our optimization problem is converted to the bi-level problem that proper techniques should be applied for solving it. Karush–Kuhn–Tucker (KKT) [36] and duality theory [37] are two common methods for converting the bi-level problem to the single one. Herein, we use the duality theory for solving the bi-level problem. In the duality theory, the optimization problem from Eqs. (6.181), (6.182), (6.183), (6.184), and (6.185) is known as the primal problem and the dual problem is reached after solving the bi-level optimization problem as follows:

$$\min_X R$$

Subject to:

$$\sum_i \left(U_i^p . X_i\right) + \min_{\Phi^{RO}, \tau_i^{RO}} \left[\left(\Theta^{RO} . \Phi^{RO}\right) + \sum_i \left(\tau_i^{RO}\right) \right] \leq R \tag{6.186}$$

$$\Phi^{RO} + \tau_i^{RO} \geq U_i^+ . X_i \tag{6.187}$$

$$AX \text{ ''} B$$

After simplifying the aforementioned problem, we can have the following final problem:

$$\min_{X, \Phi^{RO}, \tau_i^{RO}} R$$

Subject to:

$$\sum_i \left(U_i^p . X_i\right) + \left(\Theta^{RO} . \Phi^{RO}\right) + \sum_i \left(\tau_i^{RO}\right) \leq R$$

$$\Phi^{RO} + \tau_i^{RO} \geq U_i^{+}.X_i$$

$$AX \ '' \ B$$

6.3.4 *Distributionally Robust Chance Constraint (DRCC) Method*

Effective modeling of the uncertainties associated with the outputs of the stochastic producers is essential for reaching the near reality results in the hybrid energy networks incorporated with a high level of RERs. In this regard, the DRCC approach is introduced as one of the effective ways for probabilistic modeling of the system, which can provide a satisfactory amount of profit and sufficient robustness in the presence of a large number of uncertainties [12]. The DRCC method simultaneously provides some significant advantages [38] in comparison with other techniques, which are shown in Fig. 6.5.

In this research, the general mathematical formulations of the DRCC method are presented with the aim of applying it to the energy management of the MCENs' problem. At first, the vector with random variables μ is assumed to be considered with the following distribution.

$$\Theta\left(\mu,\sigma^2\right)=\left\{\vartheta : E_{\vartheta}\left[\tilde{\mu}\right]=\mu, Var_{\vartheta}\left[\tilde{\mu}\right]=\sigma^2\right\} \tag{6.188}$$

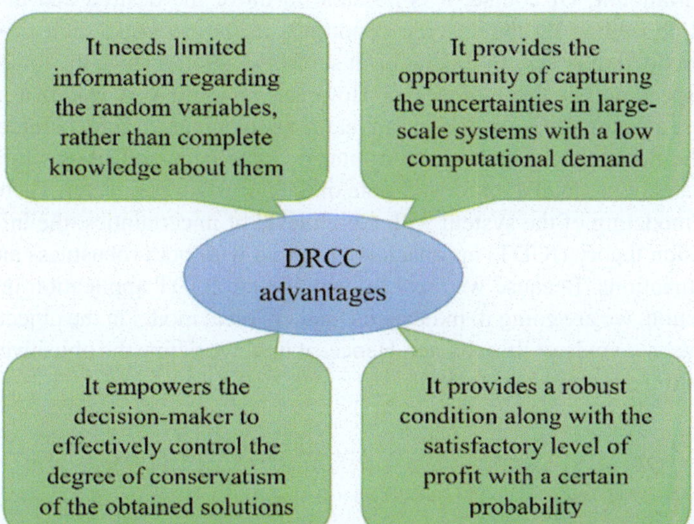

Fig. 6.5 Advantages of the DRCC method [38]

where μ and σ^2 are the mean and variance. The family of Θ contains all of the distributions with variance σ^2 and mean μ. Given the family of $\Theta(\mu,\sigma^2)$, the DRCC method for any $\psi \in (0, 1)$ can be defined as follows:

$$\underset{\vartheta \in \Theta}{Inf}\, Prob\left\{ \tilde{\mu}^T x \leq 0 \right\} \geq \psi \tag{6.189}$$

The equivalent version of this equation is the convex second-order cone constraint as follows.

$$\tilde{\mu}^T x + \sqrt{\frac{\psi}{1-\psi}} \sqrt{x^T \sigma^2 x} \leq 0 \tag{6.190}$$

where ψ states the satisfying probability of the constraints with uncertain parameters. The complete descriptions and model of the DRCC method can be fully accessed in [39].

6.3.5 *Information-Gap Decision Theory (IGDT) Method*

Regarding the uncertainty quantification, the uncertain parameters can be deviated in the desired direction or have undesirable changes during a day. For example, the desired deviation for the wind speed is that its amount has become more than our predictions and undesired deviations mean that the wind speed is lower than its predicted amount. Of course, it is possible to move the desired and undesirable concepts depending on the specific conditions in the network, such as when the energy production is very high, that is, in some special moments, the lower amount of wind speed will be favorable for us. However, we know very well that just as the worst-case scenario is likely to occur in real-time for an uncertainty parameter, there is also the possibility that the most optimistic state occurs. This is while the RO method cannot model the optimistic state of the uncertain parameter. Therefore, for realistic modeling of the system with a high level of uncertainties, the information gap decision theory (IGDT) approach is proposed with both robustness and opportunistic functions. Because we want to examine the IGDT application in the case study section, we are going to indicate its mathematical model in the objective function of the case study of this chapter. Hence, at first, we define the objective function of our optimization problem as follows.

$$OF_m = \sum_{t=1}^{T} \cos t_{m,t}^{BESS} - \sum_{t=1}^{N_T} \rho_t^{DA}.P_{m,t}^{DA}.\Delta t - \sum_{i=1}^{N_b}\sum_{t=1}^{N_T} \rho_t^{Sell}.\left(P_{i,t}^{D}\right)\Delta t \quad \forall m \tag{6.191}$$

Subject to:

$$\sum_{m=1}^{N_m}\left[\left(P_{m,t}^{PV}+P_{m,t}^{Wind}+P_{m,t}^{BESS}+P_{m,t}^{LtM}\right)-\left(P_{m,t}^{EWB}+P_{m,t}^{RC}+P_{m,t}^{MtL}\right)-P_{m,t}^{DA}\right] = \sum_{i=1}^{N_b}P_{i,t}^{D} \quad \forall t \tag{6.192}$$

$$H_{m,t}^{EWB}+H_{m,t}^{SWB}+H_{m,t}^{H,Co}+H_{m,t}^{SM,H}+H_{m,t}^{LtM} = H_{m,t}^{D}+H_{m,t}^{MS,H}+H_{m,t}^{MtL} \quad \forall m,\forall t \tag{6.193}$$

$$C_{m,t}^{RC}+C_{m,t}^{AC}+C_{m,t}^{SM,C}+C_{m,t}^{LtM}=C_{m,t}^{D}+C_{m,t}^{MS,C}+C_{m,t}^{MtL} \quad \forall m,\forall t \tag{6.194}$$

$$\text{Constraints}(6.1)\text{to}(6.3),(6.38)\text{to}(6.45),(6.69)\text{to}(6.75),(6.115)\text{to}(6.118),(6.119)\text{to}(6.131),(6.143)\text{to}(6.145),(6.149)\text{to}(6.163). \tag{6.195}$$

where OF_m is the objective function. ρ_t^{DA} and ρ_t^{Sell} are the respective indicators of day-ahead and selling energy prices.

Equations (6.192), (6.193), and (6.194) are the electrical, heating, and cooling energy balance in the MCENs, respectively.

6.3.5.1 Risk-Averse Strategy (Robustness Function)

The IGDT is one of the non-probabilistic approaches that does not rely on the information of the probability distribution of the uncertain parameter [40, 41]. The risk-averse strategy is a robustness function of the IGDT method that maximizes the horizon of uncertainty to guarantee of obtaining a certain amount of expectation for the optimization problem. This type of IGDT strategy can be defined as follows [42].

$$\max \psi^{Ro} \tag{6.196}$$

Subject to:

$$OF_m \le OF_m^{RO}=(1+\varpi)OF_m^{D} \tag{6.197}$$

$$OF_m=\max_{\Delta\rho_t^{DA}}\left\{\Pi_m-\sum_{t=1}^{N_t}\left(\rho_t^{DA,P}+\Delta\rho_t^{DA}\right).P_{m,t}^{DA}.\Delta t\right\} \quad \forall m \tag{6.198}$$

$$\Pi_m=\sum_{t=1}^{T}\cos t_{m,t}^{BESS}-\sum_{i=1}^{N_b}\sum_{t=1}^{N_T}\rho_t^{Sell}.\left(P_{i,t}^{D}\right)\Delta t \quad \forall m \tag{6.199}$$

$$-\psi^{Ro}\rho_t^{DA,P}\le\Delta\rho_t^{DA}\le\psi^{Ro}\rho_t^{DA,P} \tag{6.200}$$

All constraints in (6.195)

where ψ^{Ro} denotes the uncertain variable horizon in the risk-averse strategy. OF_m^{RO} presents the amount of robust cost level. OF_m^{D} represents the microgrids' energy cost in the deterministic problem. $\rho_t^{DA,P}$ states the amount of forecasted day-ahead energy market price at time t. $\Delta\rho_t^{DA}$ is the deviation from the forecasted price. Π_m is used for making the applying process of the IGDT method simple. The current version of the optimization problem is a bi-level programming problem. In order to simply converting the bi-level problem to the single level one without using additional techniques, we can consider the worst-state of the energy price considering the amount of it among the following states:

$$\Delta\rho_t^{DA} = \begin{cases} \psi^{Ro}\rho_t^{DA,P} & P_{m,t}^{DA} \leq 0 \\ -\psi^{Ro}\rho_t^{DA,P} & P_{m,t}^{DA} \geq 0 \end{cases} \tag{6.201}$$

After determining the worst states of the day-ahead energy price, the two following terms can be considered for reaching the single level problem.

$$\left(\Delta\rho_t^{DA} + \psi^{Ro}\rho_t^{DA,P}\right).P_{m,t}^{DA} \leq 0 \tag{6.202}$$

$$\left(\Delta\rho_t^{DA} - \psi^{Ro}\rho_t^{DA,P}\right).P_{m,t}^{DA} \leq 0 \tag{6.203}$$

According to the aforementioned formulations, when the microgrid m plays a purchaser role in the day-ahead market, that is, $P_{m,t}^{DA}$ ″ 0, the worst state for it is when $\Delta\rho_t^{DA} = \psi^{Ro}\rho_t^{DA,P}$. This is while when the microgrid is seller, that is, $P_{m,t}^{DA} \geq 0$, the worst state for it is when $\Delta\rho_t^{DA} = -\psi^{Ro}\rho_t^{DA,P}$.

6.3.5.2 Risk-Seeker Strategy (Opportunistic Function)

Contrary to the risk-averse strategy, the risk-seeker strategy is an opportunity function of the IGDT method that minimizes the horizon of uncertainty to achieve more benefits by considering the desirable changes of the uncertain parameters. This type of IGDT strategy can be defined as follows [42].

$$\min \psi^{Op} \tag{6.204}$$

Subject to:

$$OF_m \leq OF_m^{Op} = (1-\varpi)OF_m^{D} \tag{6.205}$$

$$OF_m = \min_{\Delta\rho_t^{DA}} \left\{ \Pi_m - \sum_{t=1}^{N_t} \left(\rho_t^{DA,P} + \Delta\rho_t^{DA}\right).P_{m,t}^{DA}.\Delta t \right\} \quad \forall m \tag{6.206}$$

$$-\psi^{Op}\rho_t^{DA,P} \le \Delta\rho_t^{DA} \le \psi^{Op}\rho_t^{DA,P} \tag{6.207}$$

All constraints in (6.195)

where ψ^{Op} denotes the uncertain variable horizon in the risk-seeker strategy. OF_m^{Op} presents the amount of opportunity cost level. Similar to the risk-averse strategy, we can convert the bi-level problem to the single level one by considering the optimistic state of the uncertain parameter occurrence as below.

$$\Delta\rho_t^{DA} = \begin{cases} -\psi^{Op}\rho_t^{DA,P} & P_{m,t}^{DA} \le 0 \\ \psi^{Op}\rho_t^{DA,P} & P_{m,t}^{DA} \ge 0 \end{cases} \tag{6.208}$$

After determining the optimistic states of the day-ahead energy price in both selling and purchasing modes, the two following terms can be intended for achieving the single level problem.

$$\left(\Delta\rho_t^{DA} + \psi^{Ro}\rho_t^{DA,P}\right).P_{m,t}^{DA} \ge 0 \tag{6.209}$$

$$\left(\Delta\rho_t^{DA} - \psi^{Ro}\rho_t^{DA,P}\right).P_{m,t}^{DA} \ge 0 \tag{6.210}$$

According to the opportunistic function of the IGDT, when the microgrid m plays a purchaser role in the day-ahead market, that is, $P_{m,t}^{DA}$ " 0, the optimistic state for it is when $\Delta\rho_t^{DA} = -\psi^{Op}\rho_t^{DA,P}$. This is while when the microgrid is seller, that is, $P_{m,t}^{DA} \ge 0$, the optimistic state for it is when $\Delta\rho_t^{DA} = \psi^{Op}\rho_t^{DA,P}$.

6.4 Case Study for the Multi-Carrier Energy Network with 100% RERs

As earlier mentioned, the future modern MCENs are targeted to be equipped with 100% RERs for fully clean energy production. In order to examine how this issue can be realized for a sample structure of the MCENs, the integration of the electric power system and the DHN is considered for the interconnected microgrids that are equipped with 100% RERs. In this chapter, we propose a holistic model for the optimal scheduling of the multi-microgrids in the fully deregulated environment. The related mathematical equations for the proposed model are shown as Eqs. (6.191)–(6.210). For absolute clean energy production, each microgrid is only equipped with 100% RERs including wind turbines, PV panels, BESS, TES, electric water boiler, solar water boiler, reciprocating chiller, and absorption chiller [4]. The demand-side energy management is considered to be done by applying only a price response program for just intending the shiftable loads. This is because the load shedding is not in the principles of modern MCENs. As already mentioned in the previous chapters, energy trading is the most useful way of establishing energy balance particularly in the systems with a full share of RER. In the proposed model, the new transactive multi-carrier energy technology is used for developing the local energy exchanging environment to allow for the microgrids to trade energy with each other for free to easily meet their energy demand. Herein, we just select a sample structure of the MCENs to evaluate the effectiveness of our proposed model whereas the MCENs structure is not limited to the aforementioned combination and other incorporated structures can also be developed for MCENs. In this research, the integration of the IEEE 10-bus electric power system with the 10-node DHN is considered as the case study. The schematic of this test system is illustrated in Fig. 6.6.

Because of using the non-linear equations with binary variables, the type of optimization problem is the mixed-integer non-linear programming (MINLP). For solving this problem, the general algebraic modeling system (GAMS) is used by applying its SBB solver. The problem is solved in two cases. In Case I, the deterministic state of the optimization problem is considered without uncertainty modeling. However, Case II is developed based on the proposed model that the IGDT method is used for uncertainty quantification. After solving the problem, the energy cost for each of the microgrids is depicted in Fig. 6.7.

As seen in Fig. 6.7, the amount of energy cost for microgrids in Case II is greater than Case I. Indeed, as the risk-averse strategy has made the system robust, it has been led to more energy cost due to considering the worst-state of the uncertain parameter. Since the size and scale of microgrids are different from each other, various amount of the cost increment is reached for the interconnected microgrids. In this study, the main burden of energy production is on clean energy production units. However, in order to enhance the reliability of the energy supply in the presence of the 100% RERs, the transactive multi-carrier energy technology along with the energy storage systems are used in the community-based microgrids. The

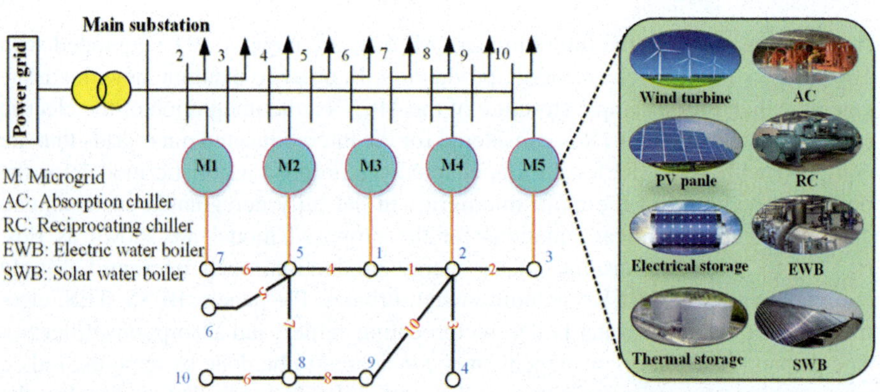

Fig. 6.6 The schematic of the integrated IEEE 10-bus and 10-node DHN with 100% renewable-based microgrids

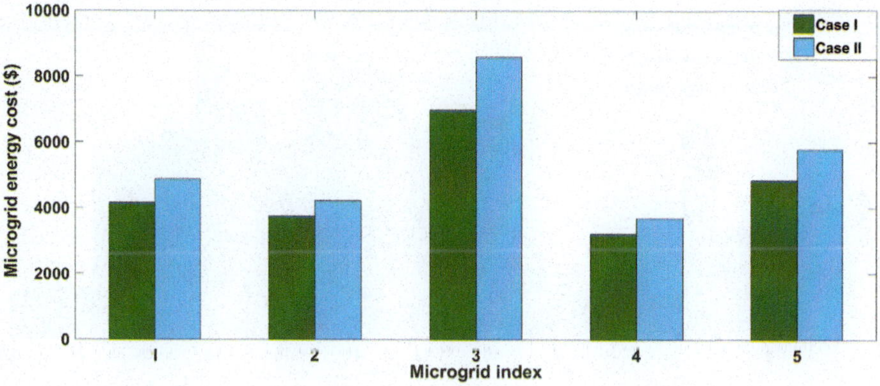

Fig. 6.7 The energy cost of microgrids in Cases I and II

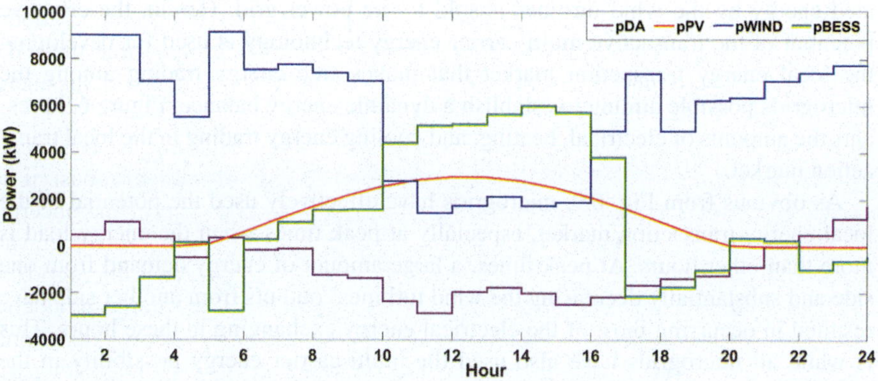

Fig. 6.8 The energy interactions between different devices in the microgrids

amounts of RERs outputs along with the BESS and energy trading in the day-ahead market are demonstrated in Fig. 6.8.

According to Fig. 6.8, a large amount of energy production by the wind turbines in the early morning (1–6 am) has been not only led to the charging of the BESS but also a few portions of the surplus of energy generation is sold to the main grid to improve the microgrids' economic benefits. This is while by hour-to-hour increasing the energy demand from 7 am, the need for more energy generation in the system has changed the BESS application from the charging mode to the discharging state. Also, the effective role of the PV panels in clean energy production is more highlighted at peak times with the highest level of energy consumption when their outputs are at the highest amount due to maximum solar radiation at noon. Moreover, the created energy trading possibility between the microgrids with the main grid is also supported the renewable-based system to reliably meet the energy demand at peak times. However, by dropping the pressure of energy consumption in the system from 5 pm, the BESS has switched to the discharging mode and the excess gener-

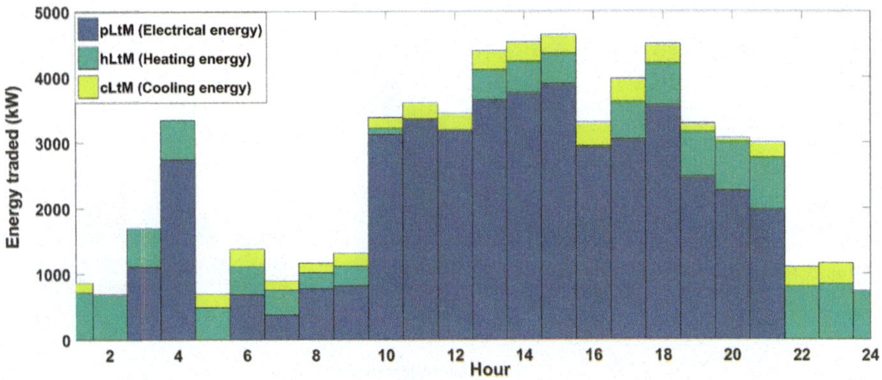

Fig. 6.9 The energy interactions between microgrids in the local transaction market

ated energy by the wind turbines is sold to the power grid. Herein, the effective potential of the transactive multi-carrier energy technology is used for developing the local energy transaction market that makes free energy trading among the microgrids possible aiming to establish a dynamic energy balance. Figure 6.9 presents the amounts of electrical, heating, and cooling energy trading in the local transaction market.

As obvious from Fig. 6.9, microgrids have effectively used the potential of the local energy transaction market, especially at peak times when the energy load is more than other hours. At peak times, a large amount of energy demand from one side and substantially decreasing the wind turbines' outputs from another side have resulted in occurring most of the electrical energy exchanging in these hours. This is while all microgrids have also used the multi-carrier energy possibility in the local transaction market for reliably meeting their heating and cooling energy demand during a day. As the energy interactions in the local transaction market transparently indicate the critical role of the transactive multi-carrier energy technology in establishing a dynamic energy balance, its presence in the future modern MCENs with a full share of RERs is essential for developing the sustainable grid. In such a deregulated environment, uncertainties have significant impacts on the realistic modeling of the system. In this study, the IGDT method is applied for uncertainty quantification and the optimization problem is solved from both risk-averse and risk-seeker strategies. The impacts of the robustness and opportunistic functions in the total energy cost of microgrids are shown in Fig. 6.10.

As illustrated in Fig. 6.10, by increasing the horizon of an uncertain parameter, the amount of objective function is increased when the robustness function is employed while its amount is decreased under the opportunistic function. This issue shows that as the degree of conservatism in the system increases, the system becomes more robust and therefore imposes more energy costs to the system.

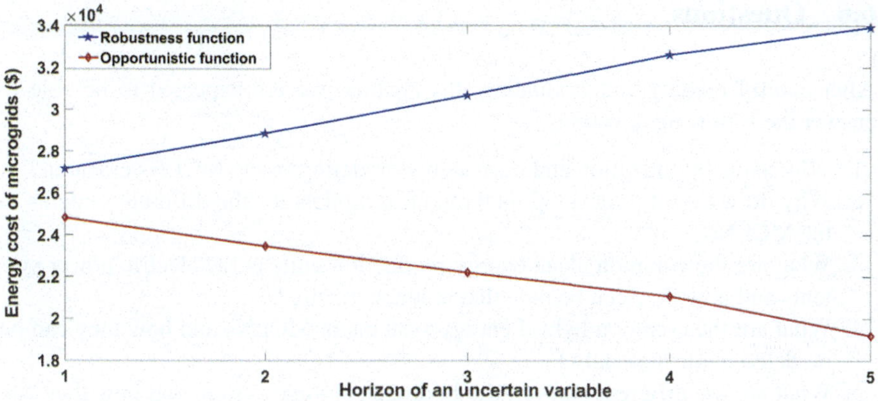

Fig. 6.10 The amount of microgrids' energy cost from the perspective of both risk-averse and risk-seeker strategies

In contrast, high optimism about the uncertain parameter's occurrence improves the economic benefits of the microgrids, but on the other hand, the system's robustness to probability changes of the uncertain parameter remains a challenge.

6.5 Summary

The energy networks with numerous intelligent systems and different entities are developing day by day and becoming fully competitive and uncertain environment with various goals. In such an environment, the optimization of the system is essential to extract the best possible set points for the different controllable devices. Therefore, in this chapter, we have prompted to present the mathematical models of most of the common systems in modern MCENs. In other words, the mathematical models are presented and in some cases are proposed for the diverse systems in the electric power system, natural gas grid, the DHN, and water distribution networks. Because the future modern MCENs are targeted to be equipped with 100% RERs, the effective modeling of the stochastic behaviors of various uncertain parameters is essential for achieving the near reality results. Thus, in the next step, different uncertainty modeling approaches are described and the related mathematical models for them are presented. Moreover, the mathematical models for demand-side energy management schemes and transactive multi-carrier energy technology are also described with the aim of facilitating their employment in future works. In the end, the case study of multiple interconnected microgrids with real data is presented to effectively analyzing the effectiveness of the proposed holistic model for a sample structure of the modern MCENs. The results indicated the applicability of the proposed model in the renewable-based structure of the modern MCENs.

6.6 Questions

After careful reading and evaluating this chapter, we are expected to be able to answer the following questions.

1. Why is the optimization and uncertainty modeling of the MCENs essential?
2. Why do we need to develop mathematical models for the different systems in the MCENs?
3. What are the common clean energy production units in the electric power systems and how they can be modeled mathematically?
4. What are the common hybrid energy systems in MCENs and how they can be modeled mathematically?
5. What are the different parts of the hydrogen storage system and how they can be modeled mathematically?
6. What are the different parts of the power to gas storage system and how they can be modeled mathematically?
7. In how many models the electricity network can be mathematically modeled in the hybrid energy networks and what are their differences?
8. What are the essential components of the natural gas grid and how they can be modeled mathematically?
9. What are the essential components of the DHN and how they can be modeled mathematically?
10. What are the essential components of the water distribution network and how they can be modeled mathematically?
11. What are the common schemes for demand-side energy management and how they can be modeled mathematically?
12. How the transactive multi-carrier energy technology can be used for developing the local energy transaction market between the interconnected microgrids in the modern MCENs?
13. Why do we need to effectively model the stochastic behaviors of the uncertain parameters in the optimization problems?
14. What are the different common uncertainty modeling approaches and for which features they are famous in the optimization world?
15. How the different common uncertainty quantification methods can be mathematically modeled?
16. What is the difference between various scenario generation techniques?
17. What are the distinguishing attributes of the stochastic programming approaches?
18. What are the superiorities of the robust-based techniques in comparison with the scenario-based methods?
19. What are the disadvantages of the robust-based approaches in the uncertainty quantification process?
20. What are the differences between the two strategies of the IGDT method with each other?

6.7 Suggestions

In this section, some suggestions are provided related to this chapter that can give a good overview of some substantial issues in the future modern MCENs.

1. Developing strong mathematical models is an essential step for effectively analyzing future modern MCENs.
2. Developing the different capable uncertainty modeling is beneficial for achieving the near reality results in the modern MCENs assessment.
3. Employing the transactive multi-carrier energy technology is necessary for developing the free energy trading environment for maintaining the sustainability of the future modern MCENs at a sufficient level.
4. Optimization of the interconnected hybrid energy networks is an inevitable step in reaching the best possible set points for the different devices of the modern MCENs.

6.8 Future Trends and Discussion Topics

Here, future trends and some discussion topics are proposed for the interested readers regarding this chapter that can be useful for developing the MCENs modernization process.

1. Developing new optimization techniques for effectively modeling of the interconnected structure of the future modern MCENs.
2. Proposing new strong models for the various hybrid energy networks and their devices to improve the energy interactions in the MCENs.
3. Developing new uncertainty modeling techniques for effectively dealing with the intermittences of the different uncertainty resources.

References

1. M. Daneshvar, B. Mohammadi-Ivatloo, M. Abapour, S. Asadi, Energy exchange control in multiple microgrids with transactive energy management. J. Modern Power Syst. Clean Energy, **8**(4), 1–8 (2020)
2. M. Daneshvar, B. Mohammadi-Ivatloo, S. Asadi, M. Abapour, A. Anvari-Moghaddam, A transactive energy management framework for regional network of microgrids, in *2019 International Conference on Smart Energy Systems and Technologies (SEST)*, (IEEE, 2019), pp. 1–6
3. C. Wang et al., Impact of power-to-gas cost characteristics on power-gas-heating integrated system scheduling. IEEE Access **7**, 17654–17662 (2019)
4. M. Daneshvar, B. Mohammadi-Ivatloo, K. Zare, S. Asadi, Two-stage robust stochastic scheduling model for transactive energy based renewable microgrids. IEEE Trans. Ind. Inform, **16**(11), 1–1 (2020)

5. M. Daneshvar, B. Mohammadi-Ivatloo, K. Zare, S. Asadi, Two-stage stochastic programming model for optimal scheduling of the wind-thermal-hydropower-pumped storage system considering the flexibility assessment. Energy **193**, 116657 (2020)
6. C. Shao, X. Wang, M. Shahidehpour, X. Wang, B. Wang, An MILP-based optimal power flow in multicarrier energy systems. IEEE Trans. Sustain. Energy **8**(1), 239–248 (2016)
7. L. Wu, M. Shahidehpour, Optimal coordination of stochastic hydro and natural gas supplies in midterm operation of power systems. IET Gener. Transm. Distrib **5**(5), 577–587 (2011)
8. J. Qiu, J. Zhao, H. Yang, Z.Y. Dong, Optimal scheduling for prosumers in coupled transactive power and gas systems. IEEE Trans. Power Syst. **33**(2), 1970–1980 (2017)
9. M. Chaudry, N. Jenkins, G. Strbac, Multi-time period combined gas and electricity network optimisation. Electr. Power Syst. Res. **78**(7), 1265–1279 (2008)
10. S. An, Q. Li, T.W. Gedra, Natural gas and electricity optimal power flow, in *2003 IEEE PES Transmission and Distribution Conference and Exposition (IEEE Cat. No. 03CH37495)*, vol. 1, (IEEE, 2003), pp. 138–143
11. A. Kabirian, M.R. Hemmati, A strategic planning model for natural gas transmission networks. Energy Policy **35**(11), 5656–5670 (2007)
12. M. Daneshvar, B.M. Ivatloo, M. Abapour, S. Asadi, R. Khanjani, Distributionally robust chance constrained transactive energy framework for coupled electrical and gas microgrids. IEEE Trans. Ind. Electron., **68**(1), 1–1 (2021)
13. Z. Li, W. Wu, J. Wang, B. Zhang, T. Zheng, Transmission-constrained unit commitment considering combined electricity and district heating networks. IEEE Trans. Sustain. Energy **7**(2), 480–492 (2015)
14. Y. Cao, W. Wei, L. Wu, S. Mei, M. Shahidehpour, Z. Li, Decentralized operation of interdependent power distribution network and district heating network: A market-driven approach. IEEE Trans. Smart Grid **10**(5), 5374–5385 (2018)
15. X. Liu, J. Wu, N. Jenkins, A. Bagdanavicius, Combined analysis of electricity and heat networks. Appl. Energy **162**, 1238–1250 (2016)
16. Y. Chen, W. Wei, F. Liu, E.E. Sauma, S. Mei, Energy trading and market equilibrium in integrated heat-power distribution systems. IEEE Trans. Smart Grid **10**(4), 4080–4094 (2018)
17. M. Daneshvar, M. Pesaran, B. Mohammadi-ivatloo, Transactive energy integration in future smart rural network electrification. J. Clean. Prod. **190**, 645–654 (2018)
18. J. Zheng, J. Chen, Q. Wu, Z. Jing, Multi-objective optimization and decision making for power dispatch of a large-scale integrated energy system with distributed DHCs embedded. Appl. Energy **154**, 369–379 (2015)
19. M.J.V. Pakdel, F. Sohrabi, B. Mohammadi-Ivatloo, Multi-objective optimization of energy and water management in networked hubs considering transactive energy. J. Clean. Prod. **126**, 121936 (2020)
20. H. Zhang, X. Xia, J. Zhang, Optimal sizing and operation of pumping systems to achieve energy efficiency and load shifting. Electr. Power Syst. Res. **86**, 41–50 (2012)
21. Z. Shu, P. Jirutitijaroen, Latin hypercube sampling techniques for power systems reliability analysis with renewable energy sources. IEEE Trans. Power Syst. **26**(4), 2066–2073 (2011)
22. R. Karki, R. Billinton, Reliability/cost implications of PV and wind energy utilization in small isolated power systems. IEEE Trans. Energy Convers **16**(4), 368–373 (2001)
23. R.Y. Rubinstein, D.P. Kroese, *Simulation and the Monte Carlo Method* (Wiley, Hoboken, 2016)
24. H. Yu, C. Chung, K. Wong, H. Lee, J. Zhang, Probabilistic load flow evaluation with hybrid latin hypercube sampling and cholesky decomposition. IEEE Trans. Power Syst. **24**(2), 661–667 (2009)
25. A. Gupta, A. Kumar, ATC calculation including wind: A probabilistic study and a comparison of MCS and LHS, in *2016 IEEE 6th International Conference on Power Systems (ICPS)*, (IEEE, 2016), pp. 1–6

26. M. Alipour, B. Mohammadi-Ivatloo, K. Zare, Stochastic risk-constrained short-term scheduling of industrial cogeneration systems in the presence of demand response programs. Appl. Energy **136**, 393–404 (2014)
27. J. Dupačová, N. Gröwe-Kuska, W. Römisch, Scenario reduction in stochastic programming. Math. Program. **95**(3), 493–511 (2003)
28. K. Bruninx, E. Delarue, W. D'haeseleer, A practical approach on scenario generation & reduction algorithms based on probability distance measures–the case of wind power forecast errors, *WP EN2014-15,* 2014
29. W.L. de Oliveira, C. Sagastizábal, D.D.J. Penna, M.E.P. Maceira, J.M. Damázio, Optimal scenario tree reduction for stochastic streamflows in power generation planning problems. Optim. Methods Softw **25**(6), 917–936 (2010)
30. J. Men, P. Jiang, H. Xu, A chance constrained programming approach for HazMat capacitated vehicle routing problem in Type-2 fuzzy environment. J. Clean. Prod. **237**, 117754 (2019)
31. J. Zhao et al., An islanding partition method of active distribution networks based on chance-constrained programming. Appl. Energy **242**, 78–91 (2019)
32. A.L. Soyster, Convex programming with set-inclusive constraints and applications to inexact linear programming. Oper. Res. **21**(5), 1154–1157 (1973)
33. A.S. Nemirovskiĭ, L. El Ghaoui, A. Ben-Tal, *Robust Optimization (Princeton Series in Applied Mathematics)* (Princeton University Press, Princeton)
34. M. Daneshvar, B. Mohammadi-Ivatloo, K. Zare, Two-stage optimal robust scheduling of hybrid energy system considering the demand response programs. J. Clean. Prod. **248**, 119267 (2020)
35. A. Soroudi, Robust optimization based self scheduling of hydro-thermal Genco in smart grids. Energy **61**, 262–271 (2013)
36. G. Gordon, R. Tibshirani, Karush-kuhn-tucker conditions. Optimization **10**(725/36), 725 (2012)
37. E. Castillo, A. Conejo, P. Pedregal, R. García, N. Alguacil, *Building and Solving Mathematical Programming Models in Engineering and Science* (Wiley, Hoboken, 2001)
38. A. Zare, C. Chung, J. Zhan, S.O. Faried, A distributionally robust chance-constrained MILP model for multistage distribution system planning with uncertain renewables and loads. IEEE Trans. Power Syst. **33**(5), 5248–5262 (2018)
39. G.C. Calafiore, L. El Ghaoui, On distributionally robust chance-constrained linear programs. J. Optim. Theory Appl. **130**(1), 1–22 (2006)
40. M. Daneshvar, B. Mohammadi-ivatloo, K. Zare, S. Asadi, A. Anvari-Moghaddam, A novel operational model for interconnected microgrids participation in transactive energy market: A hybrid IGDT/stochastic approach. IEEE Trans. Ind. Inform, 1–1 (2020). https://doi.org/10.1109/TII.2020.3012446
41. M. Majidi, B. Mohammadi-Ivatloo, A. Soroudi, Application of information gap decision theory in practical energy problems: A comprehensive review. Appl. Energy **249**, 157–165 (2019)
42. S. Shafiee, H. Zareipour, A.M. Knight, N. Amjady, B. Mohammadi-Ivatloo, Risk-constrained bidding and offering strategy for a merchant compressed air energy storage plant. IEEE Trans. Power Syst. **32**(2), 946–957 (2016)

Index

© Springer Nature Switzerland AG 2021
M. Daneshvar et al., *Grid Modernization – Future Energy Network Infrastructure*, Power Systems, https://doi.org/10.1007/978-3-030-64099-6

The manufacturer's authorised representative in the EU is Springer Nature Customer Service Centre GmbH, Europaplatz 3, 69115 Heidelberg, Germany. If you have any concerns regarding our products, please contact ProductSafety@springernature.com

Printed and bound by CPI Group (UK) Ltd, Croydon, CR0 4YY
24/09/2026
02222178-0001